BOTANICAL RESEARCH AND PRACTICES

CULTIVARS

CHEMICAL PROPERTIES, ANTIOXIDANT ACTIVITIES AND HEALTH BENEFITS

BOTANICAL RESEARCH AND PRACTICES

Additional books in this series can be found on Nova's website
under the Series tab.

Additional e-books in this series can be found on Nova's website
under the E-book tab.

AGRICULTURE ISSUES AND POLICIES

Additional books in this series can be found on Nova's website
under the Series tab.

Additional e-books in this series can be found on Nova's website
under the E-book tab.

BOTANICAL RESEARCH AND PRACTICES

CULTIVARS

CHEMICAL PROPERTIES, ANTIOXIDANT ACTIVITIES AND HEALTH BENEFITS

KATYA CARBONE
EDITOR

NOVA BIOMEDICAL

New York

For permission to use material from this book please contact us:
Telephone 631-231-7269; Fax 631-231-8175
Web Site: http://www.novapublishers.com

Additional color graphics may be available in the e-book version of this book.

Library of Congress Cataloging-in-Publication Data

Cultivars : chemical properties, antioxidant activities and health benefits / editor: Katya Carbone.
p. cm.
Includes index.
ISBN 978-1-62417-101-7 (hardcover)
1. Plants--Analysis. 2. Botanical chemistry. 3. Antioxidants. 4. Plant varieties--Research. I. Carbone, Katya.
QK865.C95 2013
333.95'34--dc23

2012038763

Published by Nova Science Publishers, Inc. † New York

CONTENTS

PREFACE

Fruits and vegetables have been the backbone of human nutrition since the beginning of time. They play a key role in the local and international trade, leading to a competition for markets, which is driving extensive research and development to produce new cultivars.

Until recent times, research has concentrated on producing new varieties that store longer, yield better, look better, taste better, suit local climates, display disease and pest resistance and suit processing technologies. In addition to this, there is a growing interest in phytonutrients or plant-derived bioactives that naturally occur in fruits and vegetables and possess curative and preventive properties above their nutritional value. In fact epidemiological data demonstrate a positive correlation between a high consumption of phytochemicals and a lower risk of several diseases, i.e. cardiovascular, degenerative and carcinogenic ones. Nutrition parameters found in plant foods vary between crops. As a consequence, many scientists as molecular biologists, biochemists, botanists and nutritionists are involved in plant breeding programs to develop new varieties of fruit and vegetables that are tailor-made to produce higher levels of health-related phytochemicals. At the same time, biotechnological tools are set up to develop biotech crops for vaccine production, while new rapid diagnostic and molecular techniques have been developed for a rapid identification of these healthy compounds.

In this contest, the aim of this book is to provide both researchers and graduated students in the areas of food technologies and related ones with accessible and up-to-date information about advances in research related to different plant cultivars and methods available to use them and their extracts. Special care is also taken to the new biotechnological and molecular tools applied in this research field, which lead to the production of vaccine as well as to development of molecular techniques able to guarantee consumers the authenticity of products.

This book was possible thanks to the valuable collaboration of many scientists who agreed to dedicate part of their time to participate in this project, with valuable contributions from their different fields of expertise. To them, I am sincerely thankful.

Katya Carbone
Agricultural Research Council – Research Unit for Enology in Central Italy (CRA-ENC),
Velletri, Rome, Italy

In: Cultivars
Editor: Katya Carbone

ISBN: 978-1-62417-101-7

Chapter 1

VEGETABLE BREEDING FOR NUTRITIONAL QUALITY AND HEALTH BENEFITS

João Silva Dias*

Technical University of Lisbon, Instituto Superior de Agronomia, Lisboa, Portugal

ABSTRACT

Vegetables are considered essential for well-balanced diets since they supply vitamins, minerals, dietary fiber, and phytochemicals. In the daily diet, they have been strongly associated with improvement of gastrointestinal health, good vision, and reduced risk of heart disease, stroke, chronic diseases such as diabetes, and some forms of cancer. Some phytochemicals of vegetables are strong antioxidants and are thought to reduce the risk of chronic disease by protecting against free radical damage by modifying metabolic activation and detoxification of carcinogens or even by influencing processes that alter the course of tumor cells. Nutrition is both a quantity and a quality issue, and vegetables in all their many forms ensure an adequate intake of most vitamins and nutrients, dietary fibers, and phytochemicals, which can bring a much-needed measure of balance back to diets contributing to solve many of these nutrition problems.

Plant breeders play a key role in determining what we eat, since every vegetable cultivar that we see on the market has benefited from plant breeding. The general objectives for producers are good yield, disease and pest resistance, uniformity, and abiotic stress resistance. Objectives for consumers are quality, appearance, shelf life, taste, and nutritional value. Quality in vegetable crops, in contrast to field crops, is often more important than yield. For producers to survive, market must accept cultivars. Thus, color, appearance, taste, and shape are usually more important than productivity. Vegetable breeding must include improving nutrition values because they are very important for human health. Historically, vegetable breeders have applied selection pressure to traits related to agronomic performance, particularly yield and quality, because these are the traits important to the producers. Growers have seldom been recompensed for nutritional factors, so there have been no economic incentives to provide significant attention to these traits. However, consumers are becoming more

* E-mail address: mirjsd@gmail.com.

aware of these traits, and selecting for nutrient-rich vegetable cultivars is becoming part of nutrition-oriented vegetable breeding programs. Successes in vegetable breeding for higher vitamin and mineral content should consider not only substance concentration but also organic components in plants that can either reduce or increase bioavailability. Enhanced nutritional content would add value for poor, malnourished populations.

Transgenic crops, commonly referred to as genetically modified (GM) crops, enable plant breeders to bring favorable genes, often previously inaccessible, into already elite cultivars, improving their value considerably and offering unique opportunities for controlling insects, viruses and other pathogens, as well as nutritional quality. Conventional plant breeding that utilizes non-transgenic approaches will remain the backbone of vegetable genetic improvement strategies. However, transgenic crop cultivars should not be excluded as products capable of contributing to more nutritious and healthy food. Many vegetable crops have been genetically modified to include improved features, such as higher nutritional status, better flavour, and reduced bitterness or can be used for vaccine delivery. Consumers could benefit further from eating more nutritious transgenic vegetables, e.g., an increase of crop carotenoids by metabolic sink manipulation through genetic engineering appears feasible in some vegetables. Genetically engineering carrots containing increased calcium (Ca) levels may boost Ca uptake, thereby reducing the incidence of Ca deficiencies such as osteoporosis. Fortified transgenic lettuce with zinc will overcome its deficiency that severely impairs organ function. Transgenic lettuce with improve tocopherol, and resveratrol composition may prevent coronary disease and arteriosclerosis and can contribute to cancer chemopreventative activity. Folates deficiency, which is regarded as a global health problem, can also be overcome with transgenic tomatoes with folate levels that provide a complete adult daily requirement. Food safety and health benefits can also be enhanced through transgenic approaches, e.g., resource-poor people in rural Asia and Africa will benefit eating cyanide-free cultivars of cassava. Biotechnology-derived vegetable crops will succeed if clear advantages and safety are demonstrated to both growers and consumers.

1. INTRODUCTION

Vegetables make up a major portion of the diet of humans in many parts of the world and play a significant role in human nutrition, especially as sources of phytonutriceuticals: vitamins (C, A, B1, B6, B9, E), minerals, dietary fiber and phytochemicals [1-4]. Some phytochemicals of vegetables are strong antioxidants and are thought to reduce the risk of chronic disease by protecting against free-radical damage, by modifying metabolic activation and detoxification of carcinogens, or even influencing processes that alter the course of tumor cells [2, 3, 5, 6]. Vegetables in the daily diet have been strongly associated with overall good health, improvement of gastrointestinal health and vision, reduced risk for some forms of cancer, heart disease, stroke, diabetes, anaemia, gastric ulcer, rheumatoid arthritis, and other chronic diseases [7-11].

Low vegetable intake, in unbalanced diets, has been estimated to cause about 31% of ischaemic heart disease and 11% of stroke worldwide. A high vegetable diet has been associated with lower risk of cardiovascular disease in humans [12]. Liu et al. [13] monitored the influence of vegetable intake on the incidence of cardiovascular disease among 15,220 male physicians without a history of heart disease, stroke, or cancer. Their results show that participants who consumed more than two servings of vegetables per day had 25% less cardiovascular disease than those who consumed less than one serving of vegetables per day. Based on this and other evidence, the American Heart Association [14] has concluded that a diet high in vegetables and fruits may reduce the risk of cardiovascular disease in humans. The International Agency for Research on Cancer (IARC) estimates that the preventable percentage of cancer due to unbalanced diets ranges from 5-12% for all cancers and 20-30% for upper gastrointestinal tract cancers. According to the 2007 World Health Report, unbalanced diets with low vegetable intake and low consumption of complex carbohydrates and dietary fiber are estimated to cause some 2.7 million deaths each year and were among the top ten risk factors contributing to mortality [15]. The American Cancer Society suggests that more than two-thirds of cancer deaths in the United States are avoidable and that a third of cancer deaths can be prevented by proper diet rich in vegetables [16, 17]. The role of balanced diet in the etiology of certain types of cancer such as colorectal cancer, which is the second leading cause of cancer deaths in the United States and in other developing countries, is gaining significant attention [17-20]. Numerous epidemiological studies conducted in the United States and in other developed countries in the last two decades have concluded that of the food groups evaluated, high vegetable intake (especially green and raw vegetables) yielded the most consistent association with decreased risk of colorectal neoplasia [18-23]. Evidence supporting high vegetable intake and decrease risk of colorectal cancer includes results from tests on adenomatous polyps, which are the precursors to colorectal cancer [20, 24-27]. Witte et al. [25] reported significantly lower incidence of colorectal polyps in men and women ages between 50 and 74 years old who consumed higher rates of vegetables including crucifers, garlic and tofu. Interestingly, this study concluded that vegetables have more beneficial effects against colorectal polyps than fruits or fiber from grains. Similarly, Steinmetz et al. [23], in a study with 41,837 women aged 55 to 69 years old, found a 20 to 40% reduction in risk of colon cancer in populations with higher vegetable consumption, especially garlic and vegetable dietary fiber. This association appears to be stronger for cancer of the distal colon than the proximal colon. Other studies have also estimated lower risk of colon cancer, ranging from three- to eightfold due to high vegetable and fruit intake [20, 28-30]. Because of their nutritional importance, vegetables have been recognized within the nutrition and medical communities, and there is a general sense that the promotion of a greater consumption of vegetables will improve health. The exact mechanisms by which vegetable consumption reduces human diseases have not yet been fully understood, however the general consensus among physicians and nutritionists is that phytonutriceuticals in vegetables are responsible for mitigating some of these diseases.

"Hidden hunger" or micronutrient deficiency is a pernicious problem around the world that is caused by a lack of vitamins and minerals such as vitamin A, iodine and iron in the human diet and affects the health of about three billion people worldwide [31, 32]. Under-consumption of vegetables and fruits is among the top ten risk factors leading to micronutrient malnutrition and is associated with the prevalence of chronic diseases [15, 33, 34]. More than 70% of malnourished children live in Asia. At least half of the preschool children and pregnant women are affected by micronutrient deficiencies in Bangladesh, Cambodia, Nepal and the Philippines [35]. Vegetables contain a range of macro- and micronutrients, including pro-vitamin A, iron and zinc, which contribute to the prevention of malnutrition disorders. Small variation in maternal diets, particularly reduction in micronutrient content, can have a significant impact on fetal growth and development. Pregnant women, in particular, benefit from good vegetable nutrition, particularly during later pregnancy and lactation. The interplay of the different micronutrients and antioxidants found in vegetables has important health impacts, explaining, for instance, the higher birth weight of children in India, when mothers consumed higher rates of green leafy vegetables and fruits during pregnancy [36]. Nutrition is both a quantity and a quality issue, and vegetables in all their many forms ensure an adequate intake of most vitamins and nutrients, dietary fibers, and phytochemicals, which can bring a much-needed measure of balance back to diets, contributing to solving many of these nutrition problems.

Vegetables are grown worldwide, on large and small farms, on good and marginal lands, in urban and rural areas, and by large commercial growers and small subsistence farmers. Short production cycle vegetables allow multiple cropping, and a significant volume of the vegetables grown worldwide are produced in small plots, which makes it difficult to estimate accurately their production, thereby preventing a clear understanding and appreciation of the value of these crops to the world food supply. According to FAO statistics, the production of vegetables in the world in 2009, was almost 1,010 million tons [37]. Asia produces and consumes more than 70% of the world's vegetables. The per capita consumption of vegetables in Asia has increased considerably from 41 kg to 141 kg between 1975 and 2003 [38], particularly in China, the largest world consumer, where the per capita consumption has increased from 43 kg (1975) to 154 kg (2003). The main reason for this increase was the rapid growth in mean per capita incomes and awareness of nutritional benefits.

A world vegetable survey showed that 402 vegetable crops are cultivated worldwide, representing 69 families and 230 genera [39,40]. Leafy vegetables, of which the leaves or young leafy shoots are consumed, are the most often utilized (53% of the total), followed by vegetable fruits (15%), while vegetables with below ground edible organs comprise 17%.

In this context, plant breeders play a key role in determining what we eat, since the cultivars they develop begin the dietary food chain. Every food product, including vegetables, that we see on the market has benefited from plant breeding. Historically vegetable breeders have applied selection pressure to traits related to agronomic performance, particularly yield and quality, because these are the traits important to the producers. Although vegetable breeding must include improving nutrition values because they are very important for human health and consumers are becoming more aware of these traits [4]. Nutritionists play a parallel role in determining nutritional requirements. Working together, plant breeders and nutritionists can play a key role in sustaining the supply of dietary nutrients in the vegetables we breed where intake is adequate and in developing strategies for increasing intake of those nutrients where intake is inadequate. Enhanced nutritional content would add value for poor, malnourished populations. Only 17% of commercial vegetables have attracted investments for crop breeding by multinational seed corporations, due to their large area of production and substantial consumption, 13% vegetables are considered minor, and other 22% species are considered rare [15]. In 2010, the global vegetable seed market was estimated at US$ 4.1 billion, of which 36% were for solanaceous, 21% for cucurbits, 13% for roots and bulbs, 12% for large seed, 11% for brassicas, and 7% for leafy and others vegetables [41]. Global commercial vegetable seed sales had an annual growth rate of 5.8% in the last decade [15]. With the increase in world population and consumption, and the advent of a high degree of added value through biotechnology, the global market of vegetable seeds is expected to expand in future years.

There is an increasing awareness among the general public of the advantages of diets rich in vegetables to ensure an adequate intake of most vitamins and micronutrients, dietary fibers, and phytochemicals that promote health. This chapter discusses the major classes of phytonutriceuticals in vegetables, their health benefits, and the importance of conventional and transgenic vegetable breeding as key for nutritional quality, increasing intake of nutrients and health benefits.

2. Major Classes of Phytonutriceuticals in Vegetables and Health Benefits

Phytonutriceuticals are all the chemical compounds derived from plants that have health-promoting properties. Most of them are found in relatively small quantities in vegetables. However, when consumed in sufficient quantities, they contribute significantly toward protecting living cells against chronic diseases. The most important phytonutriceuticals in vegetables that have biological activity against chronic diseases are: vitamins, minerals, dietary fiber, organosulfur compounds (glucosinolates and thiosulfides) and flavonoids.

2.1. Vitamins

The term vitamin is derived from the words vital and amine, because vitamins are required for life and were originally thought to be amines. Although not all vitamins are amines. They are any of various fat-soluble or water-soluble organic substances essential in minute amounts for normal growth and activity of the body and obtained naturally from plant and animal foods. A healthy body needs vitamins. All vegetables are sources of vitamins and consequently have health benefits. In this section, we will highlight and discuss only the vitamins with a high prevalence of inadequate intake in worldwide diets such as vitamins C, A, and E and folates.

2.1.1. Vitamin C

Vitamin C is an essential vitamin to humans that lack the ability to synthesize this vitamin because they are deficient in the enzyme L-gulonolactone oxidase, an enzime involved in the biosynthesis of vitamin C via the glucuronic acid pathway. The biological function of vitamin C is based on its ability to donate electrons, which provide intra- and extracellular reducing power for a variety of biochemical reactions.

The reducing power of vitamin C is capable of neutralizing most of the physiologically relevant reactive oxygen and nitrogen species in the human body [42]. Several enzymes involved in biosynthesis of collagen, carnitine and neurotransmiters require vitamin C as a cofactor [43].

Collagen is the most abundant protein in the body, and it forms into fibres, which provide strength and stability to all body tissues, including the arteries. Vitamin C serves also as cofactor for reactions that require reduced iron and/or copper metalloenzymes [43, 44]. It was also reported as an electron donor for eight enzymes involved in amino acid and hormone synthesis [45]. One of the most important indirect functions of vitamin C is its ability to regenerate other biologically important antioxidants, such as glutathione and vitamin E, into their reduced states [46-50].

As early as the 15th century, vitamin C was referred to as the "antiscorbutic factor," since it helped prevent the potentially fatal disease called scurvy. This disease was first discovered in Portuguese sailors, whose sea voyages left them far away from natural surroundings for long periods of time. Their body stores of vitamin C fell below 300 mg, and their gums and skin lost the protective effects of vitamin C. Recognizing lemon and other citrus as a good shipboard source of vitamin C, the Portuguese sailors started carrying large stores of lemons and other citrus aboard ship. Scurvy can be prevented with as little as 10 mg vitamin C per day, an amount easily obtained through consumption of vegetables [45]. Deficiency of vitamin C in humans has also been linked to reduced absorption of amino acids and lower levels of intestinal brush-border membrane proteins and phospholipids [51]. Cardiovascular diseases, several types of cancers, reduced cognitive function and memory, cataracts, risk of asthma and of the common cold, are all associated with vitamin C deficiency and can be partly prevented by optimal intake of vitamin C. Sing et al. [52] reported a twofold decrease in coronary heart disease in patients with higher plasma vitamin C compared to the lower group. Several other studies have also confirmed the association between higher plasma vitamin C concentration and lower risk of death from cardiovascular disease, ranging in benefit from 26 to 60% [53-55]. Vitamin C consumption has also been shown to reduce uterine cervical [56], colorectal [57, 58], pancreatic [59], lung [60], and gastric [61] cancers. The substantially high cellular levels of vitamin C provide antioxidant protection in the eye against photosynthetically generated free radicals avoiding cataracts [62] and protects against plasma and low-density lipoprotein oxidation [63, 64]. When vitamin C levels are low, the body also manufactures more cholesterol, especially low-density lipoprotein. Because of its ability to neutralize free radicals, it has been mentioned as having a possible therapeutic use in disorders such as ischemic stroke, Alzheimer's, Parkinson's and Huntington's diseases [65].

Near 90% of vitamin C in a typical human diet comes from vegetables and fruits (e.g., papaya, pineapple, orange, lemon, kiwifruit). Vitamin C is a water-soluble vitamin and is highly sensitive to water, air, and temperature. About 25% of the vitamin C in vegetables can be lost simply by blanching (boiling or steaming the food for a few minutes). This same degree of loss occurs in the freezing and unthawing of vegetables. Cooking of vegetables for longer periods of time (10-20 minutes) can result in a loss of over one-half the total vitamin C content. When vegetables are canned and then reheated, only one third of the original vitamin C content may be left. For these reasons, consumption of vitamin C-rich vegetables in their fresh, raw form is the best way to maximize vitamin C intake. Vitamin C has significant interactions with several key minerals in the body. Supplemental intake of vitamin C at gram-level doses can interfere with copper metabolism. Conversely, vitamin C can significantly enhance iron uptake and metabolism, even at food-level amounts. Vitamin C also has important interactions with other vitamins. Excessive intake of vitamin A, for example, is less toxic to the body when vitamin C is readily available. Vitamin C is involved in the regeneration of vitamin E, and these two vitamins appear to work together in their antioxidant effect.

2.1.2. Vitamin A

Only animal-based foods and fortified foods provide preformed vitamin A (retinol). Provitamin A carotenoids are supplied by plant-based foods and bioconverted to retinol in the body [66]. Vitamin A is a major public health problem in much of the developing world. It is estimated that in developed countries, almost 70% of vitamin A comes from animal sources, while 30% is derived from plant-based foods. In contrast, people in developing countries derive about 70% to 80% of vitamin A from plant-based foods. Vegetarians and populations with limited access to animal products depend on provitamin A carotenoids. Vitamin A deficiency affects approximately 25% of the developing world's preschoolers. It is associated with blindness, susceptibility to disease and higher mortality rates. It leads to the death of approximately one to three million children each year [67, 68].

The most abundant carotenoids in vegetables are α-carotene, β-carotene, lycopene, lutein, zeaxanthin, and β-cryptoxanthin [69]. These carotenoids account for more than 90% of the carotenoids present in the human diet. Carotenoids with molecules containing oxygen, such as lutein, zeaxanthin and β-cryptoxanthin, are known as xanthophylls. The unoxygenated carotenoids such as α-carotene, β-carotene, and lycopene are known as carotenes. The most common carotenes are β-carotene and lycopene. Lutein is the most abundant xanthophyll. Only three of the carotenoids (α-carotene, β-carotene, and β-cryptoxanthin) can be converted into the provitamin A (retinol), while lycopene, lutein, and zeaxanthin have no vitamin A activity. β-carotene is the carotenoid with the most provitamin A activity, it is the more common form, and it is also one of the most thoroughly studied. β-carotene can be found in yellow, orange, and green fruits, roots and leafy vegetables. As a rule of thumb, the greater the intensity of the orange colour of the fruit, root or leafy vegetable, the more β-carotene it contains. β-carotene can in fact be transformed to retinol, and this can in turn be transformed into retinal and retionic acid, both of which have different biological functions on many tissues [66]. β-carotene, as well as its other metabolites, affects important processes such as immunity, reproduction, growth, development and, perhaps its best known function, it holds a vital role in the visual cycle including possible inhibition of macular degeneration and cataracts [70, 71]. β-carotene has been characterized as an antioxidant and, like other carotenoids, has been shown to have inhibition mutagenesis activity contributing to decreased risk of some cancers [70]. As an antioxidant, it has been positively linked to diverse ailments related to oxidative stress, such as diabetes [72], obesity [73], low sperm motility [74], hearing loss [75] and others. All the carotenoids can act as antioxidants, although the antioxidant activity of carotenoids differs among the different compounds. Di Mascio et al. [76] reported that the singlet oxygen quenching ability of lycopene is twice that of β-carotene and ten times that of α-tocopherol. The unique nature of the lycopene molecule makes it a very potent antioxidant [77,78]. Carotenoids actions include also decreased risk of heart diseases. Gaziano et al. [79] in a study in the USA, conducted on 1,273 Massachusetts residents whose ages were greater than 65 years old, found that residents in the upper quartile of high-carotene vegetable intake had 46% lower risk of death from cardiovascular disease than those in the lowest quartile. In the eye, certain other carotenoids, as the xanthophylls lutein and zeaxanthin, apparently act directly to absorb damaging blue and near-ultraviolet light in order to protect the macula lutea [80]. Lutein and zeaxanthin reduce the risk of cataract and age-related macular degeneration [81, 82]. Vegetables that are rich in these xanthophylls can increase macular pigment density [80, 81]. Like other carotenoids, xanthophylls are found in highest quantity in the leaves of most green vegetables.

The bioavailability of β-carotene from vegetables in the human diet is limited. Studies have estimated that β-carotene bioavailability is 22 to 24% from broccoli, 19 to 34% from carrots, and 3 to 6% from leafy vegetables [83]. Several factors have been shown to influence carotenoids' bioavailability in humans, the most prominent of which is the food matrix.

Studies have shown that combination of fatty foods with carotenoid-rich vegetables enhanced carotenoid uptake. Roels et al. [84] reported a one- to fivefold increase in β-carotene uptake in boys deficient in vitamin A, when their diet was supplemented with olive oil. Lycopene was found to be more bioavailable from the thermally processed tomato products like tomato paste than from fresh tomato due to heat treatment and the presence of higher oil content in the paste [78, 85]. Lycopene exists in food primarily in the *trans* sterisomeric configuration; however, cooking and processing help convert *trans*-lycopene to *cis*-lycopene, which is more readily absorbed through the intestinal wall into the plasma [86]. Carotenoid availability is also influenced by the nature of the carotenoid. Lutein, which has no vitamin A activity, is five times more readily available in the human body than β-carotene [87].

2.1.3. Vitamin E

Vitamin E is a generic term for all tocopherols and their derivatives having biological activity. In nature, it is present under eight different forms, four tocopherols (α-, β-, γ-, δ) and four tocotrienols (α-, β-, γ-, δ) [88], of which α-tocopherol is the most predominant in nature and most bioactive form [88]. The α, β, γ, and δ isoforms of tocopherol have relative potencies of 100%, 50%, 10% and 3%, respectively [32]. Since tocopherols are of non-polar nature, their main function lies in their action in the hydrophobic environment of cell and organelle membranes protecting these structures from free radicals and also by stabilizing them [89, 90]. Tocopherols induce a protective effect against oxidative stress linked to metabolic syndrome as well as other sources [91-94]. They are also essential for normal neurological function [95]. As an antioxidant, α-tocopherol is able to prevent free-radical mediated tissue damage and thus to prevent or delay the development of degenerative and inflammatory diseases [96]; in such a role, it has been extensively investigated in many species, humans included. Compared with tocopherols, tocotrienols are sparsely studied [97-99]. Some studies have suggested that tocotrienols have specialized roles in protecting neurons from damage [99] and cholesterol reduction [100]. Tocotrienols are also thought to protect against stroke-associated brain damage *in vivo* [101].

Vitamin E deficiency is believed to be associated with the pathogenesis of cardiovascular disease including low-density lipoprotein oxidation, cytokine production, production of lipid mediators, platelet function, and smooth muscle cell proliferation, as well as interaction of the endothelium with immune and inflammatory cells [89, 96, 102-105]. Vitamin E is protective against nearly 80 cellular abnormalities [106], including cardiovascular disease [64, 89, 105, 107, 108], cancer [89, 91, 105], sterility [106], muscular dystrophy [89,109], changes in the central nervous system [110], and anemia development [106]. Since the vitamin is a liposoluble vitamin, it is found primarily in fats. The most important sources are plant based: oils and margarine, oleaginous fruits, germs of cereals. Fruits and vegetables are the second largest source of vitamin E. They do not contain high levels of vitamin E (only between 1 and 1.8 mg per 100 g for the richest sources) but consuming everyday between 100 to 200 g of vegetables can make them a significant source of Vitamin E. In a normal diet, between 12% and 18% of our vitamin E intake comes from fruits and vegetables. The current recommended dietary allowance for vitamin E intake from natural foods is about 10-15 mg/day [111, 112].

2.1.4. Folates

The terms folates (folic acid and tetrahydrofolate) are often used interchangeably for the water-soluble B-complex vitamins. Folates are involved as cofactors in carbon transfer reactions in DNA byosynthesis and the methylation cycle [113]. They are an absolute requirement for methionine, purine, and thymidylate synthesis. Tetrahydrofolate is also involved in the synthesis of S-adenosylmethionine, the universal methyl donor in all living cells, and in the control of glycine-to serine conversion [113, 114]. Ames [114] reported that folate deficiency causes chromosome breakage as a result of extensive incorporation of uracil into the human DNA. The human body needs folates to synthesize DNA, repair DNA, and methylate DNA as well as to act as a cofactor in biological reactions involving folates [115]. It is especially important in aiding rapid cell division and growth such as in infancy and pregnancy. Children and adults require folate to produce healthy red blood cells and prevent anaemia. Folate deficiency, which is regarded as a global health problem mainly in developing countries, can result in many health problems. The most notable ones are neural tube defects in developing embryos and abnormal fetal growth and risks during pregnancy. Neural tube defects produce malformations of the spine, skull, and brain including spina bifida and anencephaly. Epidemiological studies have shown that folate deficiency contributes to an accumulation of homocysteine leading to neural tube defects such as spina bifida in fetuses [116]. The risk of neural tube defects is significantly reduced when supplemental folic acid is consumed in addition to a healthy vegetable diet prior to and during the first month following conception [117, 118]. Folate deficiency during pregnancy may also increase the risk of preterm delivery, infant low birth weight and fetal growth retardation, as well as increasing homocysteine level in the blood, which may lead to spontaneous abortion and pregnancy complications, such as placental abruption and preeclampsia [119]. Women who could become pregnant are advised to eat foods fortified with folic acid or take supplements in addition to eating folate-rich vegetable foods to reduce the risk of serious birth defects [120-122].

Folate deficiency is also linked to colon cancer risk [123, 124], neurotoxicity [125], and heart attacks [126-129]. Most research studies associate high dietary folate intake with a reduced risk of prostate cancer [130, 131]. However, folate has shown to play a dual role in cancer development: low folate intake protects against early carcinogenesis, and high folate intake promotes advanced carcinogenesis [132]. Therefore, public health recommendations should be careful not to encourage too much folate intake mainly as folic acid supplements. Vegetables such as broccoli, Brussels sprouts, and potato were reported to contribute between 35 to 40% of the total intake of folate in the human diet [113].

2.2. Minerals

Minerals are inorganic substances that are originally found in rocks and soil. The vegetables we eat absorbed plenty of minerals as they were growing. When we eat those vegetables, we absorb the minerals as we digest food. All vegetables are sources of minerals and consequently have health benefits. A healthy body needs plenty of minerals. Minerals can be divided into two groups. There are some that we need only in tiny quantities (trace) and others that we need much more of (major). Major minerals, macronutrients, with a high prevalence of inadequate intake in worldwide diets include calcium, magnesium, and potassium. Trace minerals, micronutrients, identified as nutrients with great health implications and benefits, usually with a high prevalence of inadequate intake, include iron, zinc and selenium.

Calcium is an important component of a healthy diet and a mineral necessary for life. It is essential for strong bones and teeth. Calcium plays an important role in building stronger, denser bones early in life and keeping bones strong and healthy later in life [133]. Approximately 99% of the body's calcium is stored in the bones and teeth. The rest of the calcium in the body has other important uses, such as some exocytosis, especially neurotransmitter release, and muscle contraction. In the electrical conduction system of the heart, calcium replaces sodium as the mineral that depolarizes the cell, proliferating the action potential. Calcium eases insomnia and helps regulate the passage of nutrients through cell walls. If we don't get enough calcium from the food we eat, our bodies automatically take the calcium needed from our bones. If our bodies continue to tear down more bone than they replace over a period of years in order to get sufficient calcium, our bones will become weak and break easily. Calcium deficiency may result in muscle spasms and cramps in the short term and osteoporosis. Without calcium, our muscles wouldn't contract correctly, our blood wouldn't clot and our nerves wouldn't carry messages. While a lifelong deficit can affect bone and tooth formation, over-retention can cause hypercalcemia, impaired kidney function and decreased absorption of other minerals [134]. High calcium intakes or high calcium absorption were previously thought to contribute to the development of kidney stones. However, a high calcium intake has been associated with a lower risk for kidney stones in more recent research [135, 136]. Vitamin D is needed to absorb calcium. Absorption of calcium may be poor when consumed with foods rich in phytic acid (e.g., unleavened bread, raw beans, fibre rich foods, etc.) or oxalic acid (e.g., New Zealand spinach, spinach, Swiss chard, beet greens) [137]. Phytic acid (inositol hexaphosphate, Ins P_6) is the storage form of phosphorus. It contains six phosphate groups, thereby it can act as metal chelator of important minerals such as calcium, magnesium, iron, zinc and can contribute to mineral deficiencies in people. Salt of phytic acid with minerals or protein complexes are known as phytates. Compared to calcium absorption from milk, calcium from prepared dried beans and spinach is reduced by about 50% and 90%, respectively [137, 138]. Recent studies demonstrate that this "antinutrient" effect of phytic acid is only manifested when large quantities of phytic acid are consumed in combination with an micronutrient-poor diet [139].

Though most people think immediately of dairy products when it comes to calcium, there are plenty of vegetable good sources, too (e.g., kales, tronchuda cabbages, pak-choy, kailan, rutabaga, turnip, chard, rhubarb, okra, onions, amaranth, Brussels sprouts, squash, celery, parsnip, beans, etc.). For example, Galega kale contains a high content of protein, fiber, calcium, and sulfur when compared to broccoli, the reference within *Brassica* vegetables. For many years, Galega kale, based on the levels of protein and calcium, was assumed to be a substitute for milk in the poorest rural areas of Portugal where the peasants could not afford to buy milk or even raise cows [140, 141].

Magnesium is the eleventh most abundant element by mass in the human body. Its ions are essential to all living cells, where they play a major role in manipulating important biological polyphosphate compounds like ATP, DNA, and RNA [142]. Hundreds of enzymes thus require magnesium ions to function. ATP exists in cells normally as a chelate of ATP and a magnesium ion. Magnesium helps our body to release energy from food. Magnesium is needed for bone, protein, making new cells, activating B vitamins, relaxing nerves and muscles, clotting blood, and in energy production [143]. Insulin secretion and function also require magnesium [144, 145]. Magnesium also assists in the absorption of calcium, vitamin C and potassium. Deficiency may result in fatigue, nervousness, insomnia, heart problems, high blood pressure, osteoporosis, muscle weakness and cramps [146-148]. Low levels of magnesium in the body have been associated with the development of a number of human illnesses such as asthma, diabetes, and osteoporosis [149-151]. All the green leafy vegetables are good sources of magnesium as they contain chlorophyll.

Potassium is the eighth most common element by mass (0.2%) in the human body, so that a 60 kg adult contains a total of about 120 grams of potassium [152]. Potassium ions are necessary for the function of all living cells. It is essential for the body's growth and maintenance. Potassium cations are important in neuron (brain and nerve) function and in influencing osmotic balance between cells and the interstitial fluid, with their distribution mediated in all animals by the so-called Na+/K+-ATPase pump [142, 153]. Potassium ion diffusion is a key mechanism in nerve transmission and is important in preventing muscle contraction. Its depletion in humans results in various cardiac dysfunctions. Potassium plays an essential role in proper heart function [154]. A shortage of potassium in body fluids may cause a potentially fatal condition known as hypokalemia, typically resulting from vomiting, diarrhea, and/or increased diuresis [155]. Deficiency may cause muscular cramps, twitching and muscle weakness, cardiac arrhythmia, insomnia, kidney failure and decreased lung reflex response and, in severe cases, respiratory paralysis. Epidemiological studies and studies in animals subject to hypertension indicate that diets high in potassium can reduce the risk of hypertension and possibly stroke by a mechanism independent of blood pressure [154]. Potassium is found in especially high concentrations within plant cells, and in a mixed diet, it is mostly concentrated in fruits and vegetables [42].

Iron plays an important role in human health, forming complexes with molecular oxygen in hemoglobin and myoglobin [142]. These two compounds are common oxygen transport proteins in vertebrates. Iron is also the metal used at the active site of many important redox enzymes dealing with cellular respiration and oxidation and reduction in plants and animals [142]. Too little iron in human diet can lead to anaemia. Iron is found in food in two forms, heme and non-heme iron. Heme iron, which makes up 40% of the iron in meat, poultry, and fish, is well absorbed (has higher bioavailability). Non-heme iron, 60% of the iron in animal tissue and all the iron in plants (fruits, vegetables, grains, nuts), is less well absorbed. Vegan diets only contain non-heme iron. Because of this, iron recommendations are higher for vegetarians (including vegans who consume only plant foods) than for non-vegetarians. Red meat is the most obvious source of iron, but there are also good vegetable sources. Dried beans, lentils, peas, as well as all the other legumes, and dark green leafy vegetables are especially good sources of iron, even better on a per calorie basis than meat.

Iron deficiency is the most prevalent form of malnutrition worldwide, affecting an estimated two billion people. Eradicating iron deficiency can improve national productivity levels by as much as 20% [67, 156]. Most at risk of iron deficiency are infants, adolescent girls and pregnant women. Pregnant women are at special risk of low iron levels and are often advised to supplement their iron intake [157]. Iron deficiency in infants can result in impaired learning ability and behavioral problems [68]. It can also affect the immune system and cause weakness and fatigue. Vitamin C promotes the absorption of non-heme iron. It is necessary to eat foods rich in vitamin C at the same time we eat the food containing iron. Foods rich in phytic acid limit the absorption of non-heme iron [137]. The recommended dietary allowance for iron varies considerably based on age, gender, and source of dietary iron [158]. Women and teenage girls need at least 15 mg a day, whereas men can get by on 10 mg. It is important that children get about 10 to 12 mg of iron per day, preferably from their diets [158, 159].

Vegetarians need to get twice as much dietary iron as meat eaters. Surveys of vegans [160, 161] have found that iron deficiency anaemia is no more common among vegetarians than among the general population, although vegans tend to have lower iron stores [161]. The reason for the satisfactory iron status of many vegans may be that commonly eaten foods are higher in iron than animal-derived foods. For example, we would have to eat more than 1,700 calories of sirloin steak to get the same amount of iron as found in 100 calories of spinach. Another reason for the satisfactory iron status of vegans is that vegetarian diets are high in vitamin C. Vitamin C acts to markedly increase absorption of non-heme iron. Adding a vitamin C source to a meal increases non-heme iron absorption up to six fold, which makes the absorption of non-heme iron as good as or better than that of heme iron [162]. Fortunately, many vegetables, such as broccoli and pak-choy, which are high in iron, are also high in vitamin C so that the iron in these foods is very well absorbed. Commonly vegan-eaten combinations, such as beans or lentils and tomato sauce or stir-fried tofu and broccoli or pak-choy, also result in generous levels of iron absorption. Both calcium and tannins reduce iron absorption [163].

Zinc: Zinc is an essential mineral of exceptional biologic and public health importance [164]. Zinc is found in nearly 100 specific enzymes, serves as structural ions in transcription factors and is stored and transferred in metallothioneins [142]. It is the second most abundant transition metal in organisms after iron, and it is the only metal that appears in all enzyme classes [165]. In proteins, zinc ions are often coordinated to the amino acid side chains of aspartic acid, glutamic acid, cysteine and histidine [142, 166]. Zinc deficiency affects about two billion people in the developing world and is associated with many diseases [167]. In children, it causes growth retardation, delayed sexual maturation, infection susceptibility, and diarrhea, contributing to the death of about 800,000 children worldwide per year [164]. The World Health Organization advocates zinc supplementation for severe malnutrition and diarrhea [168]. Zinc supplements help prevent disease and reduce mortality, especially among children with low birth weight or stunted growth [168]. However, zinc supplements should not be administered alone, since many in the developing world have several deficiencies, and zinc interacts with other micronutrients [169].

The concentration of zinc in vegetables and plants varies based on levels of the element in soil. Zinc deficiency is crop plants' most common micronutrient deficiency. It is particularly common in high-pH soils. Plants that grow in soils that are zinc-deficient are more susceptible to disease. When there is adequate zinc in the soil, the plant sources that contain the most zinc are wheat (germ and bran) and various seeds (sesame, poppy, alfalfa, celery, mustard) [170]. Zinc is also found in vegetables such as beans, peas, pumpkin seeds, sea vegetables, and in nuts, almonds, whole grains, sunflower seeds, soy foods and black currant [112]. However, phytic acid/phytates in many whole-grains and dietary fiber in many foods may interfere with zinc absorption, and marginal zinc intake has poorly understood effects [137].Vegetarians need about 50% more zinc in their diets than meat-eaters. Despite some concerns, Western vegetarians and vegans have not been found to suffer from overt zinc deficiencies any more than meat-eaters [171].

Selenium is an essential trace mineral in humans. Its importance is underlined by the fact that it is the only trace mineral to be specified in the genetic code—as selenocysteine. Selenium is a key component of several functional selenoproteins (e.g., glutathione peroxidases, thioredoxin reductases, iodothyronine deiodinases and selenoprotein P) that protect tissues and membranes from oxidative stress and controls the cell redox status [172].

The antioxidant properties of selenoproteins help prevent cellular damage from free radicals. Free radicals are natural by-products of oxygen metabolism that may contribute to the development of chronic diseases such as cancer and heart disease. Other selenoproteins help regulate thyroid function and play a role in the immune system [172]. Selenium is found in human hair and nails and in the blood. Most selenium in animal and plant tissue is present in the unusual amino acids selenocysteine and selenomethionine. Animal tissues, however, are not capable of synthesizing selenomethionine; instead they obtain it from plant tissue. Plant-based selenomethionine accounts for nearly half of the daily requirement of selenium in humans, while the balance is obtained from water (as selenate and selenite), milk, fish, and animal-based products [173]. Selenocysteine and selenomethionine are readily absorbed by human tissue [174].

Several studies have suggested an increase risk of cancer with low selenium diet [175-180]. Helzlsouer et al. [181] indicated that women in the higher group of serum-selenium levels content were four times likely to develop ovarian cancer than women in the lower group. In another study, Yoshizawa et al. [182] reported a 33% reduction in prostate cancer in men receiving a 200 μg selenium per day supplement compared to the control. Clark et al. [183] conducted a study to test the effect of selenium supplementation on the recurrence of skin cancers on selenium-deficient men. They found that men receiving 200 μg selenium per day were 50% less likely to develop or die from carcinoma of skin than those receiving a placebo. This study did not demonstrate a reduced rate of recurrence of skin cancers but did show a reduced occurrence of total cancers, particularly for lung, colorectal and prostate cancers. Dietary selenium prevents chemically induced carcinogenesis in many rodent studies. It has been proposed that selenium may help prevent cancer by acting as an antioxidant or by enhancing immune activity. Generally, selenium functions as an antioxidant that works in conjunction with vitamin E. Tsavachidou et al. [184] report that vitamin E (400 IU) and selenium (200 μg) supplements affect gene expression and can act as a tumor suppressor in prostate cancer.

Selenium deficiency in human and animal models has also been linked to cardiovascular diseases [185] and rheumatic arthritis [186]. Keshan disease, a disease of the heart, and Kashin-Beck disease, a disease of the cartilage, have been also linked to selenium deficiency that occurs almost exclusively in children and adolescents [187, 188]. These two diseases occur mostly in Southeast Asia in areas with severe selenium deficiency. Although Moreno-Reyes et al. [189] stated that the effect of selenium deficiency on health remains uncertain, particularly in relation to Kashin-Beck disease. Selenium may inhibit Hashimotos's disease, in which the body's own thyroid cells are attacked as alien. A reduction of 21% on thyroid peroxidase antibodies was reported with the dietary intake of 0.2 mg of selenium [190]. Evidence from *in vivo* and *in vitro* studies suggests that selenium could enhance insulin sensitivity by mediating insulin-like actions [191-193]. Akbaraly et al. [194] in recent studies indicated that selenium may help inhibit the development of type 2 diabetes (diabetes at an advanced age) in men. The mechanism by which selenium may protect against dysglycemia in this study is unclear but may involve the selenoenzymes.

2.3. Dietary Fiber

Dietary fiber is a heterogeneous mixture of polysaccharide carbohydrates, such as cellulose, hemicellulose, β-glucans, pectins, mucilages and gums, plus noncarbohydrates, such as lignin, which are indigestible in the small intestine [195]. Dietary fiber is classified into water-soluble and water-insoluble fiber. Water-insoluble fiber is mainly cell wall components, such as cellulose, lignin, and hemicellulose found in the vegetables. Health benefits of insoluble fiber include shortening of the bowel transit time and bulkier and softer feces. Non-cellulosic polysaccharides, such as pectin, gums, and mucilages, are components of soluble fiber. Soluble fiber delays gastric emptying, slows glucose absorption, and lowers serum cholesterol levels, and is completely or partially fermentable in the large intestine [196]. Fermentation produces short-chain fatty acids that have many health benefits related to blood glucose levels, cholesterol synthesis, immune function, and colon health.

Dietary fiber is a major constituent of vegetables. Anderson and Bridges [197] evaluated the dietary fiber content of several food groups. They found that vegetables contain about 32% dietary fiber, based on dry weight, which they estimated to be equivalent to the amount found in many cereal products. Although vegetables possess a higher water-soluble/insoluble fiber ratio than cereals, which is considered as an indicator of good nutritional quality [198]. Pectin constitutes nearly 30% of the dietary content in vegetables [199]. Physiological and physicochemical effects of dietary fibers depend on the relative amount of individual non-digestible components [198].

The importance of dietary fiber in nutrition and its health benefits are widely recognized. Numerous clinical and epidemiological studies have addressed the role of dietary fibre in intestinal health, prevention of cardiovascular disease, some types of cancer, peptic ulcer, obesity, and diabetes [198]. Dietary fiber has many functions in diet, one of which may be to aid in energy intake control and reduced risk for development of obesity. The role of dietary fiber in energy intake regulation and obesity development is related to its unique physical and chemical properties that aid in early signals of satiation and enhanced or prolonged signals of satiety [200]. Early signals of satiation may be induced through cephalic- and gastric-phase responses related to the bulking effects of dietary fiber on energy density and palatability, whereas the viscosity-producing effects of certain fibers may enhance satiety through intestinal-phase events related to modified gastrointestinal function and subsequent delay in fat absorption. There is a strong relationship between dietary fiber intake and lower risk of type 2 diabetes. Several mechanisms have been proposed for this inhibition, including improved insulin sensitivity and/or decreased insulin requirement [201]. The mechanisms involved in cardiovascular disease are: lower blood pressure [202] and decrease total serum cholesterol and low-density lipoprotein concentrations [196]. Other effects of dietary fiber include increased stool-bulking, which dilutes chemical carcinogens, and increased production of the anticarcinogen butyric acid through fermentation of dietary fiber by the colon microflora. Le Marchand et al. [203] conducted a study among different ethnic groups and genders in Hawaii to evaluate the role of dietary fiber on the risk of colorectal cancer. Their results suggested a strong inverse association between vegetable dietary fiber (soluble and insoluble) and colorectal cancer. In spite of many years of research and nutrition education about potential benefits of dietary fiber to help prevent diseases such as cancer, heart disease, and diabetes, the consumption of complex carbohydrates and dietary fiber worldwide has decreased [204]. The recommended daily intake of dietary fibre is 25-35 g/person [198].

Dietary fiber and other phytonutriceuticals are generally addressed separately as groups of food constituents in nutritional studies. However, dietary fibre of vegetables transports a significant amount of polyphenols and carotenoids linked to the fibre matrix through the human gut [205, 206]. Therefore, associated phytochemicals can make a significant contribution to the health benefits attributed to the dietary fibre of vegetables. Moreover, phytochemicals may be considered dietary fibre constituents in view of the similarity of their properties in terms of resistance to digestive enzymes and colonic fermentability [205, 206]. In the Spanish diet, considered as Mediterranean pattern diet, fruit and vegetables provide a daily intake of 700-1,000 mg of polyphenols/person/diet; a major fraction of this (600 mg/person/day) is associated with dietary fiber [206]. These issues constitute an important qualitative difference in relation to other dietary patterns.

2.4. Organosulfur Compounds

The two major organosulfur compounds in vegetables that have been found to have biological activity against important human diseases are glucosinolates and thiosulfites.

2.4.1. Glucosinolates

Glucosinolates, or β-D-thioglucosides, are sulfur-containing secondary metabolites. Briefly, the glucosinolate molecule comprises a skeletal β-thioglucose moiety, a sulfonated oxime moiety (glucone), and an aglucone R-group that defines the structure of each glucosinolate. More than 100 glucosinolates have been identified from 16 plant families of dicotyledonous plants [207, 208]. Of the nearly 120 glucosinolates that have been identified in plants thus far, about 20 have been detected in crucifers, and of those, only three or four are present in significant amounts [208]. Based on their R-group structure, glucosinolates have been classified into aliphatic, aromatic, and indolyl. The mostly present glucosinolates in crucifers are the aliphatic, followed by the indoles, and then the aromatic. The aliphatic glucosinolates include glucoraphanin, glucoerucin, progoitrin, epi-progoitrin, sinigrin, napoleiferin, gluconapin, and glucoalysin; the indoles include glucobrassicin and 4-hydroxyglucobrassicin, 4-methoxyglucobrassicin, and glucobrassicin; and the aromatics include gluconasturtiin and sinalbin.

Many compounds occurring naturally in the human diet moderate the biotransformation of several carcinogens, resulting in reduced tumour incidence [209]. Glucosinolates and their hydrolytic products modulate the activity of xenobiotic metabolising phase I and II enzymes. Phase I enzymes responsible for bioactivation of carcinogens generally increase the reactivity of lipophilic compounds. On the other hand, phase II detoxification enzymes increase the water solubility and facilitate removal of these metabolites from the body. For the protection of cells against DNA damage by carcinogens and reactive oxygen species, inhibition of phase I and induction of phase II enzymes are required. The genes for the phase II enzyme contain a specific sequence of DNA called antioxidant response element. Activities of phase II enzyme have been reported to be enhanced by glucosinolates and their hydrolytic products [210].

Several studies have suggested that intact glucosinolates have no biological activity against cancer; however their breakdown products have been shown to stimulate mixed-function oxidases involved in detoxification of carcinogens, reducing the risk of certain cancers in humans [208, 211, 212]. In crucifers, when plant tissues are cut, chewed or otherwise damaged, glucosinolates are hydrolysed by the enzymatic action of myrosinase [thioglucoside glucohydrolase, EC 3.2.3.1[213]] releasing biologically active products including isothiocyanates, thiocyanates, indoles and nitriles [214]. However, when these tissues are cooked or longer steamed, the enzyme myrosinase is completely inactivated. It has been suggested that myrosinase activity produced by the stomach microflora is capable of breaking down glucosinolates into active products [215]. While intact glucosinolates have no biological activity against chronic diseases, certain myrosinase catalysed-breakdown products of glucosinolates have been shown to reduce chemically induced carcinogens. However, not all glucosinolate breakdown products have anticancer activity [216]. The glucosinolates glucoraphanin, glucoiberin, glucobrassicin and gluconasturtiin are involved in the anticarcinogenic activity, and glucoraphanin is known to bolster the defences of cells against carcinogens through an up-regulation of enzymes of carcinogen defence (phase II enzymes).

Isothiocyanates are an important group of breakdown products of glucosinolates and appear to act at a number of points in the tumour development by blocking the metabolism of carcinogenic compounds through biotransformation. They generally enhanced the activity of phase II enzymes and inhibited phase I enzymes [217, 218], thus reducing the carcinogenic activity and enhance the detoxification and clearance of carcinogens. Further, they serve as suppressors during the promotion phase of neo-plastic process. Induction of apoptosis and action of signal transduction pathways within the cell activities of glucosinolates has also been reported [219]. Watercress and broccoli are reported to be rich sources of phenethyl isothiocyanate, which may block the cytochrome P450-mediated metabolic activation of the common nitrosoamine to its potent carcinogenic forms [220]. Extracts of watercress and broccoli suppressed metalloproteinase-9, an enzyme closely associated with invasive potential of breast cancer [217]. It also suppressed production of proinflammatory compounds such as nitric oxide (NO) and prostaglandins [221]. Phenethyl isothiocyanate also inhibited induction of lung and oesophagus cancer in both rat and mouse tumours [222-224].

Sulpharophane (4-methylulfinylbutyl isothiocyanate) and indole-3-carbinol are the two most widely studied glucosinolate breakdown products exhibiting anticarcinogenic properties [225-227]. Sulforaphane, enzymatic hydrolysis degradation product of glucosinolate glucoraphanin [228], activates gene expression, thereby helping to clear carcinogenic substances from the body. Sulforaphane also increased levels of mammalian phase II enzymes through antioxidant response element mediated transcriptional activation [225, 226, 229-231]. Sulforaphane supported a healthy immune system by significantly enhancing the production of chemicals involved in immune response [232]. In a study in which animals were genetically bred to develop intestinal polyps, a condition that led to tumour formation, the group of animals that were fed with sulphoraphane had higher rates of apoptosis (cell suicide) and smaller tumour growing more slowly than animals not receiving sulphoraphane [233]. Indole-3-carbinol is produced from indole-3-glucosinolates like glucobrassicin through hydrolysis [234]. Under acidic conditions, indole-3-carbinol and elemental sulphur are formed. Anticarcinogenic, antioxidant and antiatherogenic activities of indole-3-carbinol have been reported [234]. Further, indole-3-carbinol modulates the activities of both phase I and II enzymes. It suppressed cancer growth and induced programmed cell deaths in tumours of breast, prostate, leukaemia, cervix and colon because of its ability to favourably influence the human body's balance of estrogens [235-238]. Indole-3-carbinol also inhibited cancer cell growth by interfering the production of proteins involved in abnormal cellular reproduction and by promoting the production of tumour suppressor proteins [238, 239]. Indole-3-carbinol has also been reported to prevent cancer by interfering with angiogenesis, process of formation of new blood vessels that tumours require for their survival and spread [240].

Glucosinolates levels are affected by methods of food preparation. Various studies have shown large effects of cooking *Brassica* vegetables, mostly resulting in substantial losses by leaching of glucosinolates into the cooking water since glucosinolates are water soluble [214, 241-243]. Goodrich et al. [241] and Rosa and Heaney [242] reported 40 to 80% leaching of glucosinolates from cabbage leaves, Brussels sprouts, and broccoli heads into the cooking water. Leaching of glucosinolates was significantly lower for Brussels sprouts than broccoli or cabbage, possibly due to the compactness of the sprout. Vallejo et al. [243] in broccoli showed large differences in glucosinolate levels among four cooking processes. Conventional and microwave cooking caused substantial losses of total glucosinolates, while steaming had minimal effects on glucosinolates [214, 243]. On the other hand, Verkerk and Dekker [244] demonstrated high retention of glucosinolates after microwave cooking of red cabbage, even up to levels exceeding the total glucosinolate content of the untreated vegetables. The authors ascribed these higher levels to increased extractability of the glucosinolates [244]. Verkerk et al. [214] also observed that steaming resulted in high retention of total aliphatic and indolyl glucosinolates in cooked broccoli and that only after extensive steaming of broccoli (30 min) substantial losses of total indolyl glucosinolates of 55% and total aliphatic glucosinolates of 8.5% were observed. These authors also observed that steaming broccoli for more than six min results in complete inactivation of the hydrolic enzyme myronase. However, steaming of broccoli for less than six min may result in a high intake of glucosinolates, in the presence of a residual active myronase, allowing the release of health-protective breakdown products of glucosinolates after consumption. Therefore, steaming for less than six min produces broccoli with high levels of glucosinolates and yet, residual active myronase enabling a more efficient conversion of glucosinolates during mastication into the health-protective isothiocyanates.

2.4.2. Thiosulfides

Thiosulfides are organosulfur compounds (OSCs) found in alliums [245]. The composition of OSCs differs depending on the *Allium* specie [246], plant cultivation and processing methods [247]. Some thiosulfides are absent in the bulbs and require mechanical exposure like cutting or chewing to be formed. According to Verma and Verma [247], whole garlic bulbs contain 16 OSCs versus 23 OSCs after crushing. In another study, Kalra et al. [248] reported 33 OSCs in fresh garlic. S-allyl-cysteine sulfoxide (alliin), S-methyl-cystein sulfoxide (methiin), and γ-glutamylcysteine are reported as the major thiosulfides in the cytoplasm of intact tissues of alliums [247, 249]. Other minor thiosulfides include S-propenil-cystein sulfoxide (isoalliin) and S-ethyl-cystein, commonly known as ethiin [247, 250]. None of the thiosulfides found in alliums have been detected in other vegetables, except methiin, which was detected in cabbage leaves and some *Brassicaceae* [251]. Several surveys of the thiosulfide content of alliums have been conducted in the last years [247, 248, 252-255]. The pathways for biosynthesis of thiosulfides in alliums have not been very clearly defined because of their diversity and instability. For example, alliin, the most abundant thiosulfide in intact garlic, is highly unstable. When garlic tissue is cut, chewed, or dehydrated, cytosolic alliin is rapidly lysed by the vacuolar enzyme alliinase or alliin lyase (E alliin alkyl-sulphenate-lyase, C 4.4.1.4) into a highly unstable diallyl thiosulfinate transient intermediate called allicin [256, 257]. Allicin then converts into several lipid-soluble alkyl alkane-thiosulfinates, including allyl sulfide, diallyl disulfide, diallyl trisulfide, allylmethyl disulfide, 2- and 3-vinyl dithiins, thioacroleines, and ajoene [258-261]. These lipid soluble compounds form the odiferous and aromatic characteristic of garlic. The type and amount of the thiosulfides in alliums varied considerably, depending on the level of alliinase activity and the nature of the substrate. The thiosulfides detected in onions are S-propenyl-cystein sulfoxide, S-methil-cysteine sulfoxide, and S-propyl-cystein sulfoxide [249, 251, 262]. Variations in the ratios of these volatile sulfur compounds are responsible for the difference in flavors and odors between *Allium* species [256, 263]. More than 80 volatile compounds of this class have been identified fresh or cooked vegetable alliums [264]. They are also major contributors to the bitter taste of some onions [263].

Consumption of *Allium* vegetables has been found to retard growth of several types of cancers. It has generally been accepted that the *in vivo* and *in vitro* anti-proliferative effect of *Allium* vegetables results from the breakdown of alliin into organosulfur compounds [265-269]. Knowles and Milner [270] listed seven allyl sulfides (alicin, diallyl sulfide, diallyl disulfide, diallyl trisulfide, ajoene, S-allyl cysteine, and S-allylmercaptocysteine) that have been reported to inhibit human tumor growth including prostate, colon, skin, breast, lung, lymphoma, erythroleukemia, and lymphocyte tumors. Examination of the mechanism of cancer prevention by organosulfur compounds indicates that they function as blocking agents. Blocking agents act during the initiation stage of carcinogenesis by either inhibiting the activation of procarcinogens, trapping reactive species, or by enhancing phase II enzymes. In addition to enhancing the activity of phase II detoxification enzymes, allyl sulfides also inhibit phase I enzymes, such as cytochrome P450 2E1, which have been shown to activate chemical carcinogens and promote apoptosis [271-273]. However, there are significant differences in the ability of allyl sulfides to inhibit tumor cell growth. For example, Knowles and Milner [270] reported that diallyl disulfides were twice as effective as S-allylmercaptocysteine in inhibiting human colon tumor growth. Diallyl trisulfide was also reported to be at least 2.5 times more effective in controlling lung tumor growth than diallyl disulfide [274-275]. Cancers of the digestive tract and prostate appear to be the most sensitive ones to the chemopreventive effects of *Allium* vegetables intake, which may reflect a problem of thiosulfides stability. Tissues/organs, which are in inner connection with the digestive/urinary systems, are obviously exposed to higher concentrations of biologically active products deriving from the degradation/catabolism of any ingested food. This encourages efforts aimed to increasing the stability of these compounds (without increasing also their toxicity). The conjugation of molecules specifically recognizing and binding to cancer cells together with the enzyme alliinase could generate high concentrations of alliin right at the tumor site and may represent an interesting attempt to overcome this limiting factor [276].

Another important health benefit of allyl sulfides is their ability to protect humans against cardiovascular diseases. Cardiovascular diseases are controlled by a multitude of factors including high serum cholesterol, high levels of low-density lipoproteins, high blood pressure, and platelet aggregation. Aggregation of platelets in the blood increases the risk of thrombosis and, as a consequence, the risk of cardiac attack. Liu and Yeh [277] examined the effect of eleven water and lipide soluble organosulfur compounds from garlic on the incorporation of [2-^{14}C]-acetate into cholesterol in rat hepatocyte cells. They found that seven of the allyl sulfides tested inhibited cholesterol syntesis by as much as 55% and that higher concentrations of diallyl disulfide, diallyl trisulfide, and dipropyl disulfide resulted in complete inhibition of cholesterol syntesis. Ide and Lau [278] reported that S-allylcyteine and other thiosulfides prevent low-density lipoprotein oxidation in vascular endothelial cells. Organosulfur compounds from garlic and onion have been suggested to reduce platelet aggregation, thereby lowering the risk of heart attack and/or stroke [279,280]. Lawson et al. [281] demonstrated that diallyl disulfide, diallyl trisulfide, and ajoene, from garlic, have *in vitro* antiplatelet activity. Similarly, Morimitsu et al. [282] and Ali et al. [283] reported that organosulfur compounds found in onions, mostly cepaenes, also have antiplatelet activity. A preliminary study conducted this time on humans showed that the consumption of the equivalent of three onions in a soup was sufficient to significantly reduce the blood platelet aggregation [284]. This activity appears to be less important after cooking [285] and more important in pungent onions [286]. It is attributed to quercetin and alkyl-propenyl cysteine sulphoxide molecules, but the exact mode of action remains elusive. It is suggested that these compounds stimulate the release of arachidonic acid from membrane phospholipids, which initiates eicosanoid metabolism in mammals leading to the inhibition of thromboxane A synthesis and a significant reduction in platelet aggregation and vasoconstriction [284].

In addition to inhibiting the two leading causes of death in humans, cancer and cardiovascular diseases, thiosulfides have also been shown to stimulate the immune system by activating T cell proliferation [287-289], to have anti-microbial effect [290], and to reduce blood glucose level in diabetics by stimulating insulin secretion by the pancreas [291-293]. Thiosulfides appear to have also anti-asthmatic properties. This anti-inflammation activity is mediated through a suppression of cyclooxygenase reaction cascades, initiating once again the eicosanoid metabolism, leading to bronchial restriction [294].

2.5. Flavonoids

Flavonoids are the largest group of phenolics present in vegetables. There are more than 4,000 secondary plant metabolites in the flavonoid family. Flavonoids are further classified into anthocyanins, flavonols, flavones, and isoflavonoides.

Anthocianins: Anthocyanins are one of the largest and most important groups of water-soluble pigments in most species in the plant kingdom. They are accumulated in cell vacuoles and are largely responsible for diverse pigmentation on vegetables. Anthocyanins give vegetable leaves and fruits their purple and/or red colour appearance, such as in eggplant, red cabbage, purple and black broccoli, red onion, beet, rhubarb, purple and red-skinned potato, purple sweet potato, purple corn, red lettuce, red endive, red radish, strawberry, etc. [295-297]. Anthocyanines exist mainly as glycosides and acylglycosides of anthocyanidins, usually as C_3 mono-, di-, and/or trisides. Anthocyanins are implicated in many biological activities that may impact positively on human health [298]. Their use for therapeutic purposes has long been supported by epidemiological evidence, but only in recent years some of the specific, measurable pharmacological properties of isolated anthocyanin pigments have been verified by controlled *in vitro*, *in vivo*, or clinical research studies [298]. These pigments may reduce the risk of coronary heart disease through inhibition of platelet aggregation [299]. Anthocyanins protect in several ways. First, they neutralize enzymes that destroy connective tissue. Second, their antioxidant capacity prevents oxidants from damaging connective tissue. Finally, they repair damaged proteins in the blood-vessel walls. Anthocyanins have also been shown to protect mammalian cell lipoproteins from damage from free radicals. For example, nasunin, an anthocyanin from eggplant, has been shown to inhibit brain cell lipid peroxidation caused by oxygen and hydroxyl free radicals [300] and to reduce intestinal absorption of cholesterol [301]. In addition, anthocyanins could exert anticarcinogenic activities, reduce inflammatory insult and also modulate immune response. All these effects might be mediated by their antioxidant activity [302]. In many other cases, the exact roles of the anthocyanins in human health versus other phytochemicals have not been completely sorted out. Some reports suggest that anthocyanin activity potentates when it is delivered in mixtures [303].

Flavonols: Flavonols are polyphenolic secondary plant metabolites and exist primarily in the glycosylated form with a hydroxil group at C_3 position. Flavonols include quercetin, kaempferol, fisetin, myricetin, and rutin. Quercetin is the most important flavonoid in vegetables. It has been detected in onions and to a lesser extent in tomato, snap and mange-tout peas, and beans [304, 305]. Kaempferol, myricetin, and fisetin have been detected in onion, lettuce, endive, and horseradish [305-308]. Several investigators have studied the effect of flavonols on cardiovascular diseases and carcinogens in human and laboratory animals, although the health benefits of flavonols have yet to be fully established. However, they have been shown to function in a way similar to antioxidant vitamins, to protect against lipoprotein oxidation *in vitro* [309-311], to have anti-platelet and anti-thrombotic actions [312-313], and to produce inhibition of estrogen binding in mammalian cells [314] and induction of the phase II enzyme quinone reductase in murine hepathoma cells [315]. Evidence obtained with an *in vitro* oxidation model for heart disease has demonstrated that several plant flavonols, such as quercetin, myricetin and rutin, are more powerful antioxidants than the traditional vitamins [316]. So there are grounds for encouraging the consumption of vegetables and other foods rich in flavonols.

Flavones: Similar to flavonols, flavones including apigenin and luteolin also exist as O-glycosides with a hydrogen atom in the C_3 position. Flavones have been detected in conjugated form in celery [304, 305], tomato, eggplant, garlic and onion [308, 317]. Celery contains flavones rather than flavonols and widely varying levels of both luteolin and apigenin in different cultivars [304, 305, 308]. In recent years, scientific and public interest in flavones has grown enormously due to their putative beneficial effects against atherosclerosis, osteoporosis, diabetes mellitus and certain cancers [318]. The mechanism of action of flavones on chronic diseases is similar to that of other flavonoids. They were proposed to function primarily as antioxidants by conserving α-tocopherol content of low-density lipoproteins and membrane lipids in the reduced state [318, 319] and may potentially be used as an anti-cancer agent for prevention and therapy of breast and other DLC1 (Deleted in Liver Cancer 1) downregulated cancers [320].

Isoflavonoids: In contrast to other flavonoids, isoflavonoids including daidzen, genistein, and glycitein have limited distribution in vegetables [321, 322]. They exist mainly in legumes such as soybean, chickpea, lentil, and alfalfa sprouts [323-326]. Smaller quantities have also been detected in other vegetables, such as broccoli, asparagus, and okra [324, 327]. Genistein and daidzein have been found in relatively high concentrations in soybean and most soy-based foods. Much of the research on the health benefits of isoflavonoids has been done on soybean isoflavonoids daidzein and genistein [328-331]. Goldwyn et al. [328] made a review of the health benefits of isoflavonoids. In this review, genistein was reported to have an estrogenic-like activity at low concentration and anti-estrogenic activity at high concentration, as well as the highest affinity for binding to the estrogen receptors and/or by blocking estrogen synthesis. In contrast, daidzein has very low estrogen activity. However, it is further metabolized in the large intestine into equol. Equol is a potent phytoestrogen and an antioxidant, but the rate of conversion varies among individuals, which could affect the health benefits of this compound [332].

Consumption of soy isoflavonoids has been suggested to have multiple beneficial effects in a number of chronic diseases and medical conditions, including certain types of cancer [333], heart disease [334], bone functions [335, 336] and most recently, prevention of obesity [337]. Estrogen-like effects have been proposed as one of the major mechanisms of action of isoflavonoids related to their health effects [338]. A second mechanism of action of isoflavonoids, particularly of genistein, was discovered—genistein is a protein tyrosine kinase inhibitor [338]. Since then, isoflavonoidses have been shown to affect a diverse array of intracellular signaling pathways [339]. As polyphenols, isoflavonoids also have antioxidant activities, which were proposed as another important mechanism of action of isoflavones. However, isoflavones are not strong antioxidants since they may not be able to scavenge oxidants directly.

Therefore, they are considered as antioxidants because of their effects on gene expression of enzymes that enhance antioxidant defenses [340]. In addition, several other mechanisms have been proposed for the activities of isoflavonoids, including stimulation/inhibition of enzyme activities involved in steroid synthesis and metabolism, targeting thyroid peroxidase, and inhibiting cancer metastasis [340]. Lee et al. [341] and Ziegler et al. [342] found that Asian women who consumed a diet rich in soy products had relatively low risk of breast cancer. Further evaluation of soy-based diets and foods has confirmed that the main anti-breast cancer ingredient is genistein [343-345]. In a most recent meta-analysis of prospective studies, it was suggested that soybean isoflavonoids intake is associated with a significantly reduced risk of breast cancer incidence in Asian populations but not in Western populations. Further studies are warranted to confirm the finding of an inverse association of soy consumption with risk of breast cancer recurrence [346]. Epidemiological evidence together with preclinical data from animal and *in vitro* studies strongly supported a correlation between soybean isoflavonoids consumption and protection towards prostate cancers [344, 347]. Besides breast and prostate cancer, isoflavonoids also showed inhibitory effects on other hormone-related (e.g., endometrial, ovarian cancer) or hormone-independent cancers (e.g., leukemia and lung cancer) [348].

3. Nutritional Quality and Health Benefits of Vegetables

There are general beliefs among nutritionists and health professionals that the health benefit of vegetables should not be linked to only one compound or one type of vegetable, but rather a balanced diet that includes more than one type of vegetable is likely to provide better protection. All the vegetables may offer protection to humans against chronic diseases. With the exception of glucosinolates and thiosulfides, which are unique to the crucifers and alliums, the phytonutriceuticals content of a number of other vegetables consist primarily of vitamin C, fiber, selenium, folate and polyphenolics (carotenoids and flavonoids). The main difference is that each vegetable group contains a unique combination and amount of these phytonutriceuticals, which distinguishes them from other groups and vegetables within their own group. For example, the *Apiaceae* family (e.g., celery, parsley, carrot) is rich in flavonoids, carotenoids, vitamin C, and vitamin E. Celery and parsley, for example, are among the best vegetable sources for the flavonoid apigenin and vitamin E [349], and carrots have an unique combination of three flavonoids: kaempferol, quercetin, and luteolin [297, 298, 350]. The *Asteraceae* or *Compositae* family (e.g., lettuce, chicory) is rich in conjugated quercetin, flavonoids, and tocopherols. Crozier et al. [305] observed sizeable variations in flavonol content with lettuce cultivars. The commonly consumed small "round" lettuce contained only 11 μg/g fresh weight of quercetin, and the levels in "iceberg" lettuce were even lower. In contrast, the outer leaves of "Lollo Rosso," a red cultivar of lettuce, contained 911 μg/g. The red color of this lettuce is due to high levels of anthocyanins, which like quercetin are products of the phenylpropanoid pathway. As one end product of the pathway has been elevated, it may well be that other related compounds, including the flavonols, are also found in higher concentrations. Romaine lettuce is richer in lutein than head lettuces; and leafy and romaine lettuces are richer in quercetin [351]. The *Cucurbitaceae* family (e.g., pumpkin, squash, melon, cucumber) is rich in vitamin C, carotenoids, and tocopherols [352]. Burger et al. [353] in a survey of 350 melon accessions from different horticultural groups of *Cucurbita melo* observed a 50-fold variation in ascorbic acid content, ranging from 0.7 mg to 35.3 mg/100g of fresh fruit weight . Ascorbic acid and β-carotene content ranged from 7.0 to 32.0 mg/100g and 4.7 to 62.2 μg/100g, respectively, in sweet melons [354]. Bitter gourd (*Momordica charantia*) has anti-diabetic properties and can be used to ameliorate the effects of type-2 diabetes. Diet is the primary therapy for this type of diabetes, and bitter gourd is particularly critical when pharmaceuticals are not available, as happens in a great part of the developing world [4]. The *Chenopodiaceae* family (e.g., spinach, Swiss chard, beet greens) is an excellent source of folate [113] and has been shown to inhibit DNA synthesis in proliferating human gastric adenocarcinoma cells [355]. The *Chenopodiaceae* vegetables are also among the most oxalate dense vegetables [356, 357]. When oxalates become too concentrated in body fluids, they can crystallize and cause health problems such as kidney calcium oxalate stones. All the legumes (*Fabaceae* or *Leguminosae* family, e.g., bean, pea, soybean, chickpea, lentils), mature and immature seeds are good sources of dietary fiber and isoflavonoids. Mallillin et al. [358] determined the total soluble and insoluble fibre and fermentability characteristics of ten legumes' mature seeds (mung bean, soybean, peanut, pole sitao, cowpea, chickpea, green pea, lima bean, kidney bean and pigeon pea) and concluded that the dietary fibre content ranged from 20.9 to 46.9 g/100g and that the best sources after *in vitro* fermentation using human faecal inoculum stimulating conditions in the human colon (as mmol/g/fibre isolate of acetate, propinate, butyrate produced after fibre fermentation measured by HPLC) were pole sitao and mung bean (acetate), kidney bean and pigeon pea, (propinate), and peanut and cowpea (butyrate). High-flavonol legumes include sugar snap peas and mange-tout, which were found to contain 98 and 145 μg quercetin/g, respectively. As mentioned, some legumes are also rich in iron. Trinidad et al. [359] determined the mineral availability *in vitro* of iron, zinc and calcium in ten local legumes (cowpeas, mung beans, pole sitao, chickpeas, green peas, groundnuts, pigeon peas, kidney beans, lima beans and soybeans). They found that the highest iron availability among legumes was for lima beans (9.5 (sem 0.1)) and mung bean, while for zinc and calcium, the highest availability was for kidney beans (49.3 (sem 4.5)) and pigeon peas (75.1 (sem 7.1)), respectively. Groundnuts have the lowest Fe (1.3 (sem 1.1)), Zn (7.9 (sem 1.3)) and Ca (14.6 (sem 2.8)) availability. They concluded that mineral availability of Fe, Zn, and Ca from legumes differs and may be attributed to their mineral content, mineral-mineral interaction and from their phytic and tannic acid content. For example, mung bean (*Vigna radiata*) either eaten as whole pod grains or grown to produce bean sprouts, is an important source of iron for women and children throughout South Asia [15].

In this section, we will highlight the health benefits of the most studied and consumed vegetables, namely crucifer, allium and solanaceous vegetables.

Crucifers: Cruciferous vegetables (*Brassicacea* or *Cruciferae family*)—which include, cabbage, broccoli, cauliflower, Brussels sprouts, kales, kailan, Chinese cabbage, turnip, rutabaga, radish, horseradish, rocket, watercress, mustards, among other vegetables—provide the richest sources of glucosinolates in the human diet. Most consumers associate cruciferous vegetable consumption with health. They have reasons for that because based on one of the largest and most detailed reviews of diet and cancer, the World Cancer Research Fund in USA [360] concluded that a diet rich in crucifers is likely to protect humans against colon, rectum, and thyroid cancers, and when consumed with vegetables rich in other phytonutriceuticals can protect against cancer in other organs. Crucifers rich in glucosinolates including broccoli, cabbage, Brussels sprouts, and kale have been shown to protect against lung, prostate cancer, breast cancer, and chemically induced cancers [361-365]. Epidemiological data show that a diet rich in crucifers can reduce the risk from several types of cancers and that the risk can be significantly reduced by an intake of at least 10 g per day [361, 362, 364-367]. Epidemiological studies have suggested that diets rich in broccoli may reduce the risk of prostate cancer, and consumption of one or more portions of broccoli per week can reduce the incidence and the progression from localized to aggressive forms of prostate cancer [364, 365].

The overwhelming evidence concerning the anticarcinogenic effect of phytonutriceuticals in crucifers were from *in vivo* studies, mainly with broccoli, using animal models and human volunteers [362, 364, 365, 367-371]. In order to establish the relationship between whole broccoli and cancer prevention, Farnham et al. [372] examined the diversity of induction of the phase II detoxification enzyme quinone reductase, in murine hepatoma cells, by 71 inbred and five hybrid lines of broccoli. They found that the rate of induction of quinone reductase in hepa 1c1c7 by the broccoli inbred lines ranged from 0 to 15,000 units and that the rate of induction was highly correlated (average r=0.85, P=0.0001) to the concentration of glucoraphanin in each broccoli inbred. These results suggest that there are significant differences in the health benefits among crucifers, which is important not only from a health point of view but also as a marketing tool to promote a certain cultivar. Comparative studies of glucosinolate profiles indicate significant quantitative and qualitative differences among accessions within each crucifer, between plant parts, developmental stage, agronomic management, and climatic conditions [208, 373-384]. Kushad et al. [378] observed in 65 cultivars of broccoli, that glucoraphanin was the major glucosinolate and that there was more than 27-fold difference between the highest concentration in cultivar "Brigadier" and the lowest concentration in cultivar "EV6-1." Hansen et al. [227] also observed in their study with 21 cultivars of red cabbage and six white cabbages that there was a considerable variation in the concentration of the individual glucosinolates between the cultivars examined. Red cabbage cultivars were found to contain significantly higher concentrations of glucoraphanin compared to white cabbage cultivars. There were also significant differences within the red cabbage cultivars. Of the red cultivars examined, "Rodima" had the highest concentration with 7.4 mg/g DW glucoraphanin, whereas "Primero" has the lowest concentration containing only 0.6 mg/g DW. The white cabbage cultivars contained significantly higher levels of glucoiberin compared to red cultivars. The white cabbage cultivar "Bartolo" contained the highest level of 7.4 mg/g dry weight, whereas the cultivar "Candela" had the lowest level of 1.7 mg/g dry weight. The red cultivars ranged from approximately 3 mg/g dry weight to 0.3 mg/g dry weight. The red cabbages were also found to contain significantly higher concentration of gluconasturtiin compared to white cabbage cultivars. The cultivar "Amager Garo" had the highest concentration, whereas "Primero" had the lowest, 1 and 0.1 mg/g dry weight, respectively. Similar differences were also observed in turnip and rutabagas [374]. Fahey et al. [218] evaluated glucosinolate content of broccoli sprouts and found that they contain nearly 20- to 50-fold higher glucosinolates than tissue from mature plants. In broccoli heads, the most significant glucosinolates are glucoraphanin, glucobrassicin, progoitrin, and gluconasturtiin [241, 376, 378, 380, 381, 383]. In cabbage, Brussels sprouts, cauliflower, kale, tronchuda and collard, the predominant glucosinolates are sinigrin, progoitrin, and glucobrassicin [227, 373, 376, 378, 384, 385]. In turnip and rutabagas, the predominant glucosinolates are glucoerucin, glucoraphanin, and glucobrassicin [374, 377]. In radish, the predominant glucosinolates are glucoerucin, glucoraphanin, and glucobrassicin [375, 379]. Each of these crucifers also contain smaller amounts of other glucosinolates. The bulk of the differences in the aliphatic glucosinolates is genetically regulated [386-388]. Differences in the indol glucosinolates, in contrast to aliphatic glucosinolates, have been attributed to environmental factors. Even though the glucosinolate content is influenced by environmental conditions, the effect of genotype is found to be greater than that of environmental factors [372].

Crucifer vegetables are also rich in vitamins, with kale rated as the second highest among 22 vegetables tested [389]. Brussels sprouts and broccoli were also ranked high in their vitamin content, containing significant amounts of vitamins C and E, and β-carotene [389]. Evaluation of α- and β-,α-, and γ-tocopherols, and vitamin C in broccoli, Brussels sprouts, cabbage, cauliflower, tronchuda, and kale showed significant variations between and within these crucifers [378, 390].Vitamin C is the most abundant vitamin in all five crucifers tested [390]. Kale had the highest amount of these vitamins, followed by broccoli, Brussels sprouts, cabbage and cauliflower. Analysis indicated that 79% of β-carotene, 82% of α-tocopherol, and 55% of vitamin C variability in broccoli were associated with genetic factors [390]. Crucifers are also excellent source of folate. Brussels sprouts and broccoli were ranked among the highest vegetable sources for folate, contributing about 110 to 135 and 70 to 90 μg/100g, respectively [113, 391].

Crucifers also contain significant amounts of dietary fiber. Dietary fiber content of cauliflower was estimated to be about 5% of the total fresh weight or about 50% of the total dry weight, consisting of about 40% non-starch polysaccharides [392]. Cellulose and lignin concentrations in Brussels sprouts were estimated to be 36% and 14.5%, while in cauliflower, they were estimated to be about 16% and 13% of the total dry matter, respectively [393].

There are plenty of crucifers (e.g., kales, tronchuda cabbages, pak-choy, kailan, rutabaga, turnip, Brussels sprouts, etc.). that are good sources of calcium. As mentioned, Galega kale contains a high content of protein, fiber, calcium, and sulfur when compared to broccoli, the reference within *Brassica* vegetables. Crucifers are capable of accumulating substantial amounts of selenium when grown on high-selenium soil. Banuelos and Meek [394] reported that broccoli grown on selenium-enriched soil accumulated sevenfold more selenium than cabbage, collards and Swiss chard. Broccoli plants grown outdoors on a sphagnum, peat moss, and vermiculite medium and fertilized with sodium-selenate and selenite accumulated 278 mg/g dry weight selenium, in the edible florets, compared to the non-fertilized control, which accumulated only 0.13 mg/g dry weight [395]. In broccoli, selenium is stored as selenocysteine [396], which is readily absorbed by human tissue [174]. Selenium-enriched broccoli has been found to reduce colon cancer and mammary tumors in animal models [395, 397]. Cabbage sprouts and fully developed heads also accumulated selenium, and the accumulation was higher in the sprouts than in the mature heads [398].

Other antioxidants in crucifers include flavonoids. Miean and Mohamed [308] examined the flavonoid content of 62 vegetables and found that broccoli, cauliflower, cabbage, Chinese cabbage, and kailan contained between 148 and 219 mg/kg of flavonoids. Broccoli contained myricetin, quercetin, and luteolin; cauliflower contained myricetin and quercetin; kailan contained quercetin and apigenin; while cabbage contained only myricetin. In a similar study, Hertog et al. [399] evaluated the methanol-extracted flavonoids from 28 vegetables and found that quercetin levels, in the edible part of most vegetables, were below 10mg/kg, except in kale (110 mg/kg), broccoli (30 mg/kg), and onion (486 mg/kg). Kale, broccoli and turnip contained 211, 72, and 48 mg/kg of kaempferol, respectively. Kaempferol had also been detected in cabbage leaves [400], but Miean and Moamed [308] did not detect kaempferol in any of the tested crucifer vegetables.

Alliums: Alliums vegetables (*Alliaceae family*) include, garlic, onion, leek, chive, Welsh onion, among other vegetables. They are rich in a wide variety of thiosulfides, which have been linked to reducing various chronic diseases. Similar to glucosinolates in crucifers, the types and amounts of thiosulfides in alliums vary significantly. Typically, they contain 1 to 5% non-protein sulfur compounds, on a dry weight basis [258]. Kubec et al. [258] reported significant variability in the total thiosulfide (0.02 to 1.3% fresh weight) content and in the relative proportion of these compounds between and within alliums, even when grown under identical conditions. They found that the total thiosulfide contents in green onion leaves, chive, and onion bulb were 0.2, 0.72, and 1.02 g/kg fresh weight, respectively. The type of thiosulphides in these vegetables was also variable. For example, onion bulbs contained 34% methiin, 5%ethiin, 6% propiin, 5% alliin, and 49% isoalliin [258], while garlic cloves contained about 92% alliin, 8% methiin, and trace amounts of ethiin, propiin, and isoalliin [401].

The second most important group of phytonutriceuticals in alliums are flavonoids. Two types of flavonoids are found in onion bulbs: anthocyanins in red onions and flavonols like quercetin and kaempferol in most yellow fresh cultivars. Miean and Mohamed [298] reported that onion leaves had the highest total flavonoid content among 62 different vegetables they tested and that total flavonoid content of onion leaves and garlic were about 2.7 and 1.0 g/kg dry weight, respectively [308]. In onion leaves, about 55% of the total flavonoids is quercetin, 31% kaempferol, and 14% luteolin. In onion bulb, more than 95% of the flavonoids is quercetin and only a trace amount of kaempferol [399]. White onion cultivars were reported to have significantly less quercetin than the red ones, and most of the quercetin is present in the outer scales [402]. Red onions contained approximately 1,350 μg/g fresh weight of total flavonols, and white onions contained only 10 μg/g. Quercetin in onion appeared mainly in the free form as the aglycone [402]. In garlic cloves, 72% of the total flavonoids are myricetin, 23% apigenin and 5% quercetin [308]. In chive, garlic chive, and leek, the predominant flavonoid is kaempferol [403].

Onion and garlic are excellent sources of calcium, potassium and manganese, providing up to 10% of the human daily requirements of these elements. Onion and garlic can also accumulate selenium if grown in selenium-rich soils in the form of selenocystein and seleno-proteins. Based on this information, Ip and Lisk [404] showed that garlic plants fertilized with high selenium and low sulfur fertilizer accumulated between 110 and 150 ppm dry weight selenium, while onion plants accumulated up to 28 ppm. Ip and Lisk [404] proposed that the selenium-enriched garlic and onion provide an ideal system to deliver selenium efficiently and safely into the human body for cancer prevention since inhibition of cancer by selenium in animal models requires between 1 to 3 ppm [405]. Most of the onions and garlic contain very low concentrations of selenium. Ip and Lisk [404] reported that "Valencia" topset garlic cloves contained 0.06 ppm selenium, while "Stuttgart" onion bulbs contained 0.02 ppm. Interestingly, onions also contain chromium and are considered a good source of this element. Chromium has been recently linked to diabetes prevention and appears to act as such by potentiating insulin receptor kinases [406] and to be a part of the glucose tolerance factor involved in cellular responses to insulin. Clinical studies on diabetic patients have shown that chromium can decrease fasting glucose levels, improve glucose tolerance, lower insulin levels and decrease total cholesterol and triglycerid levels. Two hundred grams of onions contain up to 20% of the daily requirements in chromium.

Onions are a rich source of dietary fibers and especially of inulin, a polyfructosan. The health benefits of inulin-type fructans to human health have now been studied for more than one decade [407]. It has prebiotic properties, as it is preferably fermented by beneficial bowel bacteria like *Lactobacilli* and *Bifidobacteria*, thereby altering the bacterial mycoflora of the intestine in such a way that pathogenic or harmful bacteria become less abundant [408]. Neokestose, another fructan found in onion, has recently been shown to be an excellent promoter of the growth of beneficial bacteria [409]. Frutans also promote the absorption of calcium and could thus be useful in the prevention of osteoporosis [410]. High-fructan diets have also been shown to lower concentration of cholesterol, tryacylglycerol, phospholipids, glucose and insulin in the blood of middle-aged men and women [411]. Owing to the presence of prebiotic polysaccharides (inulin), which are poorly degraded by the gut enzymes, and the presence of flavonoids, onions have been shown to possess anti-diabetic potential [412]. Sharma et al. [413] showed that onions had anti-hyperglycemic effects. Such effects were confirmed by Tjokroprawiro et al. [414], who conducted a crossover comparative study with twenty diabetic patients to assess the effect of a diet comprising onions and green beans on serum glucose levels. They showed that the consumption of 20 g fresh onion three times daily significantly reduced blood sugar levels.

The therapeutic value of onions, garlic and other *Allium* vegetables is confirmed by multiple epidemiological and experimental studies. Consumption of *Allium* vegetables has been found to retard growth of several types of cancers. For instance, there appears to be a strong link between the consumption of onions and the reduced incidence of stomach and intestine cancers [415, 416]. A number of epidemiological studies show inverse correlations between the consumption of alliums like onions and garlic and the reduced incidence of cancers. A synthesis of case control studies carried in Italy and Switzerland reveals that consumption of one to seven portions of onions per week reduces the risks of colon, ovary, larynx, and mouth cancers [417]. Other epidemiological studies clearly show the correlation between moderate garlic intake and a low esophageal and stomach tract cancers incidence [417-419]. These studies have been performed on different geographical areas/continents and countries: China [420, 421], Japan [422], Netherlands [423], Italy [424], Hawaii [425], Venezuela [426], Uruguay [427]. Overall, a protective effect was reported, despite the obvious genetic variance existing among the populations examined in the different studies. A regular consumption of garlic has been associated with the reduction in the incidence of preneoplastic lesions occurring in the gastric mucosa of individuals infected by *Helicobacter pylori* [420]. A reduced cancer risk has been widely documented also for colorectal and prostate cancers [417, 419, 428, 429]. Other studies analyzing the preventive effect of garlic extracts on colorectal cancer have evidenced their suppressive potential on the development and progression of colorectal adenomas [430, 431]. A population-based study analyzing the impact of a diet rich in *Allium* vegetables on the incidence of prostate cancer showed that the anti-cancer effects were more pronounced in men presenting localized rather than advanced forms [429]. Mortality due to prostate cancer also appears to be reduced by a diet making a large place for onions [432]. A limited number of studies explored the impact of a regular intake of *Allium* vegetables on the incidence of cancers affecting breast, endometrium and lungs [433-435]. The risk of breast cancer was shown to decrease as consumption of alliums was increased in a French case-control study [433]. It was found that garlic and some of its constituents prevent tumor initiation by inhibiting the activation of pro-carcinogens and by stimulating their elimination [436-438]. Some studies say that onion extracts can inhibit the mutation process [439] and reduce the proliferation of cancer cells [440]. This effect is being attributed to quercetin in particular.

Prevention of cardiovascular diseases has been attributed to regular garlic consumption. Onions also contain a number of bioactive molecules that can presumably reduce the risks for cardiovascular diseases [286]. A preliminary study conducted in humans showed that the consumption of the equivalent of three onions in a soup was sufficient to significantly reduce the blood platelet aggregation [284].

Solanaceous vegetables: The use of solanaceous vegetables in traditional medicine is ancient. There are significant differences in the phytonutreucical content between solanaceous vegetables, and therefore each vegetable will be examined separately.

Tomato: Tomato is the second most consumed and widely grown vegetable in the world after potato. Tomato is popular fresh and in many processed forms (e.g., ketchup, canned whole or in pieces, puree, sauce, soup, juice, or sun-dried). In addition to their culinary role in the diet, tomatoes represent a low energy dense food with unique constituents that may positively affect health. Compositionally, the tomato has a unique nutritional and phytochemical profile. The major phytochemicals in tomato are the carotenoids consisting of 60 to 64% lycopene, 10 to 12% phytoene, 7 to 9% neurosporene, and 10 to 15% carotenes [441]. Based on a fresh weight basis, tomato (on average) contains about 35 mg/kg of lycopene, with red cultivars containing in average 90 mg/kg of lycopene and yellow ones only 5 mg/kg [442]. Processed tomatoes (sauce, paste, juice, and ketchup) contain two- to 40-fold higher lycopene than fresh tomatoes [441, 443, 444].

Tomatoes and tomato-based foods are the world richest sources of lycopene. The average daily intake of lycopene in the human diet is about 25 mg/day; nearly 85 percent is obtained from fresh and processed tomato products [444, 445]. Tomato contains significant amounts of α-, β-, γ-, δ-carotene ranging in concentrations form 0.6 to 2.0 mg/kg [446, 447], which ranks tomato as the fourth leading contributor of provitamin A and vitamin A in the American diet [112, 448].

In addition to lycopene, tomatoes are one of the top contributors of potassium in developed countries and in the American diet [449]. Based on a 1999-2000 USA National Health and Nutrition Examination Survey food intake data, tomatoes rank seventh after milk, potatoes, beef, coffee, poultry, and orange/grapefruit juice as a potassium source. Besides, in the USA, potassium is a nutrient of concern, as most Americans consume amounts well below the Dietary Reference Intake (DRI). In 2004, the new adult DRI for potassium (4,700 mg) was substantially higher than the amount previously reported in the 1989 Recommended Dietary Allowance (3,500 mg) [450, 451]. The increased recommendation was based on evidence indicating that 4,700 mg potassium should help lower blood pressure, reduce the adverse effect of excess sodium intake on blood pressure, reduce the risk of kidney stones, and possibly reduce age-related bone loss. Tomatoes provide at least twice the potassium per 100 kcal compared with other common sources, except coffee, a non-significant calorie source of potassium. Consuming potassium from fruits and vegetables is ideal because it occurs with a biologically advantageous ratio of bicarbonate or citrate, important for bone health. Increasing potassium intake through increased tomato intake is a healthful, calorically sensible strategy for world-developed countries.

Tomato fruits are also an excellent source of ascorbic acid, about 200 mg/kg, and are the major source of vitamin C next to citrus [452]. Tomato contains small but significant amounts (1-2 mg/kg) of lutein, α-, β-, and γ-tocopherols, and conjugated flavonoids [446, 447, 453]. In a study of 20 tomato cultivars, total flavonoids' content ranged from 1.3 to 22.2 mg/kg, with about 98% present in the skin [454]. Flavonoids in fresh tomato are present only in the conjugated form as quercetin and kaempferol [304], but processed tomato products contain significant amounts of free flavonoids [454]. Flavonoid content is affected by cultivar and culture. For example, cherry tomatoes have a markedly higher flavonoid content than standard or beef tomato cultivars, and field-grown fruits have higher flavonoids content than greenhouse-grown tomatoes [305, 454]. Tomato cultivars are available containing twice the normal vitamin C (Doublerich), forty times normal vitamin A (97L97), high levels of anthocyanin (resulting in purple tomatoes), and two to four times the normal amount of lycopene (numerous available cultivars with the high crimson gene).

Tomato lycopene is found in appreciable levels in human serum and tissues when tomatoes and tomato products are consumed frequently [455-457]. Several research investigations have shown an inverse relationship between plasma/serum lycopene concentrations and risk of some cancers [458-464]. Similar associations have been reported for markers of cardiovascular disease, osteoporosis, cognitive function, and body weight [465-472]. Moreover, relationships between dietary intakes of tomato products or tomato extract supplements have been observed in epidemiological studies and clinical trials examining markers of some cancers, cardiovascular disease, and ultraviolet light–induced skin erythema [473-488]

The majority of research conducted in the area of tomato and lycopene intake and cancer risk has been observational. In a recent review of the literature, 178 original research articles were compiled reporting findings in humans on the relationship between lycopene, tomatoes and tomato-based products and cancer risk [489]. Among these publications, nearly 90% were observational, highlighting the paucity of cause and effect investigations in this area. Reports on 13 cancer types were identified, of which breast, colorectal, gastric/upper gastrointestinal, and prostate cancers have the most original research published in humans, ranging from 17 to 60 publications. For breast, colorectal, and gastric cancers, the data support a neutral, although potentially protective, relationship between tomato/lycopene intake and cancer risk, although the data are limited for gastric and lung cancers; the protective association is strongest with tomato intake verses dietary lycopene intake. Among the cancers investigated relative to lycopene and tomato intake, prostate cancer is the most widely researched. Although randomized controlled trial data are less available than observation data, a small number of dietary intervention trials using processed tomato products have been conducted. The results have been relatively successful as measured by improvements in prostate-specific antigen concentrations [477, 479, 490, 491, 492] or increased apoptotic cell death in carcinomas [481]. A prostate cancer risk reduction of nearly 35% was observed when the test subjects consumed ten or more servings of tomato products per week, and the effect was much stronger for subjects with more aggressive and advanced stages of cancer [493]. People consuming diets rich in tomato and tomato-based products, which are rich in the carotenoid lycopene, were found to be less likely to develop stomach and rectal cancers than those who consume lesser amounts of lycopene-rich vegetables [494].

Tomatoes, with their distinctive nutritional attributes, may play also an important role in reducing the risk of cardiovascular and associated diseases through their bioactivity in modulating disease process pathways. In 2004, Sesso et al. [468] reported an inverse association for women consuming greater intakes (>7 servings per week) of tomato-based products and cardiovascular disease—an association not observed with lycopene intake alone. Several hypotheses are being tested related to the antioxidant properties of lycopene and a combination of carotenoids with coexisting water-soluble constituents delivered by tomatoes, such as vitamin C. The antioxidant capacity of plasma decreases when tomatoes and tomato products are removed from the diet and increased when they are added back [480]. Consuming tomato products daily for two to four weeks increases antioxidant enzyme defenses and has been shown to reduce plasma lipid peroxides and the susceptibility of low-density lipoprotein to oxidation [475, 482, 495]. Oxidative modification of low-density lipoprotein is a key step in the development of atherosclerotic lesions [496]. Consuming diets with appreciable amounts of antioxidants from plant foods, such as tomatoes, to inhibit the oxidative process of low-density lipoprotein may be one way to reduce the risk of cardiovascular atherosclerotic disease. One study (n = 60) in relatively healthy individuals [480] and two studies (n = 40 and n = 57) in individuals with type 2 diabetes, who are in a relatively pro-oxidant state, showed decreased lipid peroxidation rates [497] and decreased susceptibility of low-density lipoproteins to oxidation [475] after daily consumption of tomatoes or tomato juice.

Others have reported less susceptibility to oxidation of DNA [498, 499] and low-density lipoprotein [500] after tomato product consumption delivering approximately half (or more) of the lycopene dose typically used in lycopene supplementation studies. These data suggest that the health benefits of tomato/tomato product consumption are not solely because of lycopene content but rather the result of the combination of nutrients and bioactive constituents delivered when the whole food is consumed.

Tomatoes and tomato products are also being investigated for possible anti-inflammatory, antithrombotic, and lipid-lowering effects. Supplementation of a low tomato diet with tomato products produces mixed results as measured by changes in inflammatory markers such as C-reactive protein (CRP), interleukin-6, and tumor necrosis factor-a. Jacob et al. [487] reported decreased CRP after a two-week tomato juice supplementation containing approximately 21 mg lycopene and two levels of vitamin C (45.5 mg and 435 mg, respectively). Both juices reduced CRP as well as total cholesterol concentration. Natural antithrombotic agents that influence platelet function or fibrinolytic activity are of interest as primary and secondary cardio-preventive strategies. Aqueous extracts from tomatoes have been shown to display anti-platelet activity *in vitro* [501]. Subsequent research in humans shows significant reductions in *ex vivo* platelet aggregation three hours after supplementation with tomato extract from the yellowish membrane surrounding seeds in amounts equivalent to two or six fresh tomatoes [501].

For skin protection, tomato intake (40 g tomato paste corresponding to a lycopene dose of approximately 16 mg) for more than eight weeks reduced ultraviolet light–induced erythema [484, 485, 502]. Epidemiological studies suggest a beneficial relationship between dietary sources of lycopene and bone mass [503-505]. Likewise, lower serum lycopene concentrations have been documented in osteoporotic women compared with controls [469, 472]. Rao et al. [471] have also reported an inverse association between serum lycopene and markers of oxidative stress and bone turnover in 33 postmenopausal women aged 50 to 60 years. Research for a possible role of tomatoes in brain health has largely been limited to case–control studies investigating the relationship between plasma/serum lycopene and oxidative stress markers in people with documented Alzheimer's disease, Parkinson's disease, vascular dementia, and mild cognitive impairment compared with control/non-cognitively impaired individuals. In general, plasma/serum lycopene concentrations are lower in cognitively impaired compared with control individuals [465-467, 506], and oxidative stress markers are elevated and inversely correlated with plasma carotenoids concentrations [466, 467, 507].

Potato: Potato ranks as the third most important food crop after wheat and rice. Potatoes yield on average more food energy on a per-hectare and a per-day basis than either cereals or cassava. In general, potato is perceived only as a source of carbohydrates but is also an excellent source of essential amino acids. The predominant form of this carbohydrate is starch. A small but significant portion of this starch is resistant to digestion by enzymes in the stomach and small intestine and so reaches the large intestine essentially intact. This resistant starch is considered to have similar physiological effects and health benefits as dietary fiber: it provides bulk, offers protection against colon cancer, improves glucose tolerance and insulin sensitivity, lowers plasma cholesterol and triglyceride concentrations, increases satiety, and possibly even reduces fat storage [508-510]. The amount of resistant starch in potatoes is highly dependent on preparation methods. Cooking and then cooling potatoes significantly increases resistant starch. For example, cooked potato starch contains about 7% resistant starch, which increases to about 13% upon cooling [511]. Due to carbohydrate content, potatoes are considered to make a person obese if used in excess, i.e., more than RDA of carbohydrates and fats. Recent research by the University of California, Davis and the National Center for Food Safety and Technology, Illinois Institute of Technology demonstrates that people can include potatoes in their diet and still lose weight [512]. Potatoes are also often broadly classified as high on the glycemic index (GI) and so are often excluded from the diets of individuals trying to follow a low-GI diet. In fact, the GI of potatoes can vary considerably depending on type (such as red, russet or white), origin (where it was grown), preparation methods (i.e., cooking method, whether it is eaten hot or cold, whether it is mashed or cubed or consumed whole, etc.), and with what it is consumed (i.e., the addition of various high-fat or high-protein toppings) [513]. Potato contains a small amount of protein (less than 6%), but the biological value of potato protein is the best among vegetable sources and comparable to cow's milk. Human feeding trials suggested that potato proteins are of a very high quality, possibly because they are rich in essential amino acids, such as lysine, and other metabolites, which may enhance protein utilization [514-516]. The lysine content of potato complements cereal-based diets, which are deficient in this amino acid.

In addition to high-quality proteins, potato tubers accumulate significant amounts of vitamins and minerals, as well as an assortment of phytochemicals including phenolics, phytoalexins, and protease inhibitors. Chlorogenic acid constitutes up to 90% of the potato tuber natural phenols. Others found in potatoes are 4-O-caffeoylquinic acid (crypto-chlorogenic acid), 5-O-caffeoylquinic (neo-chlorogenic acid), 3,4-dicaffeoylquinic and 3,5-dicaffeoylquinic acids [517]. A medium-size 150 g potato with the skin provides 27 mg of vitamin C (45% of the daily allowance), 620 mg of potassium (18% of daily allowance), 0.2 mg vitamin B6 (10% of daily allowance) and trace amounts of thiamin, riboflavin, folate, niacin, magnesium, phosphorus, iron, and zinc. The fiber content of a potato with skin (2 g) is equivalent to that of many whole grain breads, pastas, and cereals.

Potato contributes a small but significant amount of phytochemicals. Several yellow, red, and purple fleshed types with high phytochemical content have recently been introduced into the market. The purple potato has purple skin and flesh, which becomes blue once cooked. A mutation in the varieties' P locus causes production of the antioxidant anthocyanin [518]. Total phenolics in potato tubers range in concentration from 0.5 to 1.7 g/kg [519, 520]. Al-Saikhan et al. [521] reported significant differences in total phenolics among cultivars, with flesh color having no significant effect on total phenolics. Nearly 50% of the total phenolic compounds in potato are located in the peel and adjoining tissue but decrease toward the center of the tuber [521, 522], with chlorogenic acid representing about 90 percent of the total polyphenolic content [517].

Potato tubers contain a moderate amount of vitamin C, in the range of about 10 to 104 mg/kg, depending on the cultivar and the growing season, but it declined rapidly (30 to 50 percent) during storage and cooking [515, 523, 524]. Other antioxidants found in potato include 0.5 to 2.8 mg/kg α-tocopherol, 0.13 to 0.6 mg/kg lutein, and 1 mg/kg β-carotene [525-527]. Cao et al. [389] estimated the total antioxidant capacity of potato to be in the medium range among 22 commonly consumed vegetables. Potato also contributes a small amount of selenium (0.01 mg/kg) and folate (0.35 mg/kg) to the human diet [391, 528].

Sweet and Hot Peppers: Peppers come in a beautiful array of colors and shapes. They add flavor, color, and crunch to many low-calorie dishes. All fresh peppers are excellent sources of vitaminsC, K, carotenoids, and flavonoids [529]. Antioxidant vitamins A and C help to prevent cell damage, cancer, and diseases related to aging, and they support immune function. They also reduce inflammation like that found in arthritis and asthma. Vitamin K promotes proper blood clotting, strengthens bones, and helps protect cells from oxidative damage. Red peppers are a good source of lycopene, which is earning a reputation for helping to prevent prostate cancer as well as cancer of the bladder, cervix, and pancreas. Beta-cryptoxanthin, another carotenoid in red peppers, is holding promise for helping to prevent lung cancer related to smoking and secondhand smoke. Besides being rich in phytochemicals, peppers provide a decent amount of fiber. Significant differences in vitamin C were observed between cultivars but not between species. On average, fruits contain between 1 to 2 g/kg vitamin C, which is equivalent to 200 to 300% of the recommended daily allowance for adult men and women [530]. The level of provitamin A carotenoids (α- and β-carotene) is cultivar specific.

Some cultivars of hot peppers have as much as 12 mg/kg total carotenoids, while others are below the detectable level [530, 531]. Major flavonoids in the peppers are quercetin and luteolin. They are present in conjugated form and their content varies among cultivars ranging from not detectable to 800 mg/kg [532]. When comparing the nutrient values of the different bell peppers, studies have shown that red bell peppers have significantly higher levels of nutrients than green. Red bell peppers also contain lycopene, which helps to protect against cancer and heart disease. Possibly due to their vitamin C and beta-carotene content, bell peppers have been shown to be protective against cataracts. Just like other nutrient-dense vegetables, bell peppers contain many different powerful phytochemicals. Bell peppers have also been shown to prevent blood clot formation and reduce the risk of heart attacks and strokes probably due to their content of substances such as vitamin C, capsaicin, and flavonoids. Although chili hot peppers contain a higher amount of those substances, bell peppers should still be promoted, especially for individuals with elevated cholesterol levels.

Hot peppers don't have that spicy image for nothing. The major phytochemicals in hot peppers are capsaicinoids. More than 20 capsaicinoids, belonging to two groups, capsaicin and dihydrocapsaicin, have been identified in pepper. Capsaicin was discovered in 1846, and its structure, as an acid amide, was elucidated by Nelson [533]. Capsaicin constitutes about 70% of the pungent flavour in hot pepper, while its analogue dihydrocapsaicin represents 30% [534]. The two groups differ in the presence or absence of double bonds in the fatty acid side-chain, and within each group, they differ in the length and branching point in the fatty acid side-chain [535]. Significant variations in the profile of capsaicinoids are found between and within pepper species, ranging from about 220 ppm (3,400 Scoville Heat Units, SHU) in *Capsicum annum* to 20,000 ppm (320,000 SHU) in *Capsicum chinense* [536]. Hot peppers' fire comes from capsaicin, which acts on pain receptors, not taste buds, in our mouths. Capsaicin predominates in the white membranes of peppers, imparting its "heat" to seeds as well. The capsaicin in hot peppers has been shown to decrease blood cholesterol and triglycerides, boost immunity, and reduce the risk of stomach ulcer. It used to be thought that hot peppers aggravated ulcers. Instead, they may help kill in the stomach that can lead to ulcers. Capsaicin has also analgesic, anti-bacterial, and anti-diabetic properties. Capsaicin is an ingredient in several commercial formulations for the treatment of muscle pains, toothaches, burning-mouth syndrome, gastric ulceration, painful diabetic neuropathy, post-mastectomy pain syndrome, and osteo- and rheumatoid-arthritis. It is also prescribed for bladder hypersensivity, vasomotor rhinitis, and hyperreflexia of spinal origin [537].

Chili hot peppers have amazingly high levels of vitamins and minerals. Just 100 g provides (in % of recommended daily allowance), 240% of vitamin C (ascorbic acid), 39% of vitamin B6 (pyridoxine), 32% of vitamin A, 13% of iron, 14% of copper, 7% of potassium [42]. Fresh chili hot peppers, red or green, are rich source of vitamin-C. One hundred g fresh chilies provide about 143.7 μg or about 240% of recommended daily allowance. Chilies are also good in B-complex group of vitamins such as niacin, pyridoxine (vitamin B6), riboflavin and thiamin (vitamin B1). These vitamins are essential in the sense that body requires them from external sources to replenish. Chili hot peppers contain a good amount of minerals like potassium, manganese, iron, and magnesium. As mentioned, potassium in an important component of cell and body fluids that helps controlling heart rate and blood pressure. The body as a co-factor for the antioxidant enzyme superoxide dismutase uses manganese.

Both hot and sweet peppers contain substances that have been shown to increase the body's heat production and oxygen consumption for about 20 minutes after eating. This is great news; it means our body is burning extra calories, which helps weight loss.

Eggplant: The eggplant, known as aubergine in Europe and brinjal in south Asia, is a popular vegetable cop grown in many countries throughout the subtropics, tropics and Mediterranean area, since it requires a relatively long season of warm weather to give good yields. In addition to featuring a host of vitamins and minerals, eggplant also contains important phytochemicals, which have antioxidant activity. Phytochemicals contained in eggplant include phenolic compounds, such caffeic and chlorogenic acid, and flavonoids, such as nasunin. Nasusin or delphinidin-3-(coumaroylrutinoside)-5-glucoside is the major phytochemical in eggplant. Nasunin is part of the anthocyanin purple pigment found in the peel of eggplant, purple radish, red turnip, and red cabbage [300, 301, 538]. Matsuzoe et al. [539] examined the profile of anthocyanins in several eggplant cultivars and found that nasunin represents between 70 to 90% of the total anthocyanins in the peel; however, the actual concentration of nasunin has not been reported. Nasunin is an antioxidant that effectively scavenges reactive oxygen species, such as hydrogen peroxide, hydroxyl and superoxide, as well as inhibits the formation of hydroxyl radicals, probably by chelating ferrous ions in the Fenton reaction [300, 538].

By chelating iron, nasunin lessens free radical formation with numerous beneficial results, including protecting blood cholesterol from peroxidation; preventing cellular damage that can promote cancer; and lessening free radical damage in joints, which is a primary factor in rheumatoid arthritis. The predominant phenolic compound found in all culticars tested is chlorogenic acid, which is one of the most potent free radical scavengers found in plant tissues [539].

Benefits attributed to chlorogenic acid include antimutagenic (anti-cancer), antimicrobial, anti-low-density lipoproteins (bad cholesterol) and antiviral activities. Whitaker and Stommel [539] studied seven eggplant cultivars grown commercially in the United States and a diverse collection of exotic and wild eggplants from other counties. In addition to chlorogenic acid, they found 13 other phenolic acids present at significantly varying levels in the commercial cultivars, although chlorogenic acid was the predominant phenolic compound in all of them. “Black Magic’s” commercial eggplant cultivar representative of US market types was found to have nearly three times the amount of antioxidant phenolics as the other eggplant cultivars that were studied. In addition to their nutritive potential, the phenolic acids in eggplant are responsible for some eggplants' bitter taste and the browning that results when their flesh is cut. The enzyme polyphenol oxidase triggers a phenolic reaction that produces the brown pigments. Breeders have begun work on developing eggplant cultivar with an optimal balance of phenolics to ensure both optimal nutritional value and pleasing taste.

Eggplant fruits also contain several other antioxidants including the carotenoids lycopene, lutein, and α-carotene, as well as the flavonoids myricetin and kaempferol [308, 540]. Total antioxidant activity of eggplant was estimated to be about 190 μmol Trolox equivalent per 40g serving size, which ranks it in the middle among 22 commonly consumed vegetables [389]. Eggplant is an excellent source of digestion-supportive dietary fiber and bone-building manganese. It is very good source of enzyme-catalyzing molybdenum and heart-healthy potassium. Eggplant is also a good source of bone-building vitamin K and magnesium as well as heart-healthy copper, vitamin C, vitamin B6, folate, and niacin [541, 542].

Studies have shown that eggplant is effective in the treatment of high blood cholesterol [543]. Guimarães et al. [544] showed a significant decrease in blood levels of low-density lipoproteins and total cholesterol in human volunteers who were fed with eggplant powder. A study from Heart Institute of the University of São Paulo (Brazil) found no effects at all and does not recommend eggplant as a replacement to statins [545]. Another study showed that eggplant extract inhibits human fibrosarcoma HT-180 cell invasiveness [546]. Kwon et al. [547] presented eggplant phenolics as inhibitors of key enzymes relevant for type 2 diabetes and hypertension.

Eggplant is richer in nicotine than any other edible vegetable, with a concentration of 100 ng/g (or 0.01 mg/100g). However, the amount of nicotine from eggplant or any other food is negligible compared to passive smoking [548]. On average, 9 kg of eggplant contains about the same amount of nicotine as a cigarette. Eggplant is also among a small number of vegetables that contain measurable amounts of oxalates. When oxalates become too concentrated in body fluids, they can crystallize and cause health problems such as kidney calcium oxalate stones. For this reason, individuals with already existing and untreated kidney or gallbladder problems may want to avoid eating eggplant [549,550]. Laboratory studies have shown that oxalates may also interfere with absorption of calcium from the body [551].

4. Vegetable Breeding for Nutritional Quality, Increasing Intake of Nutrients and Health Benefits

Farming and plant breeding have been closely associated since the early days when crops were first domesticated. Plant breeders play a key role in determining what we eat, since the cultivars they develop begin the dietary food chain. Every food product, including vegetables that we see on the market, has benefited from plant breeding. Without understanding the science, early farmers saved the seed from the best portion of their crop each season. Over the years, they selected the traits that they liked best and were most useful for the crop, transforming and domesticating the vegetable species they grew. The first vitamins were described with the discovery of vitamin A in 1913 [552], and several others thereafter, only a decade after Mendel's work was rediscovered to provide a more scientific foundation for plant breeding. Differences in corn colour due to carotenoids (yellow versus white corn) were soon associated with differences in vitamin A nutrition [553], and in fact yellow corn virtually replaced white corn by 1970, because it was recognized to be a better animal feed [554]. With this fact, it might be expected that plant breeders would include the improvement of nutritional quality as a basic breeding goal for all food and feed crops, including vegetables. Although, as we will see, numerous challenges in producing the vegetable and delivering an acceptable commodity to the consumers must be met before nutritional quality comes to the attention of plant breeders.

Vegetable breeding has to address and satisfy the needs of both the consumer and the grower. The general objectives for growers are good yield, disease and pest resistance, uniformity, and abiotic stress resistance. Objectives for consumers are quality, appearance, shelf life, taste, and nutritional value. Quality in vegetable crops, in contrast to field crops, is often more important than yield. For growers to survive, the market must accept cultivars. Thus, color, appearance, taste, and shape are usually more important than productivity. For example, tomatoes to be used either fresh or in processing must have distinct quality characteristics. Fresh tomatoes must have acceptable flavor, color, texture, and other taste parameters to satisfy consumer demands and handling requirements. Processing tomatoes, on the other hand, must have intrinsic rheological characteristics that make them suitable for various processing applications, such as juice, ketchup, or sauce production. Traditional breeding requires the selection of a tomato genotype or a related wild species that has a desirable trait, such as early ripening or disease resistance, and crossing it with another tomato cultivar that has a good genetic background. The desired result is an earlier ripening tomato that makes it to the market sooner or cultivars that resist pathogen attack. In this way, several thousands of tomato cultivars have been developed over the years. The final goal of vegetable breeding programs is then to release new cultivars having elite combinations of many desirable horticultural characteristics. Plant breeding for improved taste, convenience, and consumer appeal has already contributed to increased per capita vegetable consumption with the development of products such as baby carrots, yellow and orange peppers, cherry and pear tomatoes, non-bitter cucumbers, mild-tasting eggplants, seedless watermelons, and lettuces with different colors, textures and flavors for baby leaf and precut salads.

Since vegetables are rich in vitamins, minerals, dietary fiber and other phytochemicals, and therefore vital for health, breeding objectives should include improving their nutritional value. Historically, vegetable breeders have applied selection pressure to traits related to agronomic performance, particularly yield and quality, because these are the traits important to the producer. Rarely have growers been paid for nutritional factors, so there have not been economic incentives to pay much attention to these traits. However, consumers are becoming more aware of these traits.

Vegetable breeding for nutritional quality was not mentioned as a primary goal in plant breeding textbooks through the mid-20th century (e.g., Hayes and Garber [555]; Allard [556]; Fehr [557]). However vegetable breeding efforts targeting improved micronutrient content and composition had begun in the 1940s and 1950s, with research describing the inheritance and development of tomato breeding stocks and lines high in provitamin A carotenoids and vitamin C [558-560]. Lincoln et al. [558] noted a fourfold variation in vitamin C among commercial cultivars and up to 1,194 ppm in red-fruited tomato interspecific crosses with *Solanum pimpinellifolium*. Similar research leading to the development of darker orange, and consequently high provitamin A, carrots began in the 1970s [561]. Yellow core color occurs only in older open-pollinated carrot cultivars since uniform orange storage root color has been a trait of interest in carrot for over a century [562]. Similar studies were made in squash, where rapid gains in carotenoid content have been made with phenotypic selection for orange color versus green and cream [563]. Genetic improvement to increase levels of specific micronutrients has been pursued in several other vegetables such as melon, spinach, sweet potato, potato, lettuce, broccoli, pepper, watermelon, collard, kale, peas, and bean. This field of study is relatively new and also complex because of mineral interactions with each other and numerous other compounds in the soil and in the plant [564]. There is usually a large environmental effect when the component is present in tiny amounts, such as for some micronutrients and phytochemicals. As mentioned the concentration of zinc and selenium in vegetables varies based on levels of these elements in soil. Success in vegetable breeding for higher vitamin and mineral content must consider not only substance concentration but also organic components in plants that can be abundant and either reduce or increase bioavailability [564]. Breeders should also be aware that by increasing certain phytochemicals in any vegetable, they might be compromising flavor, palatability, quality, or other nutritional components. For example, breeders and researchers who are attempting to increase glucosinolates in crucifers or thiosulfides in alliums need to be aware that some of these compounds have off-flavor characteristics. Another concern is related to the potential change in the balance of these compounds. For example, selenium and sulfur use the same transport channels into the plant and the same pathway to synthesize their organic forms. It is possible that by increasing selenium in the plant, sulfur uptake will be reduced, which may limit glucosinolate and thiosulfide synthesis.

Even when nutrient or phytochemical content is increased with plant breeding, it does not necessarily follow that more nutrient or phytochemical will be absorbed and reach its target. Diets are complex, and various dietary factors can promote or inhibit absorption of nutrients and phytochemicals. For example, carotenoids are best absorbed when consumed along with dietary fat. This means that after consuming a salad with fat-free salad dressing, carotenoid absorption can be negligible; but with full-fat dressing or olive oil, carotenoid absorption will be substantially greater than with fat-free or reduced fat dressing [565]. Furthermore, absorption differs for different carotenoids. Absorption of calcium may be poor when consumed with foods rich in phytic acid (e.g., whole grains, unleavened bread, raw beans) or oxalic acid (e.g., New Zealand spinach, spinach, Swiss chard, beet greens). Phytate also limits absorption of non-heme iron and zinc, although, as mentioned, vitamin C promotes absorption of non-heme iron. Magnesium also assists in the absorption of calcium, vitamin C and potassium. Vitamin D is needed to absorb calcium. Selenium as an antioxidant works in conjunction with vitamin E. It is also worth noting that there can be large inter-individual variability in absorption of nutrients and phytochemicals and in nutrient and phytochemical requirements. So consuming more of a nutrient or a phytochemical does not necessarily produce a proporcional increase in blood levels of the nutrient or phytochemical. This means that consuming 60 mg rather than 30 mg of lycopene, or 600 mg rather than 300 mg of specific anthocyanins from raw and cooked purple carrots, does not double the concentration of these phytochemicals in blood circulation as observed by Diwadkar- Navsariwala et al. [566] and Kurilich et al. [567], respectively.

With these numerous considerations, breeding vegetable plants for improved nutritional value is a complicated goal that needs expertise in many disciplines such as plant breeding, nutrition, and soil science. When a vegetable compound (micronutrient or phytochemical) is found to be important for human health, and growers, vegetable markets and seed companies can capitalize on the value of the compound, there may be an opportunity for vegetable breeders to increase the amount of this compound. Breeders can be successful in reaching this goal, if the vegetable crop contains genetic variability for the compound, if selection is effective without detrimental pleiotropic effects, and if there is an easy method to measure the compound.

Enhanced nutritional content would add value for poor, malnourished populations. Breeding for provitamin A carotenoids, iron, and zinc is of keen interest as a strategy to alleviate nutrient deficiencies in developing countries [568-572]. An example is the "golden tomato," which contains three to six times more provitamin A carotenoids than standard tomatoes [573]. Developed at AVRDC with conventional breeding techniques, these improved nutritionally rich tomato lines could help prevent many children of developing countries from going blind, since vegetarians and populations with limited access to animal products depend on provitamin A carotenoids for vitamin A. One "golden tomato" can provide a person's full daily vitamin A requirements. Tomato fruit and its processed products are the principal dietary sources of carotenoids such as lycopene. Lycopene is a potent antioxidant with the potential to prevent epithelial cancers and improve general human health. Therefore, there is considerable interest in elevating the levels of carotenoids in tomato fruit and thereby improve the nutritional quality of the crop [574, 575]. The B gene from *Solanum hirsutum* shifts tomato carotenoid accumulation from lycopene almost entirely to β-carotene and results in orange fruit color [576]. This consequently dramatically increases the provitamin A carotenoid content. β-carotene content of commercial cultivars, as mentioned, is of interest [577], and several high β-carotene orange cherry tomato breeding lines have been bred [578]. Rainbow carrots, which are super sweet and crunchy, have multi-pigmented roots that naturally contain several antioxidants, such as lycopene, lutein, and anthocyanin. Similarly, yellow sweet potatoes are much more nutritious than white ones since they are high in provitamin A carotenoids. Unfortunately, the popularity of white-fleshed sweet potato cultivars in many tropical regions may complicate the acceptance of more nutritious orange ones. Recent studies across a range of Andean potato cultivars show wide variation in calcium, iron and zinc content [579] as well as anthocyanins [580, 581] due to the existence of red-, blue- and purple-fleshed potatoes.

In spite of many years of research and nutrition education about potential benefits of dietary fiber to help prevent diseases such as cancer, heart disease, and diabetes, the consumption of complex carbohydrates and dietary fiber worldwide has decreased [204]. Due to this fact, an increasing of the amount of soluble or insoluble fiber is a logical breeding goal to improve the nutritional quality of vegetables. Although traditional methods to measure total dietary fiber and components of dietary fiber are expensive and time consuming, thus unsuitable for plant breeding, where many samples need to be measured in a short time frame. Whole grains and bran are concentrated sources of fiber, compared to vegetables with their high water content [195].

A vegetable, in order to have impact for its nutrient content, must be appealing to consumers. Sensory appeal, including color, is an attribute important to consumers when selecting many vegetables. Enhanced pigmentation of carrot, potato, tomato, and pepper, for example, is considered a quality factor [562, 582]. In peppers, carotenoid content of green, yellow, orange and particularly red peppers can be relatively high. Selection for high pigment is an important goal because these carotenoids are important for visual appeal in many markets but until now didn't have any specific nutritional impact [583, 584]. In the past, yellow or orange tomatoes could not compete with red tomatoes because they were unfamiliar to consumers, but now they are commercialized and are challenging the market. This fact and recent development of orange-colored cauliflower [585] and orange flesh cucumbers [586] reflect a new direction in vegetable breeding: nutritional quality. The white versions of these two vegetables lack carotenoids present in high enough concentrations to alter their appearances. As nutritional quality becomes a more common breeding goal, and the novelty of unusual colors brings added value to seed companies and vegetable growers, unusual colors will quite certainly become more available and perhaps more widely consumed. But consumer requirements for quality—appearance, shelf life, and taste—must be met. Breeding to increase consumer appeal by improving convenience and the quality factors of a moderately nutritious crop often can be a more effective approach to increase intake of shortfall nutrients [587]. Nutritional quality identifiable by the consumer and available at a moderate price might induce increased consumption and thus confer an important marketing incentive for breeding activity.

5. Transgenic Vegetables and Nutritional Health Benefits

Biotechnology is a new, and potentially powerful, tool that has been added by all the major seed corporations to their vegetable breeding research programs. It can augment and/or accelerate conventional cultivar development programs through time saved, better products, and more genetic uniformity, or achieve results not possible by conventional breeding. Transgenic crops, commonly referred to as genetically modified (GM) crops, enable plant breeders to bring favorable genes, often previously inaccessible, into already elite cultivars, improving their value considerably and offering unique opportunities for controlling insects, viruses and other pathogens, as well as nutritional quality. Conventional plant breeding that utilizes non-transgenic approaches will remain the backbone of vegetable genetic improvement strategies. However, transgenic crop cultivars should not be excluded as products capable of contributing to more nutritious and healthy food. At present, many vegetable crops have been genetically modified to include resistance to insects, other pathogens (including viruses), and herbicides and for improved features, such as slow ripening, higher nutritional status, seedless fruit, and increased sweetness. Recently, Dias and Ortiz [32] did an analysis of the status (until 2010) of transgenic tomato, eggplant, potato, cucurbits, brassicas, lettuce, alliums, sweet corn, cowpea, cassava, sweet potato, and carrots. From this analysis, we can see several benefits of transgenic vegetables to nutritional quality and health benefits very important for consumers.

Consumers could benefit further from eating more nutritious transgenic vegetables, e.g., an increase of crop carotenoids by metabolic sink manipulation through genetic engineering appears feasible in some vegetables. Tomato fruit and its processed products are the principal dietary sources of carotenoids such as lycopene and β-carotene. Therefore, there is considerable interest in elevating the levels of carotenoids in tomato fruit by genetic manipulation and thereby improving the nutritional quality of the crop. To enhance the carotenoid content and profile of tomato fruit, Romer et al. [588] have produced transgenic lines containing a bacterial carotenoid gene (crtI) encoding the enzyme phytoene desaturase, which converts phytoene into lycopene. Expression of this gene in transgenic tomato plants of the cultivar "Ailsa Cray" did not elevate total carotenoid levels. However, the β-carotene content increased about threefold, up to 45% of the total carotenoid content. The alteration in carotenoid content of these transgenic tomatoes did not affect growth and development. Levels of noncarotenoid isoprenoids were unchanged in the transformants. The phenotype has been found to be stable and reproducible over at least four generations. Fraser et al. [589] reported also an increase in tomato fruit carotenoids phytoene, lycopene, β-carotene and lutein in cultivar "Ailsa Craig." Phytoene synthase from the bacterium *Erwinia uredovora* (crtB) has been overexpressed in tomato cultivar. Fruit-specific expression was achieved by using the tomato polygalacturonase promoter, and the CRTB protein was targeted to the chromoplast by the tomato phytoene synthase-1 transit sequence. Total fruit carotenoids of primary transformants (T(0)) were two- to fourfold higher than the controls, whereas phytoene, lycopene, β-carotene, and lutein levels were increased 2.4-, 1.8-, and 2.2-fold, respectively. The biosynthetically related isoprenoids, tocopherols plastoquinone and ubiquinone were unaffected by changes in carotenoid levels. The progeny (T(1) and T(2) generations) inherited both the transgene and phenotype. Ripe tomato fruits accumulate large amounts of lycopene and small amounts of β-carotene (pro-vitamin A). Lycopene is transformed into β-carotene by the action of lycopene beta-cyclase (beta-Lcy). Rosati et al. [590] introduced, via *Agrobacterium*-mediated transformation, DNA constructs aimed at upregulating (OE construct) or downregulating (AS construct) the expression of the beta-Lcy gene in a fruit-specific fashion. Three tomato transformants containing the OE construct show a significant increase in tomato fruit β-carotene content. The tomato fruits from these plants display different colour phenotypes, from orange to orange-red, depending on the lycopene/beta-carotene ratio. Fruits from AS transformants show up to 50% inhibition of beta-Lcy expression, accompanied by a slight increase in lycopene content. Leaf carotenoid composition is unaltered in all transformants. In most transformants, an increase in total carotenoid content is observed with respect to the parental line. This increase occurs in the absence of major variations in the expression of endogenous carotenoid genes.

Current advances in genetic engineering of brassicas have enabled the production of plants with alterations in a range of vitamins or amino acids for improved human nutrition. Lu et al. [591] developed transgenic cauliflower with high levels of β-carotene accumulation, and Wahlroos et al. [592] produced oilseed *Brassica rapa* .with increased histidine content. It is likely in the future that more transgenic vegetable brassicas with altered vitamin or amino acid content also will be developed.

As mentioned, vitamin E, which includes tocopherols, is a lipid-soluble antioxidant. There are α, β, γ, and δ isoforms of tocopherol with relative vitamin E potencies of 100%, 50%, 10%, and 3%, respectively. Conversion of γ-tocopherol to α-tocopherol in vegetable crops could increase their value and importance in human health because vitamin E reduces the risk of several serious disorders (e.g., cardiovascular diseases and cancer), slows aging and enhances the function of the immune system. Cho et al. [593] developed transgenic lettuce plants of the cultivar "Chungchima" expressing a cDNA encoding γ-tocopherol methyltransferase from *Arabidopsis thaliana* to improve tocopherol composition. Transgene inheritance and expression in transformed plants increased enzyme activity and conversion of γ-tocopherol to the more potent α form.

As mentioned, folate deficiency, regarded as a global health problem, causes neural tube defects and other human diseases. Folates are synthesized from pteridine, *p*-aminobenzoate (PABA), and glutamate precursos. Díaz de la Garza et al. [594, 595] developed trangenic tomatoes by engineering fruit-specific overexpression of GTP cyclohydrolase I that catalyzes the first step of pteridine synthesis and amynodeoxychorismate synthase that catalyzes the first step of PABA synthesis. Vine-ripened fruits contained on average 25-fold more folate than controls by combining PABA- and pteridine-overproduction traits through crossbreeding of transgenic tomato plants. The achieved folate level provides a complete adult daily requirement with less than one standard serving.

Vegetables also offer consumers a diverse mixture of nutrients that promote human health more beneficially than dietary supplements. However, the ingestion of plant-based diets rather than diets that rely primarily on animal products could limit the intake of essential nutrients such as calcium (Ca). Consequently, genetically engineering vegetables containing increase Ca levels may boost Ca uptake, thereby reducing the incidence of Ca deficiencies such as osteoporosis. In this regard, Park et al. [596] modified carrots to express increased levels of the plant Ca transporter sCAX1. These carrot lines were fertile and displayed no adverse phenotypes. Further, mice and human feeding trials demonstrated increased Ca absorption from *sCAX1*-expressing transgenic carrots vis-à-vis controls [597]. This research supports alternative means of biofortifying vegetables with bioavailable Ca. Zinc is also an essential element in human nutrition, as its deficiency severely impairs organ function. In experiments to fortify lettuce with this element, Zuo et al. [598] used *Agrobacterium*-mediated gene delivery of a mouse metallothionein mutant β-cDNA in the lettuce cultivar "Salinas 88." The concentration of zinc in the lettuce transgenic plants increased to 400 μg/g dry weight, which is considerably higher than in wild-type plants.

As mentioned, flavonoids are polyphenols whose dietary intake has the potential to prevent chronic diseases. Schijlen et al. [599] introduced heterologous, flavonoid pathway genes —stilbene synthase, chalcone synthase, chalcone reductase, chalcone isomerase, and flavone synthase— to produce novel flavonoids in tomato fruit. These novel flavonoids-flavones and flavonols increased threefold, mostly in the Q12 peel, which had higher total antioxidant capacity. These findings add further support to the potential of engineering tomato fruit for accumulation of high levels of beneficial nutrients. Similarly, the poliphenol resveratrol, a stilbene, shows cancer chemo-preventative activity and may prevent coronary heart disease and arteriosclerosis. Liu et al. [600] in a quantitative analysis showed that resveratrol in transgenic lettuce plants was 56.0 ± 5.52 μg/g leaf fresh weight, which is comparable to that in the skin of grape fruit (*Citrus* × *paridisi* Macfad.). Flavonoids such as anthocyanins are known as antioxidants *in vitro* and can reduce the risk of many diseases related to aging. However, some vegetable brassicas, such as cauliflower, are low in anthocyanins.

In an attempt to manipulate pigment biosynthesis to increase the health benefits of brassica vegetables, the effect of a regulatory locus of flavonoid content was assessed. *Agrobacterium tumefaciens*–mediated transformation of a *Brassica oleracea* line, selected for high transformation ability by Sparrow et al. [601], was used to produce plants transgenic for the maize lc (leaf color) locus. Lc is a regulatory gene in the anthocyanin pathway, and it is expected that its presence will increase the flavonoid content. Seedling explants were co-cultivated with *Agrobacterium tumefaciens* strain LBA4404 containing a binary vector Q27 with a neomycin phosphotransferase II (NPTII) gene. Under tissue culture conditions, lc-containing plants were green with no visible increase in anthocyanin production. However, after transfer to the greenhouse, the exposure to high light intensity led to visible signs of pigmentation within one week. Increased pigmentation was apparent in stems, petioles, main leaf veins, and sepals. Lc-containing lines had 10 to 20 times higher levels of total anthocyanins than controls. In addition, antioxidant activity of lc-containing lines was 1.5 times higher than that of controls [602].

As mentioned, the unique flavor and odor of alliums is derived from the hydrolysis of organosulfur compounds, which produces pyruvate, ammonia, and volatile sulfur compounds [263]. This reaction is catalyzed by the enzyme alliinase, which is contained in vacuoles within cells and released upon disruption of the tissue [603]. Variations in the ratios of these volatile sulfur compounds are responsible for the difference in flavors and odors between *Allium* species [263]. Along with health and nutritional benefits associated with these compounds, these thiosulfides are also major contributors to the bitter taste of some onions [263, 351]. Three sets of transgenic onion plants containing antisense alliinase gene constructs (a CaMV 35S-driven antisense root alliinase gene, a CaMV 35S-driven antisense bulb alliinase, and a bulb alliinase promoter-driven antisense bulb alliinase) have been recently produced [604]. Results from the antisense bulb alliinase lines have been much more encouraging, and three lines were produced with barely detectable bulb alliinase levels and activity. Progress has been confounded by the poor survival of transgenic plants. Crossing a non-transgenic open-pollinated parental line with a transgenic parental plant carrying a single transgene in the hemizygous state has conducted to a transgenic hybrid onion seed from these transgenic lines. Some resulting seed produced by the non-transgenic parents will be hemizygous for the transgene and can be selected to give F1 heterozygous individuals containing the transgene. Self-fertilization of these individuals produces homozygous, hemizygous, and null F2 progeny for the transgene locus. These homozygous individuals can then be used to generate the bulk seed required for the production of commercial transgenic onion lines with less bitter taste. In attempts to reduce bitterness in lettuce, Sun et al. [605] cloned the gene for the sweet and taste, modifying protein miraculin from the pulp of berries of *Richadella dulcifica*, a West African shrub. This gene, with the CaMV 35S promoter, was introduced into the lettuce cultivar "Kaiser" using *A. tumefaciens* GV2260. Expression of this gene in transgenic plants led to the accumulation of significant concentrations of the sweet-enhancing protein. People suffering diabetes may use Miraculin as a food sweetener, which is active at extremely low concentrations. The first successful study conducted to engineer the taste of tomato fruit involved transformation of tomato with the thaumatin gene from the African plant katemfe (*Thaumatococcus daniellii*) [606]. Thaumatin is a sweet-tasting protein. Fruit fromT2 transgenic plants tasted sweeter than the control plants, leaving a unique and sweet-specific aftertaste.

Food safety can also be enhanced through transgenic approaches. The first transgenic cassava plants became available in the mid-1990s [607-609] as plants with reduced cyanogenic content [610-612], which can benefit resource-poor people in rural Asia and Africa where this starchy root crop is the base of their diet.

Some vegetables, mainly tomato, are also been genetically modified to be used as vaccine delivery. Plant delivery of oral vaccines has attracted much attention because this strategy offers several advantages over vaccine delivery by injection [613-615]. Oral vaccines also offer the hope of more convenient immunization strategies and a more practical means of implementing universal vaccination programs worldwide. Tomato has been tested for expression of vaccines that can address human health issues of the developing world. Transgenic tomato plants potentially can bring several positive effects and improve human health. McGarvey et al. [616] engineered tomato plants of cultivar "UC82b" to express a gene encoding a glycoprotein (G-protein), which coats the outer surface of the rabies virus. The recombinant constructs contained the G-protein gene from the environmental risk assessment strain of rabies virus.

The G-protein was expressed in leaves and fruit of the transgenic plants, and it was found localized in Golgi bodies, vesicles, lasmalemma, and cell walls of vascular parenchyma cells. Ma et al. [617] overexpressed hepatitis E virus (HEV) open reading frame 2 partial gene in tomato plants to investigate its expression in transformants, the immunoactivity of expressed products, and explore the feasibility of developing a new type of plant-derived HEV oral vaccine. The recombinant protein was produced at 61.22 ng/g fresh weight in tomato fruits and 6.37–47.9 ng/g fresh weight in the leaves of the transformants. It was concluded that the HEV-E2 gene was correctly expressed in transgenic tomatoes and that the recombinant antigen derived had normal immunoactivity. These transgenic tomato plants are a valuable tool for the development of edible oral vaccines. Chen et al. [618] developed an effective antiviral agent against enterovirus 71 (EV71), which causes seasonal epidemics of hand, foot, and mouth disease associated with fatal neurological complications in young children, by transforming the gene for VP1 protein—a previously defined epitope and also a coat protein of EV71—in tomato plant.VP1 protein was first fused with sorting signals to enable it to be retained in the endoplasmic reticulum of tomato plant, and its expression level increased to 27 mg/g in fresh tomato fruit. Transgenic tomato fruit expressing VP1 protein was then used as an oral vaccine, and the development of VP1- specific fecal IgA and serum IgG were observed in BALB/c mice. Additionally, serum from mice fed transgenic tomato could neutralize the infection of EV71 to rhabdomyosarcoma cells, indicating that tomato fruit expressing VP1was successful in orally immunizing mice. Moreover, the proliferation of spleen cells from orally immunized mice was stimulated by VP1 protein and provided further evidence of both humoral and cellular immunity. Results of this study not only demonstrated the feasibility of using transgenic tomato as an oral vaccine to generate protective immunity in mice against EV71 but also the probability of enterovirus vaccine development. The Gram-negative bacterium *Yersinia pestis* causes plague, which has affected human health since ancient times. It is still endemic in Africa, Asia, and the American continent. There is the urgent need for a safe and cheap vaccine due to the increasing reports of the incidence of antibiotic-resistant strains and concern with the use of *Y. pestis* as an agent of biological warfare. Out of all the *Y. pestis* antigens tested, only F1 and Vinduce present a good protective immune response against a challenge with the bacterium [619]. Alvarez et al. [620] reported the expression in tomato of the *Y. pestis* F1-Vantigen fusion protein. The immunogenicity of the F1-V transgenic tomatoes was confirmed in mice that were injected subcutaneously with bacterially produced F1-V fusion protein and boosted orally with transgenic tomato fruit. Expression of the plague antigens in the tomato fruit allowed producing an oral vaccine candidate without protein purification and with minimal processing technology, offering a good system for a large-scale vaccination programs in developing countries. The future of edible plant-based vaccines through transgenic approaches will depend on producing them safely on sufficient amounts.

Despite these examples, horticulture remains in its infancy regarding the use of transgenic crop technology because vegetables are considered minor crops (compared to field crops), due to the lower resources invested (especially by the multinational private seed corporations) and derived of the high costs for deregulation. Most biotech research is done in field crops because multinational seed corporations expect the highest rate of return for their investments. While it is becoming less expensive to create transgenic crops for pest management, developing a marketable product and a regulatory package remains costly. Development and regulatory costs can be recouped more readily if the product is grown on an extensive area (as happens with field crops), which is not generally the case for individual vegetable crops. The large multinational seed corporations have, for the most part, abandoned the development of transgenic vegetable crops because of the high costs associated with product development and deregulation.

There are many cultivars of the same vegetable crop, and their expected life can be relatively short. Moreover, introducing a transgene into a breeding program can be complicated and cost prohibitive, especially in crops with difficulty for using backcrossing (e.g., cassava, potato or sweet potato). Furthermore, deregulation of a transgenic trait is event-specific in many countries. It may not, therefore, be possible to develop a single transgenic event that can be converted into many different cultivars of a single or closely related group of vegetable species through conventional breeding.

Although transgenic cultivars have proven to be a powerful tool for nutritional health benefits, many countries are still engaged in discussions about their potential food safety, and consumer antagonism has precluded many farmers and consumers from sharing the benefits that these crops can provide. These unfounded concerns are present despite the highly successful and rapid adoption of transgenics in maize, soybeans, canola and cotton in many countries of the world. Transgenic crops must pass a rigorous assessment for potential risks based on scientific data. The objective of this appraisal is to determine whether the transgenic crop is as safe as its conventional counterpart before transgenic modification. For this purpose, scientific data are provided to be reasonably confident that it will not damage the health of consumers.

The World Health Organization [621], the Food and Agriculture Organization of the United Nations [622], the Royal Society of London, the U.S. National Academy of Sciences, the Brazilian Academy of Sciences, the Chinese Academy of Sciences, the Indian National Science Academy, the Mexican Academy of Sciences and the Third World Academy of Sciences [623], the American College of Nutrition [624], the Society of Toxicology [625], the British Medical Association [626], and the Union of German Academies of Sciences and Humanities [627], among others, have stated that GM crops approved for commercialization do not pose more risk to human health than conventional crops, and they should be considered as safe as conventional ones.

The world has witnessed a steady increase of transgenic crop area in the last 16 years. The potential impacts on human and animal health have been subject of extensive research, and no evidence has been found against transgenic crops. Some people, however, continue to argue the potential long-term risks but without indicating what those risks may be. The Federal Office of Consumer Protection and Food Safety of Germany and partners [628] issued the report "Biological and Ecological Evaluation towards Long-term Effects" (also known as the BEETLE report) with the aim of providing scientific data to the European Commission. The BEETLE report reviewed in excess of 100 publications and consulted 52 experts in health issues to assess the possible long-term effect of GM crops on the health of consumers and the environment. This report concluded that so far, no adverse effects to human health from eating GM food have been found. The report further stated that although unexpected negative effects are known in conventional crops, none has yet been detected in GM crops. The report concludes that there is a negligible probability for adverse effects to consumers' health in the long term.

CONCLUSIONS

Regular consumption of a vegetable-rich diet has undeniable positive effects on health since phytonutriceuticals of vegetables can protect the human body from several types of chronic diseases. The mechanism by which vegetables decrease risk of disease is complex and largely unknown. Various components of the whole food are likely to contribute to the overall health benefit. Various phytonutriceuticals with antioxidant properties may work directly by quenching free radicals or indirectly by participating in cell-signaling pathways sensitive to redox balance. Nutrients such as potassium contribute to blood pressure regulation. The dietary fiber content and type of different vegetables may also contribute to the overall health benefit, such as improving bowel transit, lowering cholesterol, helping manage blood glucose concentrations, and by transporting a significant amount of minerals and phytochemicals linked to the fibre matrix through the human gut. Finally, increasing vegetables in the diet may reduce the intake of saturated fats, trans fats, and foods with higher caloric density, all of which may be related to a healthier overall diet. Because each vegetable contains a unique combination of phytonutriceuticals (vitamins, minerals, dietary fiber and phytochemicals), a great diversity of vegetables should be eaten to ensure that individual's diet includes a combination of phytonutriceuticals and to get all the health benefits. The availability of a large diversity of vegetables year round, and the knowledge of vegetable health benefits, has enabled consumers to include a variety of health-promoting phytonutriceuticals in their diets.

Historically, vegetable breeders have applied selection pressure to traits related to agronomic performance, particularly yield and quality, because these are the traits important to the producers. Growers have seldom been recompensed for nutritional factors, so there have been no economic incentives to provide significant attention to these traits. However, consumers are becoming more aware of these traits, and selecting for nutrient-rich vegetable cultivars is becoming part of nutrition-oriented vegetable breeding programs. Overall, there is great genetic and phenotypic diversity for types and amounts of nutrients and phytochemicals in the various vegetables.

Consequently, there is a good potential for increasing nutrient and phytochemical content and thus enriching the diet of the average consumer. Genetic improvement to increase levels of specific nutrients and phytochemicals is complex because there is often a large environmental effect when the component is present in tiny amounts. Success in vegetable breeding for higher vitamin, mineral and phytochemical content should consider not only substance concentration but also organic components in plants that can either reduce or increase bioavailability. More breeding research is needed with the goals of providing benefit to poor and malnourished populations.

Nutritional quality as understood by the consumers and available at a moderate price may encourage enhanced consumption, thereby conferring an important marketing incentive to plant breeding. Research on the health benefits of vegetables, from a horticultural and breeding perspective, needs to focus on key areas in the near future such as: i) to continue the evaluation of phytonutriceutical content among older versus newer major cultivars; ii) to identify the genetic mechanisms that regulate the syntesis of their key phytochemicals, such as the glucosinolates, thiosulfides and flavonoids, in order to develop cultivars rich in a variety of phytochemicals and in order to ensure that a mixture of phytochemicals enters into the human diet; iii) to study the potential change in the balance of these compounds (and their eventual synergisms or interactions); and iv) to identify the optimum conditions for maintaining these phytochemicals after harvest and processing since studies have shown that the bioavailability of some of the phytochemicals increases dramatically after storage and processing, while the bioavailability of others degrades rapidly.

Molecular tools will be useful for increasing quality, nutritional value, and yields. Transgenic crop cultivars should not be excluded as products capable of contributing to more nutritious and healthy vegetables. Biotechnology-derived vegetable crops will succeed if clear advantages and safety are demonstrated to both growers and consumers. Consumers are not sufficiently informed about transgenics and about which combination of vegetables and phytonutriceuticals to eat.

Progress in breeding for human nutrition and health benefits will necessitate forming teams that besides plant breeders and molecular biologists also include nutritionists, health professionals and food scientists and processors. Seed corporations with associated biotechnologies must become concerned not only with the immediate effects of profits of those using their biotech products but also with the far-reaching effects on society health benefits. If world consumers take advantage of healthier vegetables and make dietary improvements, the changes could impact world agriculture for a better future.

REFERENCES

[1] Quebedeaux, B. and Eisa, H.M. (1990). Horticulture and human health. Contributions of fruits and vegetables. Proc. 2nd Intl. Symp. Hort. and Human Health. *Hort. Science* 25:1473-1532.

[2] Craig, W. and Beck, L. (1999). Phytochemicals: health protective effects. *Can. J. Diet. Pract. Res.* 60:78-84.

[3] Wargovich M.J. (2000). Anticancer properties of fruits and vegetables. *Hort. Science* 35: 573-575.

[4] Dias, J.S. and Ryder, E. (2011). World vegetable industry: production, breeding, trends. *Hort. Rev.* 38:299–356.

[5] Southon S. (2000). Increased fruit and vegetable consumption within the EU: Potential health benefits. *Food Res. Int.* 33:211–217.

[6] Herrera, E., Jimenez, R., Aruoma, O.I., Hercberg, S., Sanchez-Garcia, I., and Fraga, C. (2009). Aspects of antioxidant foods and supplements in health and disease. *Nutr. Rev.* 67(1):S140-S144.

[7] Prior, R.L. and Cao, G. (2000). Antioxidant phytochemicals in fruit and vegetables, diet and health implications. *Hort. Science* 35:588-592.

[8] Hyson, D. (2002). The health benefits of fruit and vegetables. *A scientific overview for health professionals.* Produce for Better Health Foundation. Wilmington DE.

[9] Golberg, G. (ed.) (2003). Plants: diet and health. *The Report of a British Nutrition Foundation Task Force.* Blackwell Science, Oxford.

[10] IFAVA (2006). *Fruit, vegetables and health: a scientific overview.* International fruit and vegetable Alliance, Ottawa, Canada. www.5aday. co.nz/5adaywebsite/cms_ resources/file/Scientific20%Overview.

[11] Keatinge, J.D.H., Waliyar F., Jammadass R.H., Moustafa A., Andrade M., Drechsel P., Hughes J.d'A., Kardivel P., and Luther K. (2010). Re-learning old lessons for the future of food: By bread alone no longer—diversifying diets with fruit and vegetables. *Crop Sci.* 50(1):51-62.

[12] Mullie P. and Clarys, P. (2011). Association between cardiovascular disease risk factor knowledge and lifestyle. *Food Nutr. Sci.* 2:1048-1053.

[13] Liu, S., Lee, I.M., Ajani, U., Cole, S.R., Buring, J.E. and Manson, J.E. (2001). Intake of vegetables rich in carotenoids and risk of coronary heart disease in men: the physicians' health study. *Int. J. Epidemiol.* 30:130-135.

[14] American Heart Association (2000). *Heart and stroke statistical update.* American Heart Association. Dallas, Texas.

[15] Dias, J.S. (2011). World importance, marketing and trading of vegetables. *Acta Horticulturae* 921:153-169.

[16] Fergunson, L.R. (1999). Prospect of cancer prevention. *Mutat. Res.* 428:329-338.

[17] AICR (American Institute of Cancer Research) (2007). Food, nutrition, physical activity, and the prevention of cancer: a global perspective. AIRC/World Cancer Research Fund. Washington, DC.

[18] Potter, J.D., Slattery, M.L., and Bostick, R.M. (1993). Colon cancer: a review of the epidemiology. *Epidemiol. Rev.* 15:499-545.

[19] Van Duijnhoven, F.J.B., Bueno-de-Mesquita, H.B., Ferrari, P., Jenab, M., et al. (2009). Fruit, vegetables and colorectal cancer risk: the European prospective investigation into cancer and nutrition. *Am. J. Clin. Nutr.* 89:1441-1452.

[20] Trinidad, T.P., Mallillin, A.C., Loyola, A.S., Sagum, R.S. and Encabo, R.R. (2010). The potential health benefits of legumes as a good source of dietary fibre. *Br. J. Nutr.* 103(4):569-574.

[21] Trock, B., Lanza, E. and Greenwald, P. (1990). Dietary fiber, vegetables, and colon cancer: critical review and meta-analysis of the epidemiologic evidence. *J. Natl. Cancer Inst.* 82:650-661.

[22] Neugut, A.I., Jacobson, J.S. and DeVivo, L. (1993). Epidemiology of colorectal adenomas: macronutrients, cholesterol and fiber. *J. Natl. Cancer Inst.* 85:884-891.

[23] Steinmetz, K.A., Kushi, L.H., Bostick, R., Folisom, A.R. and Potter, J.D. (1994). Vegetable, fruit and colon cancer in the Iwoa women's health study. *Am. J. Epidemiol.* 139:1-15.

[24] Peipins, L.A. and Sandler, R.S. (1994). Epidemiology of colorectal adenomas. *Epidemiol. Rev.* 16:273-279.

[25] Witte, J.S., Longnecker, M.P., Bird, S.L., Lee, E.R., Frakl, H.D. and Haile, R.W. (1996). Relation of vegetable, fruit and grain consumption to colorectal adenomatous polyps. *Am. J. Epidemiol.* 144:1015-1025.

[26] Lin, H.J., Probst-Hensch, N.M., Louie, A.D., Kau, J.H., Witte, J.S., Ingles, S.A., Frankl, H.D., Lee, E.R. and Haile, R.W. (1998). Glutathione transferase null genotype, broccoli, and lower prevalence of colorectal adenomas. *Cancer Epidemiol. Biomakers Prev.* 7:647-652.

[27] Hu, J., Liu, Y., Yu, Y., Zhao, T.Z., Liu, S.D. and Wang, Q.Q. (1991). Diet and cancer of the colon and rectum: a case-control study in China. *Int. J. Epidemiol.* 20:362-367.

[28] Iscovich, J.M., L'Abbe, K.A., Castelleto, R., Calzona, A., Bernaedo, A., Chopita, N.A., Jmelnitzsky, A.C. and Kaldor, J. (1992). Colon cancer in Argentina. Risk from intake of dietary items. *Int. J. Cancer* 51:851-857.

[29] Zaridize, D., Filipchenco, V. and Kustov, V. (1993). Diet and colorectal cancer: results of two case-control studies in Russia. *Eur. J. Cancer.* 29A:113-115.

[30] Tanaka, S., Haruma, K., Yoshihara, M., Kajiyama, G., Kira, K., Amagase, H. and Chayama, K. (2006). Aged garlic extract has potential suppressive effect on colorectal adenomas in humans. *J. Nutr.* 136(3): 821S-826S.

[31] Pfeiffer, W., and B. McClafferty. 2007. HarvestPlus: Breeding crops for better nutrition. *Crop Sci.* 57:s88–s105.

[32] Dias, J.S. and Ortiz, R. (2012). Transgenic vegetable crops: progress, potentials and prospects. *Plant Breed. Rev.* 35:151–246.

[33] Ezzati, F. (2002). Selected major risk factors and global and regional burden of disease. *Lancet* 360:1347-1360.

[34] WHO (World Health Organization) (2003). Diet, nutrition and the prevention of chronic diseases. Report of a joint WHO/FAO expert consultation. WHO Tech. Report Ser. 916 World Health Organization, Geneva, Switzerland.

[35] Helen Keller International. (2010). Homestead food production model contributes to improved household food security, nutrition and female empowerment–experience from scaling-up programs in Asia (Bangladesh, Cambodia, Nepal and Philippines). *Nutrition Bulletin* 8:1-8.

[36] Rao, S., Yajnik, C.S., Kanade, A., Fall, D.H.D., Margetts, B.M. and Jackson, A.A. (2001). Intake of micronutrient-rich foods in rural Indian mothers is associated with the size of their babies at birth: Pune maternal nutrition study. *J. Nutr.* 131:1217–1224.

[37] FAO. 2011. FAOSTAT data. Food and Agriculture Organization of the United Nations, Rome, Italy. http://www.fao.org.

[38] FAO. (2009). FAOSTAT data. Food and Agriculture Organization of the United Nations, Rome, Italy. http://www.fao.org.

[39] Kays, S.J. and Dias, J.S. (1995). Common names of commercially cultivated vegetables of the world in 15 languages. *Economic Botany* 49:115-152.

[40] Kays, S.J. (2011). *Cultivated vegetables of the world: a multilingual onomasticon.* Wageningen Academic Publishers, Wageningen, The Netherlands.

[41] Monsanto. 2011. *Supplemental toolkit for investors*. Updated April 2011. Monsanto.

[42] Frei, B. and Lawson, S. (2008). Vitamin C and cancer revisited. *Proc. Natl. Acad. Sci. (USA)* 105(32):11037-11038.

[43] Packer, L., and Fuchs, J. (eds.). (1997). *Vitamin C in health and disease*. Marcel Dekker, New York.

[44] Englard, S. and Seifter, S. (1986). The biochemical functions of ascorbic acid. *Annu. Rev. Nutr.* 6:365-406.

[45] Levine, M., Rumsey, S., Wang, Y., Park, J., Kwon, O., Xu, W. and Amano, N. (1996). Vitamin C. In: Ziegler, E.E. and Filer, L.J., Jr. (eds.). *Present knowledge in nutrition.* 7th ed. ILSI Press, Washington DC. Pp.146-149.

[46] Jacob, R.A. (1995). The integrated antioxidant system. *Nutr. Res.* 15:755-766.

[47] Halpner, A.D., Handelman, G.J., Belmont, C.A., Harris, J.M. and Blumberg, J.B. (1998). Protection by vitamin C of oxidant-induced loss of vitamin of vitamin E in rat hepatocytes. *J. Nutr. Biochem.* 9:355-359.

[48] Cuddihy, S.L., Parker, A., Harwood, D.T., Vissers, M.C. and Winterbourn, C.C. (2008). Ascorbate interacts with reduced glutathione to scavenge phenoxyl radicals in HL60 cells. *Free Radic. Biol. Med.* 44(8):1637-1644.

[49] Valero, E., Gonzalez-Sanchez, M.I., Macia, H. and Garcia-Carmona, F. (2009). Computer simulation of the dynamic behavior of the glutathione-ascorbate redox cycle in chloroplasts. *Plant Physiol.* 149(4):1958-1969.

[50] Foyer, C.H. and Noctor, G. (2011). Ascorbate and glutathione: the heart of the redox hub. *Plant Physiol.* 155(1):2-18.

[51] Dulloo, R.M., Mjundar, S. Chakravarti, R.N., Mehta, R.N. and Mahmood, A. (1981). Effect of vitamin C deficiency in guinea pigs on intestinal functions and chemical composition of brush border membrane. *Annu. Nutr. Metab.* 25:213-220.

[52] Singh, R.B., Ghosh, S., Niaz, M.A., Singh, R., Beegum, R., Chibo, H., Shoumin, Z. and Postiglione, A. (1995). Dietary intake, plasma levels of antioxidant vitamins, and oxidative stress in relation to coronary artery disease in elderly subjects. *Am. J. Cardiol.* 76:1233-1238.

[53] Gale, C.R., Martyn, C.N., Winter, P.D. and Cooper, C. (1995). Vitamin C and risk of death from stroke and coronary heart disease in cohort of elderly people. *Brit. Med. J.* 310: 1563-1566.

[54] Nyyssonen, K., Parviainen, M.T., Salonen, R., Toumilehto, J. and Salonen, J.T. (1997). Vitamin C deficiency and risk of myocardial infarction: prospective population study of men from eastern Finland. *Brit. Med. J.* 314:634-638.

[55] Simon, J.A., Hudes, E.S. and Brown, W.S. (1998). Serum ascorbic acid and cardiovascular disease prevalence in US adults. *Epidemiology* 9:316-321.

[56] Wassertheil-Smoller, S., Romney, S.L., Wylie-Rosett, J., Slagle, S., Miller, G., Lucido, D., Duttagupta, C. and Palan, P.R. (1981). Dietary vitamin C and uterine cervical dysplasia. *Am. J. Epidemiol.* 114: 714-724.

[57] Freudenheim, J.L., Graham, S., Marshall, J.R., Haughey, B.P. and Wilkinson, G. (1990). A case-control study of diet and rectal cancer in western New York. *Am. J. Epidemiol.* 131:612-624.

[58] Howe, G.R., Benito, E., Castelleto, R., Cornee, J., Esteve, J., Gallagher, R.P., Iscovich, J.M., Deng-ao, J., Kaaks, R. and Kune, G.A. (1992). Dietary intake of fiber and decreased risk of cancers of the colon and rectum: evidence from the combined analysis of 13 control studies. *J. Natl. Cancer Inst.* 84:1887-1896.

[59] Howe, G.R., Jain, M. and Miller, A.B. (1990). Dietary factors and risk of pancreatic cancer: results of a Canadian population-based case-control study. *Int. J. Cancer* 45:604-608.

[60] Fontham, E.T., Pickle, W., Haenszel, W., Correa, P., Lin, Y.P. and Falk, R.T. (1988). Dietary vitamins A and C and lung cancer risk in Louisiana. *Cancer* 62:2267-2273.

[61] O'Toole, P, and Lombardi, M. (1996). Vitamin C and gastric cancer: supplements for some or fruit for all? *Gut* 39:345-347.

[62] Delamere, N.A. (1996). Ascorbic acid and the eye. *Subcell. Biochem.* 25:313-329.

[63] Frei, B.T., England, L. and Ames, B.N. (1989). Ascorbat on an outstanding antioxidant in human blood plasma. *Proc. Nat. Acad. Sci. (USA).* 86:6377-6381.

[64] Jialal, I., Vega, G.I. and Grundy, S.M. (1990). Physiologic levels of ascorbate inhibit the oxidative modification of low-density lipoprotein. *Atherosclerosis* 82:185-191.

[65] Harrison, F. E. and J. M. May. (2009). Vitamin C function in the brain: vital role of the ascorbate transporter SVCT2. *Free Radic. Biol. Med.* 46(6):719-730.

[66] Bremner, J.D. and McCaffery, P. (2008). The neurobiology of retinoic acid in affective disorders. *Prog. Neuropsychopharmacol Biol. Psychiatry* 32(2):315-331.

[67] WFP (World Food Programme) 2007. Annual Report. WFP- UN, Washington.

[68] UNICEF 2010. Vitamin and Mineral Deficiency. *A Global Progress Report,* UNICEF. 2 pp.

[69] Britton, G. and Khachik, F. (2009). Carotenoids in Food. In: Britton, G., Pfander, H. and Liaaen-Jensens, S. (eds.). *Carotenoids.* Birkhäuser, Basel. pp. 45-66.

[70] Olson, J.A. (1999). Carotenoids. In: Shils, M.E., Olson, J.A., Shick, M. and Ross, A.C. (eds.). *Modern nutrition in health and disease*. 9th ed. Williams and Wilkins. Baltimore, MD. Pp. 525-541.

[71] Pasquali, M.A., Gelain, D.P., Zanotto-Filho, A., Souza, L.F., Oliveira, R.B., Klamt, F. and Moreira, J.C. (2008). Retinol and retinoic acid modulate catalase activity in Sertoli cells by distinct and gene expression-independent mechanisms. *Toxicol. In Vitro* 22(5):1177-1183.

[72] Ramakrishna, V. and Jailkhani, R. (2008). Oxidative stress in non-insulin-dependent diabetes mellitus (NIDDM) patients. *Acta Diabetol.* 45(1):41-46.

[73] Botella-Carretero, J.I., Balsa, J.A., Vazquez, C., Peromingo, R., Diaz-Enriquez, M. and Escobar-Morreale, H.F. (2010). Retinol and alpha-tocopherol in morbid obesity and nonalcoholic fatty liver disease. *Obes. Surg.* 20(1):69-76.

[74] Kao, S.H., Chao H.T., Chen, H.W., Hwang, T.I., Liao, T.L. and Wei, Y.H. (2008). Increase of oxidative stress in human sperm with lower motility. *Fertil. Steril.* 89(5):1183-1190.

[75] Michikawa, T., Nishiwaki, Y., Kikuchi, Y., Hosoda, K., Mizutari, K., Saito, H., Asakura, K., Milojevic, A., Iwasawa, S., Nakano, M. and Takebayashi, T. (2009). Serum Levels of Retinol and Other Antioxidants for Hearing Impairment among Japanese Older Adults. *J. Gerontology Series-Biol. Sci. Med. Sci.* 64(8):910-915.

[76] Di Mascio, P., Kaiser, S. and Sies, H. (1989). Lycopene as the most effective biological carotenoid singlet oxygen quencher. *Arch. Biochem. Biophys.* 274:532-538.

[77] Agarwal, S. and Rao, A.K. (2000). Tomato lycopene and its role in human health and chronic diseases. *CAMJ* 163(6):739-744.

[78] Rao A.V. (ed.) (2006). *Tomatoes, Lycopene and Human Health Preventing Chronic Diseases*. Caledonian Science Press. Stranraer, Scotland.

[79] Gaziano, J.M., Manson, J.E., Branch, L.G., Colditz, G.A., Willet, W.C. and Buring, J.E. (1995). A prospective study of consumption of carotenoids in fruits and vegetables and decreased cardiovascular mortality in elderly. *Ann. Epidemiol.* 5:55-60.

[80] Johnson, E.J., Hammond, B.R., Yeum, K.J., Qin, J., Wang, X.D., Castaneda C., Snodderly, D.M. and Russel, R.M. (2000). Relation among serum and tissue concentrations of lutein and zeaxanthin and macular pigment density. *Am. J. Clin. Nutr.* 71(6):1555–1562.

[81] Niizu, P.Y. and Rodriguez-Amaya D.B. (2005). Flowers and leaves of *Tropaeolum majus* L. as rich sources of lutein. *J. Food Science* 70 (9): S605-S609.

[82] SanGiovanni J.P., Chew E.Y., Clemons T.E., Ferris F.L., Gensler G., Lindblad A.S., Milton R.C., Seddon J.M. and Sperduto R.D. (2007). The relationship of dietary carotenoid and vitamin A, E, and C intake with age-related macular degeneration in a case control study: AREDS Report No. 22. *Arch. Ophthalmol.* 125(9): 1225–1232.

[83] Van het Hof, K.H., West, C.E., Weststrate, J.A. and Hautvast, J.G. (2000). Dietary factors that affect the bioavailability of carotenoids. *J. Nutr.* 130:503-506.

[84] Roels, O.A., Trout, M., and Dujacquier, R. (1958). Carotene balances on boys in Ruanda where vitamin A deficiency is prevalent. *J. Nutr.* 65: 115-127.

[85] Gartner, C., Stahl, W. and Sies, H. (1997). Lycopene is more bioavailable from tomato paste than from fresh tomato. *Am. J. Clin. Nutr.* 66:116-122.

[86] Unlu N.Z., Bohn T., Francis D.M., Nagaraja H.N., Clinton S.K. and Schwartz S.J. (2007). Lycopene from heat-induced cis-isomer-rich tomato sauce is more bioavailable than from all-trans-rich tomato sauce in human subjects. *Brit. J. Nutr.* 98:140-146.

[87] Van het Hof, K.H., Brouwer, I.A., West, C., Haddeman, E., Steegers-Theunissem, R., van Dusseldorp, M., Weststrate, J., Eskes, T. and Hautvast, G. (1999). Bioavailability of lutein from vegetables in 5 times higher than that of β-carotene. *Amer. J. Clin. Nutr.* 70:261-268.

[88] Ravaglia, G., Forti, P., Lucicesare, A., Pisacane, N., Rietti, E., Mangialasche, F., Cecchetti, R., Patterson, C. and Mecocci, P. (2008). Plasma tocopherols and risk of cognitive impairment in an elderly Italian cohort. *Am. J. Clin. Nutr.* 87(5):1306-1313.

[89] Atkinson, J., Epand, R.F. and Epand, R.M. (2008). Tocopherols and tocotrienols in membranes: a critical review. *Free Radic. Biol. Med.* 44(5):739–764.

[90] Naumowicz, M., Petelska, A.D. and Figaszewski, Z.A. (2009). Impedance spectroscopic investigation of the interactions between phosphatidylethanolamine and [alpha]-tocopherol in bilayer membranes. *Electrochimica Acta* 54(3):1089-1094.

[91] Devaraj, S., Leonard, S., Traber, M.G. and Jialal, I. (2008). Gamma-tocopherol supplementation alone and in combination with alpha-tocopherol alters biomarkers of oxidative stress and inflammation in subjects with metabolic syndrome. *Free Radic. Biol. Med.* 44(6): 1203-1208.

[92] Grattagliano, I., Palmieri V.O., Portincasa, P., Moschetta, A. and Palasciano, G. (2008). Oxidative stress-induced risk factors associated with the metabolic syndrome: a unifying hypothesis. *J. Nutr. Biochem.* 19(8):491-504.

[93] Roberts, C.K. and Sindhu, K.K. (2009). Oxidative stress and metabolic syndrome. *Life Sci.* 84(22):705-712.

[94] Chung, M.Y., Yeung, S.F., Park, H.J., Volek, J.S. and Bruno R.S. (2010). Dietary alpha- and gamma-tocopherol supplementation attenuates lipopolysaccharide-induced oxidative stress and inflammatory-related responses in an obese mouse model of nonalcoholic steatohepatitis. *J. Nutr. Biochem.* 21(12):1200-1206.

[95] Sen C.K, Khanna, S, and Roy, S (2006a).Tocotrienol: the natural vitamin E to defend the nervous system? *Ann. New York Acad. Sci.* 1031: 127–142.

[96] Traber, M.G. and Atkinson, J. (2007). Vitamin E: antioxidant and nothing more. *Free Radic. Biol. Med.* 43 (1):4–15.

[97] Traber, M.G. and Packer, L. (1995). Vitamin E: beyond antioxidant function. *Am. J. Clin. Nutr.* 62(6):1501S–1509S.

[98] Traber, M.G. and Sies, H. (1996). Vitamin E in humans: demand and delivery. *Ann. Rev. Nutr.* 16: 321–347.

[99] Sen C.K.; Khanna, S. and Roy, S. (2006b). Tocotrienols: vitamin E beyond tocopherols. *Life Sci.* 78(18):2088–2098.

[100] Das, S., Lekli, I., Das, M., Szabo, G., Varadi, J., Juhasz, B., Bak, I., Nesaretam, K, et al. (2008). Cardioprotection with palm oil tocotrienols: comparison of different isomers." *Am. J. Physiol. Heart and Circulatory Physiol.* 294(2):970–978.

[101] Khanna, S., Roy, S., Slivka, A., Craft, T., Chaki, S., Rink, C., Notestine, M., DeVries, A.C., Parinandi N. and Sen, C. (2005). Neuroprotective properties of the natural vitamin E α-tocotrienol. *Stroke* 36(10):2258–2264.

[102] Ozer, N.K., Boscoboniki, D., and Azzi, A. (1995). New roles of low-density lipoproteins and vitamin E in the pathogenesis of atherosclerosis. *Biochem. Mol. Biol. Int.* 35:117-124.

[103] Azzi, A., Boscoboinik, D., Marilley, D., Ozer, N. K., Stauble, B., and Tasinato, A. (1995). Vitamin E: a sensor and an information transducer of the cell oxidation state. *Am. J. Clin. Nutr.* 62:1337S-1346S.

[104] Kowdley, K.V., Mason, J.B., Meydani, S.N., Cornwall, S. and Grand, R.J. (1992). Vitamin E deficiency and impaired cellular immunity related to intestinal fat malabsorption. *Gastroenterology* 102(6):2139–2142.

[105] Brigelius-Flohé, R. and Davies, K.J. (2007). Is vitamin E an antioxidant, a regulator of signal transduction and gene expression, or a “junk” food? Comments on the two accompanying papers: "Molecular mechanism of alpha-tocopherol action" by Azzi, A. and "Vitamin E, antioxidant and nothing more" by M. Traber and J. Atkinson.” *Free Radic. Biol. Med.* 43(1):2–3.

[106] Gey, K.F., and Puska, P. (1989). Plasma vitamins E and A inversely correlated to mortality from ischemic heart disease in cross-cultured epidemiology. *Ann. New York Acad. Sci.* 570:268-282.

[107] Dowd, P. and Zheng, Z.B. (1995). On the mechanism of the anti-clotting action of vitamin E quinone. *Proc. Natl. Acad. Sci. (USA)* 92(18):8171–8175.

[108] Devaraj, S., Hugou, I. and Jialal, I. (2001). Tocopherol decreases CD36 expression in human monocyte-derived macrophages. *J. Lipid Res.* 42(4):521–527.

[109] Srivastava, U.S., and Goswami, T. (1988). Biochemical changes in progressive muscular dystrophy XV. Distribution of radioactive glutamate and proximate composition of various components of skeletal muscle and liver in vitamin E deficient dystrophy rabbits and 129/RJ(dy/dy) mice. *Expt. Biol.* 47:185-193.

[110] Muller, D.P. (2010). Vitamin E and neurological function. Review. *Mol. Nutr. Food Res.* 54 (5):710–718.

[111] National Research Council (USA). (1989). Committee on Diet and Health, National Academy of Sciences. Diet and Health. National. Academies Press, Washington, DC.

[112] USDA (United States Department of Agriculture) (2009). *USDA National Nutrient Database for Standard Reference, Release 20*. USDA. http://www.nal.usda.gov/fnic/foodcomp/Data/SR20/nutrlist/sr20w309.pdf.

[113] Scott, J., Releille, F., and Fletcher, J. (2000). Folic acid and folates: the feasibility for nutritional enhancement in plant foods. *J. Sci. Food Agr.* 80:795-824.

[114] Ames, B.N. (2001). DNA damage from micronutrient deficiencies is likely to be major cause of cancer. *Muta. Res.* 475:7-20.

[115] Weinstein, S.J., Hartman T.J. and Stolzenberg-Solomon R. (2003). Null association between prostate cancer and serum folate, vitamin B6, vitamin B12, and homocysteine. *Cancer Epidem. Biomarkers Prev.* 12(11): 1271–1272.

[116] Oakley, G.P., Adams, M.J. and Dickinson, C.M. (1996). More folic acid for everyone, now. *J. Nutr.* 126:751S-755S.

[117] Mulinare, J., Cordero, J.F., Erickson, J.D. and Berry, R.J. (1988). Periconceptional use of multivitamins and the occurrence of neural tube defects. *J. Am. Med. Assoc.* 260 (21):3141–3145.

[118] Milunsky, A., Jick, H., Jick, S.S., Bruell, C.L., MacLaughlin, D.S., Rothman, K.J. and Willett, W. (1989). Multivitamin/folic acid supplementation in early pregnancy reduces the prevalence of neural tube defects. *J. Am. Med. Assoc.* 262(20):2847–2852.

[119] Scholl, T.O. and Johnson, W.G. (2000). Folic acid: influence on the outcome of pregnancy. *Am. J. Clin. Nutr.* 71(5):1295S–1303S.

[120] Quinn, C.T., Griener, J.C., Bottiglieri, T., Hyland, K., Farrow, A., and Kamen, B.A. (1997). Elevation of homocysteine and excitatory amino acid neurotransmitters in the CSF of children who receive methotrexate for the treatment of cancer. *J. Clin. Oncol.* 15:2800-2806.

[121] Stewart, C.P., Christian, P., Schulze, K.J., Leclerq, S.C., West Jr, K.P. and Khatry, S.K. (2009). Antenatal micronutrient supplementation reduces metabolic syndrome in 6- to 8-year-old children in rural Nepal. *J. Nutr.* 139(8):1575–1581.

[122] Wilton, D.C. and Foureur, M.J. (2010). A survey of folic acid use in primigravid women. *Women and Birth: J. Austral. College of Midwives* 23(2):67–73.

[123] Mason, J.B. (1994). Folate and colonic carcinogenesis: searching for a mechanistic understanding. *J. Nutr. Biochem.* 5:170-175.

[124] Sanjoaquin, M.A., Allen, N., Couto, E., Roddam, A.W. and Key, T.J. (2005). Folate intake and colorectal cancer risk: a meta-analytical approach. *Int. J. Cancer* 113 (5): 825–828.

[125] Blount, B.C., Mack, M.M., Wehr, C., MacGregor, J., Hiatt, R., Wang, G., Wickramasinghe, S., Everson, R., and Ames, B.N. (1997). Folate deficiency causes uracil mis-incorporation into human DNA and chromosome breakage: implications for cancer and neuronal damage. *Proc. Natl. Acad. Sci. (USA)* 94:3290-3295.

[126] Imasa, M.S., Gomez, N.T. and Nevado Jr, J.B. (2009). Folic acid-based intervention in non-ST elevation acute coronary syndromes. *Asian Cardiov. and Thoracic Ann.* 17(1):13–21.

[127] Terwecoren, A., Steen, E., Benoit, D., Boon, P. and Hemelsoet, D. (2009). Ischemic stroke and hyperhomocysteinemia: truth or myth? *Acta Neurologica Belgica* 109(3):181–188.

[128] Bazzano, L.A. (2009). Folic acid supplementation and cardiovascular disease: the state of the art. *Am. J. Med. Sci.* 338(1):48–49.

[129] Bazzano, L.A. (2011). No effect of folic acid supplementation on cardiovascular events, cancer or mortality after 5 years in people at increased cardiovascular risk, although homocysteine levels are reduced. *Evidence-Based Med.* 16(4):117–118.

[130] Johansson, M., Appleby, P.N., Allen, N.E., Travis, R.C., Roddam, A.W., Egevad, L., Jenab, M., et al. (2008). Circulating concentrations of folate and vitamin B12 in relation to prostate cancer risk: results from the European Prospective Investigation into Cancer and Nutrition study. *Cancer Epidem. Biomar. and Prev.* 17(2):279–285.

[131] Figueiredo, J.C., Grau, M.V., Haile, R.W., Sandler, R.S., Summers, R.W., Bresalier, R.S., Burke, C.A., McKeown-Eyssen, G.E. and Baron, J.A. (2009). Folic acid and risk of prostate cancer: results from a randomized clinical trial. *J. Natl. Cancer Inst.* 101(6):432–435.

[132] Ulrich, C.M. (2007). Folate and cancer prevention: a closer look at a complex picture. *Am. J. Clin. Nutr.* 86(2):271–273.

[133] Dawson-Hughes, B., Harris, S.S., Krall, E.A. and Dallal, G.E. (1997). Effect of calcium and vitamin D supplementation on bone density in men and women 65 years of age or older. *N. Engl. J. Med.* 337(10):670–676.

[134] Ross, A.C., Taylor, C.L., Yaktine, A.L., and Del Valle H.B. (eds.) (2011). *Dietary Reference Intakes for Calcium and Vitamin D*. Institute of Medicine. Food and Nutrition Board. Committee to Review Dietary Reference Intakes for Vitamin D and Calcium. National Academies Press. Washington, DC.

[135] Bihl, G. and Meyers A. (2001). Recurrent renal stone disease-advances in pathogenesis and clinical management. *Lancet* 358:651–656.

[136] Hall, W.D., Pettinger, M. and Oberman, A. (2001). Risk factors for kidney stones in older women in the Southern United States. *Am. J. Med. Sci.* 322(1):12–18.

[137] Raboy, V. (2001). Seeds for a better future: low phytic grains help to overcome malnutrition and reduce pollution. *Trends Plant Sci.* 6:458-462.

[138] Institute of Medicine (2006). *Dietary reference intakes: the essential guide to nutrient requirements*. National Academies Press. Washington, DC.

[139] Grases, F., Simonet, B.M., Vucenik, I., Prieto, R.M., Costa-Bauza, A., March, J.G. and Shamsuddin, A.M. (2001). Absorption and excretion of orally administered inositol hexaphosphate (IP(6) or phytate) in humans. *Biofactors* 15:53-61.

[140] Dias, J.S. (1995). The Portuguese tronchuda cabbage and galega kale landraces: a historical review. *Genet. Resour. Crop Evol.* 42:179-194.

[141] Monteiro, A.A. and Rosa E. (2008). Caldo verde: a story of Portuguese brassicas. *Chronica Horticulturae* 38(1):9-11.

[142] Nelson, D.L. and Cox, M.M. (2005). *Lehninger, Principles of Biochemistry*. 4th Ed. W.H. Freeman and Company, New York.

[143] Guerrera, M.P., Volpe, S.L., Mao, J.J. (2009). Therapeutic uses of magnesium. *Am Fam. Physician* 80(2):157-162.

[144] Bartlett, H.E. and Eperjesi, F. (2008). Nutritional supplementation for type 2 diabetes: a systematic review. *Ophthalmic Physiol. Opt.* 28(6): 503-523.

[145] Kim, D.J., Xun, P., Liu, K., Loria, C., Yokota, K., Jacobs, D.R. Jr. and He K. (2010). Magnesium intake in relation to systemic inflammation, insulin resistance, and the incidence of diabetes. *Diabetes Care* 33(12): 2604-2610.

[146] Johnson, S. (2001).The multifaceted and widespread pathology of magnesium deficiency. *Med. Hypotheses* 56(2):163-170.

[147] Champagne, C.M. (2008). Magnesium in hypertension, cardiovascular disease, metabolic syndrome, and other conditions: a review. *Nutr. Clin. Pract.* 23(2):142-151.

[148] Eby, G.A. and Eby, K.L. (2010). Magnesium for treatment-resistant depression: a review and hypothesis. *Med. Hypothesis* 74(4):649-660.

[149] Hijazi, N., Abalkhail, B. and Seaton, A. (2000). Diet and childhood asthma in a society in transition: a study in urban and rural Saudi Arabia. *Thorax* 55:775-779.

[150] Ciarallo, L., Brousseau, D. and Reinert, S. (2000). Higher-dose intravenous magnesium therapy for children with moderate to severe acute asthma. *Arch. Ped. Adol. Med.* 154(10):979-983.

[151] Bo, S. and Pisu, E. (2008). Role of dietary magnesium in cardiovascular disease prevention, insulin sensitivity and diabetes. *Curr. Opin. Lipidol.* 19(1):50-56.

[152] Abdelwahab, M., Youssef, S., Aly, A., Elfiki, S., Elenany, N. and Abbas, M. (1992). A simple calibration of a whole-body counter for the measurement of total body

potassium in humans. *Int. J. Radiat. Applic. Instrumen. Part A. Appl. Radiat. Isotopes* 43(10):1285–1289.

[153] Campbell, N. (1987). *Biology*. Benjamin/Cummings Pub. Co. Menlo Park, California.

[154] D'Elia, L., Barba, G., Cappuccio, F. and Strazzullo, P. (2011). Potassium intake, stroke, and cardiovascular disease: a meta-analysis of prospective studies. *College Cardiology* 57(10):1210–1219.

[155] Slonim, A.D. and Pollack, M.M. (2006). *Potassium. Pediatric critical care medicine*. Williams and Wilkins, Lippincott.

[156] WHO (World Health Organization) (2008). *Worldwide prevalence of anaemia 1993–2005 WHO global database on anaemia.* Edited by Benoist, B., McLean, E., Ines Egli, I. and Cogswell, M. WHO, Geneva,, Switzerland.

[157] Milman, N. (1996). Serum ferritin in Danes: studies of iron status from infancy to old age, during blood donation and pregnancy. *Int. J. Hematology* 63 (2): 103–135.

[158] Hoppe, M., Hulthén, L. and Hallberg, L. (2005). The relative bioavailability in humans of elemental iron powders for use in food fortification. *Eur. J. Nutr.* 45(1):37–44.

[159] Pineda, O. and Ashmead, H.D. (2001). Effectiveness of treatment of iron-deficiency anemia in infants and young children with ferrous bis-glycinate chelate. *Nutrition* 17(5):381–384.

[160] Haddad, E.H., Berk, L.S., Kettering, J.D., Hubbard, R.W. and Peters, W.R. (1999). Dietary intake and biochemical, hematologic, and immune status of vegans compared with nonvegetarians. *Am. J. Clin. Nutr.* 70: 586S-593S.

[161] Obeid, R., Geisel, J. and Schorr, H. (2002). The impact of vegetarianism on some haematological parameters. *Eur. J. Haematol*. 69:275-279.

[162] Hallberg, L. (1981). Bioavailability of dietary iron in man. *Ann. Rev. Nutr.* 1:123-147.

[163] Gleerup, A., Rossander Hulthen, L. and Gramatkovski, E. (1995). Iron absorption from the whole diet: comparison of the effect of two different distributions of daily calcium intake. *Am. J. Clin. Nutr.* 61:97-104.

[164] Hambidge, K.M. and Krebs, N.F. (2007). Zinc deficiency: a special challenge. *J. Nutr.* 137(4):1101–1105.

[165] Broadley, M.R., White, P.J., Hammond, J.P., Zelko I. and Lux A. (2007). Zinc in plants. *New Phytologist* 173(4):677–702.

[166] Brandt, E.G., Hellgren, M., Brinck, T., Bergman, T. and Edholm, O. (2009). Molecular dynamics study of zinc binding to cysteines in a peptide mimic of the alcohol dehydrogenase structural zinc site. *Phys. Chem. Chem. Phys.* 11(6):975–983.

[167] Prasad, A.S. (2003). Zinc deficiency. *Brit. Med. J.* 326:409–410.

[168] WHO (2007). The impact of zinc. World Health Organization. http://www.who.int/child_adolescent_health/ documents/zinc_mortality/en/index.html.

[169] Shrimpton, R., Gross, R., Darnton-Hill, I. and Young, M. (2005). Zinc deficiency: what are the most appropriate interventions? *Brit. Med. J.* 330: 347–349.

[170] Ensminger, A.H. and Konlande, J.E. (1993). *Foods and Nutrition Encyclopedia*. 2nd ed. CRC Press. Boca Raton, Florida.

[171] Freeland-Graves, J.H., Bodzy, P.W. and Epright, M.A. (1980). Zinc status of vegetarians. *J. Am. Dietetic Assoc.* 77:655–661.

[172] Rayman, M.P. (2000). The importance of selenium to human health. *Lancet* 356:233-241.

[173] Esaki, N., Nakamura, T., Tanaka, H., and Soda, K. (1982). Selenocysteine lyase, a novel enzyme that specifically acts on selenocyteine. Mammalian distribution and purification and properties of pig liver enzyme. *J. Biol. Chem.* 257:4386-4391.

[174] Swanson, C.A., Patterson, B.H., Levander, O.A., Veillon, C., Taylor, P.R., Helzlsouer, K., McAdams, P.A., and Zech, L.A. (1991). Human [74Se]selenomethionine metabolism: a kinetic model. *Am. J. Clin. Nutr.* 54:917-926.

[175] Lane, H.W., and Medina, D. (1985). Mode of action of selenium inhibition of 7, 12dimethylbenz[a]anthrancene-induced mouse mammary tumorogenesis. *J. Natl. Cancer Inst.* 75:674-679.

[176] Ip, C. (1986). Chemopreventive role of selenium in carcinogenesis. *J. Am. Coll. Toxicol.* 5:7-20.

[177] Burguera, J.L., Burguera, M., Gallignani, M., Alarcón, O.M., Burguera, J.A. (1990). Blood serum selenium in the province of Mérida, Venezuela, related to sex, cancer incidence and soil selenium content. *J. Trace Elements and Electrolytes in Health and Disease* 4(2):73–77.

[178] Russo, M.W., Murray, S.C., Wurzelmann, J.I., Woosley, J.T. and Sandler, R.S. (1997). Plasma selenium levels and the risk of colorectal adenomas. *Nutr. and Cancer* 28(2): 125–129.

[179] Knekt, P., Marniemi, J., Teppo, L., Heliövaara, M. and Aromaa, A. (1998). Is low selenium. *Am. J. Epidem.* 148(10):975–982.

[180] Young, K.J. and Lee, P.N. (1999). Intervention studies on cancer. *Eur. J. Cancer Prev.* 8(2): 91–103.

[181] Helzlsouer, K.J., Alberg, A.J., Norkus, E.P., Morris, J.S., Hoffman, S.C., and Comstock, G.W. (1996). Prospective study of serum micronutrients and ovarian cancer. *J. Natl. Cancer Inst.* 88:32-37.

[182] Yoshizawa, K., Willett, W.C., Morris, S.J., Stampfer, M.J., Spiegelman, D., Rimm, E.B., and Giovannucci, E. (1999). Study of pre-diagnostic selenium level in toenails and the risk of advanced prostate cancer. *J. Natl. Cancer Inst.* 90:1219-1224.

[183] Clark, L.C., Combs, G.F. and Turnbull, B.W. (1996). Effects of selenium supplementation for cancer prevention in patients with carcinoma of the skin. A randomized controlled trial. Nutritional Prevention of Cancer Study Group. *JAMA* 276(24):1957–1963.

[184] Tsavachidou, D, McDonnell, T.J. and Wen S. (2009). Selenium and Vitamin E: Cell Type– and Intervention-Specific Tissue Effects in Prostate Cancer. *Nat. Cancer Institute* 101(5):306–320.

[185] Overcast, J.D., Ensley, A.E., Buccafusco, C.J., Cundy, C., Broadnax, R.A., He, S., Yoganathan, A.P., Pollock, S.H., Hartley, C.J., and May, S.W. (2001). Evaluation of cardiovascular parameters of a selenium-based antihypertensive using pulsed Doppler ultrasound. *J. Cardiovasc. Pharmacol.* 38:337-346.

[186] Knekt, P., Heliovaara, M., Aho, K., Alfthan, G., Marniemi, J., and Aromaa, A. (2000). Serum selenium, serum alpha-tocopherol, and the risk of rheumatoid arthritis. *Epidemiology* 11:402-405.

[187] Keshan Disease Research Group. (1979). Observation on effect of sodium selenite in prevention of Keshan disease. *China Med. J.* 92:471-476.

[188] Yang, G.Q., Ge, R.K., Chen, J., and Chen, X. (1988). Selenium-related endemic diseases and the daily selenium requirement of humans. *World Rev. Nutr. Diet* 55:98-152.

[189] Moreno-Reyes, R., Mathieu, J., Vanderpas, M., Begaux, F., Suetens, C., Rivera, M.T., Nève, J., Perlmutter, N. et al. (2003). Selenium and iodine supplementation of rural Tibetan children affected by Kashin-Beck osteoarthropathy. *Am. J. Clin. Nutr.* 78(1):137–144.

[190] Mazokopakis, E.E., Papadakis, J.A., Papadomanolaki, M.G., Batistakis, A.G., Giannakopoulos, T.G., Protopapadakis, E.E. and Ganotakis, E.S. (2007). Effects of 12 months treatment with L-selenomethionine on serum anti-TPO Levels in Patients with Hashimoto's thyroiditis. *Thyroid* 17 (7): 609–612.

[191] Mueller, A. and Pallauf, J. (2006): Compendium of the anti-diabetic effects of supra-nutritional selenate doses. *In vivo* and *in vitro* investigations with type II diabetic db/db mice. *J. Nutr. Biochem.* 17: 548-560.

[192] Bleys, J., Navas-Acien, A. and Guallar, E. (2007): Serum selenium and diabetes in U.S. adults. *Diabetes Care* 30:829-834.

[193] Laclaustra M., Navas-Acien, A., Stranges, S., Ordovas, J.M. and Guallar, E. (2009). Serum selenium concentrations and diabetes in U.S. adults: National Health and Nutrition Examination Survey (NHANES) 2003-2004. *Environ. Health Perspect.* 117:1409-1413.

[194] Akbaraly, T.N., Arnaud, J., Rayman, M.P., Hininger-Favier, I., Roussel, A.M., Berr, C. and Fontbonne, A. (2010). Plasma selenium and risk of dysglycemia in an elderly French population: Results from the prospective Epidemiology of Vascular Aging Study. *Nutr. and Metabolism* 7:21-28.

[195] Dreher, M.L. (2001). Dietary fiber overview. In: Cho, S.S. and Dreher, M.L. (eds.). *Handbook of dietary fiber*. Marcel Dekker, New York.

[196] Anderson, J.W., Deakins, D.A., Floore, T.L., Smith, B.M., and Whitis, S.E. (1990). Dietary fiber and coronary heart disease. *Crit. Rev. Food Sci. Nutr.* 29:95-147.

[197] Anderson, J.W. and Bridges, S.R. (1988). Dietary fiber content of selected foods. *Am. J. Clin. Nutr.* 47:440-447.

[198] Buttriss, J.L. and Stokes, C.S. (2008). Dietary fibre and health: An overview. *Nutr. Bull.* 33(3):186-200.

[199] Marlett, J.A. (1992). Content and composition of dietary fiber in 117 frequently consumed foods. *J. Am. Diet. Assoc.* 92:175-186.

[200] Blundell, J.E. and Burley, V.J. (1987). Satiation, satiety and the action of fibre on food intake. *Int. J. Obes.* 11:9-25.

[201] Harold, M., Reeves, R., Blze, M., Guthrie, R., and Guthrie, D. (1985). Effect of dietary fiber in insulin-dependent diabetics: insulin requirements and serum lipids. *J. Am. Diet. Assoc.* 85:1455-1461.

[202] Rouse, L.L., Beilin, L.J., Armstrong, B.K., and Vandongen, R. (1983). Blood pressure lowering effect of a vegetarian diet: controlled trial in normotensive subjects. *Lancet* 1:5-10.

[203] Le Marchand, L., Hankin, J.H., Wilksens, L.R., Kolonel, L.N., Englyst, H.N., and Lyu, L.C. (1997). Dietary fiber and colorectal cancer risk. *Epidemiology* 8:658-665.

[204] Slavin, J.L. (20019. Resistant oligosaccharides. In: Cho, S.S. and Drecher, M.L. (eds.), *Handbook of dietary fiber*. Marcel Dekker, New York. Pp.125-145.

[205] Saura-Calixto, F. and Goñi, I. (2006). Antioxidant capacity of the Spanish Mediterranean diet. *Food Chem.*94(3):442-447.

[206] Saura-Calixto, F., Serrano, J. and Goñi, I. (2007). Intake and bioaccessibility of total polyphenols in a whole diet. *Food Chem.* 101(2): 492-501.

[207] Rohdman, E.E. (1991). A taxonomic analysis of glucosinolate producing plants, part 1. Phenetics. *Syst. Bot.* 16:598-618.

[208] Fahey, J.W., Zalcmann, A.T. and Talalay, P. (2001). The chemical diversity and distribution of glucosinolates and isothiocyanates among plants. *Phytochemistry* 56:5-51.

[209] Wright, A.S. (1980). The role of metabolism in chemical mutagenesis and chemical carcinogenesis. *Mutation Res.* 75:215-223.

[210] Holst, B. and Williamson, G. (2004). A critical review of the bioavailability of glucosinolates and related compounds. *Natural Product Reports* 21:425-447.

[211] Zhang, Y. and Talalay, P. (1994) Anticarcinogenic activities of organic isothiocyanate: chemistry and mechanisms. *Cancer Res.* 54:1976s-1981s.

[212] Mithen, R., Faulkner, K., Magrath, R., Rose, P., Williamson, G. and Marquez, J. (2003). Development of isothiocyanate enriched broccoli and its enhanced ability to induce phase 2 detoxification enzymes in mammalian cells. *Theor. Appl. Gen.* 106: 727-734.

[213] Fenewick, G.R., Heaney, R.K. and Mullin, J. (1983). Glucosinolates and their break down products in food and food plants. *CRC Critical Rev. Food Sci. Nutr.* 18: 123-201.

[214] Verkerk, R., Knol, J.J. and Dekker, M. (2010). The effect of steaming on the glucosinolate content in broccoli. *Acta Hort.* 867:37-45.

[215] Campbell, L.D., Slominski, B.A., Nugon-Baudon, L., Rabot, S., Lory, S., Quinsac, A., Krouti, M. and Ribaillier, D. (1995). Studies on intestinal tract glucosinolate content, xenobiotic metabolizing enzymes and thyroid status in germ-free and conventional rats fed rapeseed meal. *Proc. 9th Int. Rapeseed Congr.* 1:212-214.

[216] Rosa, E.A.S., Heaney, R.K., Fenewick, G.R. and Portas, C.A.M. (1997). Glucosinolates in crop plants. *Hort. Rev.* 19: 99-215.

[217] Conaway, C.C., Jiao, D. and Chung, F.L. (1996). Inhibition of rat liver cytochrome p 450 isozymes by isothiocyanates and their conjugates: a structure activity relationship study. *Carcinogenesis* 17: 2423-2427.

[218] Fahey, J.W., Zhang, Y.S. and Talalay, P. (1997). Broccoli sprouts: an exceptionally rich source of inducers of enzymes that protects against chemical carcinogens. *Proc. Natl. Acad. Sci. (USA)* 94:10367-10372.

[219] Smith, T.K., Lund, E.K., Parker, M.L., Clarke, R.J. and Johnson, I.T. (2004). Allyl isothiocyanate causes mitotic block, loss of cell adhesion and disrupted cytoskeletal structure in HT 29 cells. *Carcinogenesis* 25: 1409-1415

[220] Palaniswamy, U.R., McAvoy, R.J., Bible, B.B. and Stuart, J.D. (2003). Ontogenic variations of ascorbic acid and phenethyl isothiocyanate concentrations in watercress (*Nasturtium officianale* R.Br.) leaves. *J. Agri. Food Chem.* 51:5504-5509.

[221] Ribnicky, D.M., Poulev, A., Henry, E. and Raskin, J. (2001). Seed of *Barbarea verna* as a rich source of phenethyl isothiocyanate to provide natural protection from environmental and dietary toxins. *J. Nutrac. Funct. Medic. Foods* 3:43-65.

[222] Morse, M.A., Zu, H., Galati, A.J, Schmidt, C.J. and Stoner, G.D. (1993). Dose-related inhibition by dietary phenethyl isothiocyanate of esophageal tumorigenesis and DNA

methylation induced by N-nitrosomethylbenzylamine in rats. *Cancer Letters* 72: 103-110.

[223] Stoner, G.D., Morse, M.A. (1996). Isothiocyanates as inhibitors of esophageal cancer. *Adv. Exp. Med. Biol.* 401:13-23.

[224] Hecht, S.S. (1996). Chemoprevention of lung cancer by isothiocyanates. *Adv. Exp. Med. Biol.* 401:1-11.

[225] Hwang, E.S. and Jeffery, E.H. (2005). Induction of quinone reductase by sulforaphane N-acetyl cystein conjugate and in murine hepatoma cells. *J. Medic. Food* 8: 198-203.

[226] Khor, T.O., Hu, R., Shen, G., Jeong, W.S., Hebbar, V., Chen, C., Xu, C., Nair, S., Reddy, B., Chada, K. and Kong, A.N.T. (2006). Pharmacogenomics of cancer chemo preventive isothiocyanate compound sulforaphane in the intestinal polyps of Apc Min/+ mice. *Biopharm. Drug Dispos.* 27:407-420.

[227] Hansen, M., Bengtsson, G.B., Borge, G.I., Berge, L. and Wold, A.B. (2010). Red cabbage, a vegetable rich in health-related glucosinolates. *Acta Hort.* 867:61-65.

[228] Fimognari, C., Nusse, M., Cesari, R., Iori, R., Cantelli-Forti, G. and Hrelia, P. (2002). Growth inhibition, cell cycle arrest and apoptosis in human T-cell leukaemia by the isothiocyanates sulphoraphane. *Carcinogenesis* 23: 581-586.

[229] Zhang, Y., Kensler T.W., Cho, C.G., Posner, G.H and Talalay, P. (1994). Anti-carcinogenic activities of sulforaphane and structurally related synthetic norbornyl isothiocyanate. *Proc. Natl. Acad. Sci. (USA)* 91:3147-3150.

[230] [230] Talalay, P., Fahey, J.W., Holtzclaw, W.D., Prestera, T. and Zhang, Y. (1995). Chemo protection against cancer by phase 2 enzyme induction. *Toxicol. Lett.* 82:173-179.

[231] Talalay, P. and Zhang, Y. (1996). Chemoprotection against cancer by isothiocyanate and glucosinolates. *Biochem. Soc. Transactions* 24:806-810.

[232] Thejass, P. and Kuttan, G. (2006). Augmentation of natural killer cell and antibody dependent cellular cytotoxicity in BALB/c mice by sulforphane, a naturally occurring isothiocyanate from broccoli 1L-2 and IFN-gamma. *Immunnopharmacology and Immunnotoxicology* 28:443-457.

[233] Wang, L.J., Giovannucci, E.L., Hunter, D., Neuberg, D., Su, L. and Christiani, D.C. (2004). Dietary intake of Cruciferous vegetables glutathione -S- transferase (GST) polymorphisms and lung cancer risk in a Caucasian population. *Cancer Causes Control* 15:977-985.

[234] Jongen, W.M.F. (1996). Glucosinolates in brassica: occurrence and significance as cancer modulating agents. *Proc. Nutr. Soc.* 55:433-446.

[235] Yuan, F., Chend, D.Z., Liu, K., Sepkovic, D.W., Bradlow, H.L. and Auborn, K. (1999). Anti estrogenic activities of indole-3-carbinol in cervical cell: Implication for prevention of cervical cancer. *Anticancer Res.* 19:1673-1680.

[236] Ashok, B.T., Chen, Y., Liu, X., Bradlow, H.L., Mittelman, A. and Tiwari, R.K. (2001). Abrogation of estrogen mediated cellular and biochemical effects by indole -3-carbinol. *Nutr. and Cancer* 41:180-187.

[237] Ashok, B.T, Chen ,Y., Liu ,X., Garikapaty, V.P., Seplovitz, R., Tschorn, J., Roy, K., Mittelman, A. and Tiwari, R.K. (2002). Multiple molecular targets of indole-3-carbinol: a chemo preventive anti estrogen in breast cancer. *Eur. J. Cancer Prev.* 11:S86-S93.

[238] Aggarwal, B.B. and Ichikawa, H. (2005). Molecular targets and anti cancer potential of indole-3-carbinol and its derivative. *Cell Cycle* 4:1201-1215.

[239] Firestone, G.L. and Bjeldanes, L.F. (2003). Indole-3-carbinol and 3',3'-diindolylmethane anti-proliferative signaling pathways control cell cycle gene transcription in human breast cancer cells by regulating promoter sp-1 transcription factor interactions. *J. Nutr.* 133: 2448S-2455S.

[240] Wu, H. T., Lin, S.H. and Chen, Y.H. (2005). Inhibition of cell proliferation and in vitro markers of angiogenesis by indole-3-carbinol a major indole metabolite present in Cruciferous vegetables. *J. Agric. Chem.* 53:5164-5169.

[241] Goodrich, R.M., Anderson J.L. and Stoewsand, G. (1989). Glucosinolate changes in blanched broccoli and Brussels sprouts. *J. Food Proc. Preserv.* 13:275-280.

[242] Rosa, E.A.S. and Heaney, R.K. (1993). The effect of cooking and processing on the glucosinolate content: studies on four varieties of Portuguese cabbage an hybrid white cabbage. *J. Sci. Food Agric.* 62:259-265.

[243] Vallejo, F., Tomás-Barberán, F. A. and Garcia-Viguera, C. (2002b). Glucosinolates and vitamin C content in edible parts of broccoli inflorescences after domestic cooking. *Eur. Food Res. Technol.* 215:310-316.

[244] Verkerk, R. and Dekker, M. (2004). Glucosinolates and myrosinase activity in red cabbage (*Brassica oleracea* L. "*capitata*" f. *rubra* DC.) after various microwave treatments. *J. Sci. Food Chem.* 52:7318-7323.

[245] Desjardins, Y. (2008). Onions as a nutraceutical and functional food. *Chronica Horticulturae* 48(2):8-14.

[246] Nencini, C., Cavallo, F., Capasso, A., Franchi, G.G., Giorgio, G. and Micheli, L. (2007). Evaluation of antioxidative properties of Allium species growing wild in Italy. *Phytotherapy Research* 21(9):874-878.

[247] Verma, S.K. and Verma, D. (2008). Garlic - "The spice of life": composition, cooking, chemistry and preparations. *J. Herbal Med. Toxicol.*, 2(2):21-28.

[248] Kalra, N., Arora, A. and Shukla, Y. (2006). Involvement of multiple signaling pathways in diallyl sulfide mediated apoptosis in mouse skin tumors. *Asian Pacific J. Cancer Prev.* 7(4):556-562.

[249] Rose, P., Whiterman, M., Moore, P.K. and Zhu, Y.Z. (2005). Bioactive S-alk(en)yl cysteine sulfoxide metabolites in the genus *Allium*: the chemistry of potential therapeutic agents. *Nat. Prod. Rep.* 22:351-368.

[250] Rubec, R., Svobodovaand, M. and Velisek, J. (1999). Gas chromatographic determination of *S*-alk(en)yl-L-cysteine sulfoxides. *J. Chromatogr.* 862:85-94.

[251] Thomas, D.J. and Parkin, K.L. (1994). Quantification of alk(en)yl-L-cysteine sulfoxide and related amino acids in alliums by high-performance liquid chromatography. *J. Agr. Food. Chem.* 42:1632-1638.

[252] Block, E., Naganathan, S., Putnamand, D. and Zhao, S.H. (1992a). *Allium* chemistry: HPLC-analysis of thiosulfinates from onion, garlic, wild garlic (ramsons), leek, scallion, shallot, elephant (great-headed) garlic, chiveand Chinese chive: uniquely high allyl-to methyl ratios in some garlic samples. *J. Agr. Food Chem.* 40:21418-21430.

[253] Breu, W. (1996). *Allium cepa* (onion). Part 1: Chemistry and analysis. *Phytomedicine* 3:293-306.

[254] Calvey, E.M., Matusik, J.E., White, K.D., DeOrazio, R., Shaand, D. and Block, E. (1997). *Allium* chemistry: Supercritical fluid Extraction and LD-APCI-MS of thiosulfinates and related compounds from homogenates of garlic, onion, ramp.

Identification in garlic and ramp and synthesis of 1-propanesulfinothionic acid *S-allyl* ester. *J. Agr. Food Chem.* 45:4406-4413.

[255] Yin, M., Hwang, S. and Chan, K. (2002). Nonenzymatic antioxidant activity of four organosulfur compounds derived from garlic. *J. Agric. Food Chem.* 50: 6143-6147.

[256] Lanzotti, V. (2006). The analysis of onion and garlic. *J. Chromatography. A* 1112(1-2):3-22.

[257] Weiner, L., Shin, I., Shimon, L.J., Miron, T., Wilchek, M., Mirelman, D., Frolow, F. and Rabinkov, A. (2009). Thiol-disulfide organization in alliin lyase (alliinase) from garlic (*Allium sativum*). *Protein Sci.* 18(1):196-205.

[258] Kubec, R., Svobodovaand, M. and Velisek, J. (2000). Distribution of *S*-alk(en)ylcysteine sulfoxides in some *Allium* species. Identification of a new flavour precursor: *S*-Ethylcysteine sulfoxide (ethiin). *J. Agr. Food Chem.* 48:428-433.

[259] Kamel, A. and Saleh, M. (2000). Recent studies on the chemistry and biological activities of the organosulfur compounds of garlic (*Allium sativum*). *Studies Natural Products Chem.* 23:455-485.

[260] Amagase, H. (2006). Clarifying the real bioactive constituents of garlic. *J. Nutr.* 136(3):716S-725S.

[261] Munchberg, U., Anwar, A., Mecklenburg, S. and Jacob, C. (2007). Polysulfides as biologically active ingredients of garlic. *Organic Biomo. Chem.* 5(10):1505-1518.

[262] Yoo, K.S. and Pike, L.M. (1998). Determination of flavor precursor compound *S*-alk(en)yl-L-cysteine sulfoxides by HPLC method and their distribution in *Allium* species. *Scientia Hort.* 75:1-10.

[263] Randle, W.M., and Lancaster, J.E. (2002). Sulphur compounds in Alliums in relation to flavour quality. In: Brewster, J.L. (ed.). *Onions and other vegetable alliums.* CAB International, Wallingford, Oxfordshire, UK. Pp. 329–356.

[264] Brewster, J.L. (ed.) (2008). *Onions and other vegetable alliums.* Crop Production Science in Horticulture 15. 2nd Ed. CAB International, Wallingford, Oxfordshire, UK.

[265] Siegers, C., Steffen, B., Robkeand, A. and Pentz, R. (1999). The effect of garlic preparations against tumor cell proliferation. *Phytomedicine* 6:7-11.

[266] Harris, J.C., Cottrell, S.L., Plummer, S. and Lloyd, D. (2001). Antimicrobial properties of *Allium sativum* (garlic). *Appl. Microb. Biotech.* 57: 282-286.

[267] Jackson, R., McNeil, B., Taylor, C., Holl, G., Ruf, D. and Gwebu, E.T. (2002). Effect of aged garlic extract in casepase-3 activity in vitro. *Nutr. Neuroscience*, 5:287-290.

[268] Davenport, D.M. and Wargovich, M.J. (2005). Modulation of cytochrome P450 enzymes by organosulfur compounds from garlic. *Food Chemical Toxicol.* 43(12):1753-1762.

[269] Wargovich, M.J. (2006). Diallylsulfide and allylmethylsulfide are uniquely effective among organosulfur compounds in inhibiting CYP2E1 protein in animal models. *J. Nutr.* 136(3):832S-834S.

[270] Knowles, L.M. and Milner, J.A. (2001). Possible mechanisms by which allyl sulphides suppress neoplastic cell proliferation. *J. Nutr.* 131:1061S-1066S.

[271] Brady, J.F., Ishizaki, H., Fukuto, J.M., Lin, M.C., Fadel, A., Gupacand, J.M. and Yang, C.S. (1991). Inhibition of cytochrome P-450 2E1 by diallyl sulfide and its metabolites. *Chem. Res. Toxicol.* 4:642-647.

[272] Kwak, M.K., Kim, S.G., Kwak, J.Y., Novakand, R.F. and Kim, N.D. (1994). Inhibition of cytochrome P450 2E1 expression by organosulfur compounds. *Biochem. Pharmacol.* 47:531-539.

[273] Reicks, M.M. and Crankshaw, D.L. (1996). Modulation of rat hepatic cytochrome P450 activity by garlic organosulfur compounds. *Nutr. Cancer* 25:241-248.

[274] Sundaram, S.G. and Milner, J.A. (1993). Impact of organosulfur compounds in garlic on canine mammary tumor cells in culture. *Cancer Lett.* 74:85-90.

[275] Sakamoto, K., Lawsonand, L.D. and Milner, J.A. (1997). Allyl sulfides from garlic suppresses the in vitro proliferation of human A549 lung tumor cells. *Nutr. Cancer* 29:152-156.

[276] Appel, E., Rabinkov, A., Neeman, M., Kohen, F. and Mirelman, D. (2011). Conjugates of daidzein-alliinase as a targeted pro-drug enzyme system against ovarian carcinoma. *J. Drug Targeting* 19(5):326-335.

[277] Liu, L. and Yeh, Y.Y. (2000). Inhibition of cholesterol biosynthesis by organosulfur compounds derived from garlic. *Lipids* 35:197-203.

[278] Ide, N. and Lau, B.H. (1997). Garlic compounds protect vascular endothelial cells from oxidized low-density lipoprotein-induced injury. *J. Pharm. Pharmacol.* 49:908-911.

[279] Patrono, C. (2001). Aspirin: new cardiovascular uses for an old drug. *Am. J. Med.* 11:62S-65S.

[280] Rahman, K. (2001). Historical perspective on garlic and cardiovascular disease. *J. Nutr.* 131:977S-979S.

[281] Lawson, L.D., Ransomand, D.K. and Hughes, B.G. (1992). Inhibition of whole blood platelet aggregation by compounds in garlic clove extracts and commercial garlic preparations. *Thromb. Res.* 65:141-156.

[282] Morimitsu, Y., Moriokaand, Y. and Kawakishi, S. (1992). Inhibitors of platelet aggregation generated from mixtures of *Allium* species and/or S-alk(en)y-L-cysteine sulfoxides. *J. Agr. Food Chem.* 40:368-372.

[283] Ali, M., Thomson, M. and Afzal, M. (2000). Garlic and onions: their effect on eicosanoid metabolism and its clinical relevance. *Prost. Leuko. Ess. Fatty Acids* 62: 55-73.

[284] Hubbard, G.P., Wolffram, S., Lovegrove, J.A. and Gibbins, J.M. (2006). Ingestion of quercetin inhibits platelet aggregation and essential components of the collagen-stimulated platelet activation pathway in man: a pilot study. *J. Thromb. Haem.* 2:2138-2145.

[285] Janssen, K., Mensink, R.P., Cox, F.J., Harryvan, J.L., Hovenier, R., Hollman, P. C.H. and Katan, M.B. (1998). Effects of flavonoids quercetin and apigenin on hemostasis in healthy volunteers: results from an *in vitro* and a dietary supplement study. *Am. J. Clin. Nutr.* 67:255-262.

[286] Osmont, K.S., Arnt, C.R. and Goldman, I.L. (2003). Temporal aspects of onion-induced antiplatelet activity. *Plant Food Hum. Nutr.* 58:27-40.

[287] Lau, B.H., Yamasakiand, T. and Gridley, D.S. (1991). Garlic compounds modulate macrophage and T-lymphocyte functions. *Mol. Biother.* 3:103-107.

[288] Feng, Z.H., Zhangand, G.M. and Hao, T.L. (1994). Effect of dially trisulfide on the activation of T cell and macrophage-mediated cytotoxicity. *J. Tongji Med. Univ.* 14:142-149.

[289] Amagase, H., Petesch, B.L., Matsuura, H., Kasuga, S. and Itakura, Y. (2001). Intake of garlic and its bioactive components. *J. Nutr.* 131(3):955S-962S.

[290] Goncagul, G. and Ayaz, E. (2010). Antimicrobial effect of garlic (*Allium sativum*). *Recent Patents on Anti-infective Drug Discovery* 5(1):91-93.

[291] Sheela, C.G., Kumud, K. and Augusti, K.T. (1995). Anti-diabetic effect of onion and garlic sulfoxide amino acids in rats. *Planta Medica* 61:356-357.

[292] Augusti, K.T. and Sheela, C.G. (1996). Antiperoxide effect of S-allyl cysteine sulfoxide, an insulin secretagogue, in diabetic rats. *Experientia* 52:115-120.

[293] Agarwal, K.C. (1996). Therapeutic actions of garlic constituents. *Med. Res. Rev.* 16(1):111-124.

[294] Wagner, H., Dorsch, W., Bayer, T., Breu, W. and Willer, F. (1990). Anti-asthmatic effects of onions: inhibition of 5-lipoxygenase and cyclooxygenase in vitro by thiosulfinates and "cepaenes." *Prost. Leuko. Ess. Faty Acids* 39:59-62.

[295] Cao, G., Muccitelli, H.U., Sanchez-Morenoand, C. and Prior, R. (2001). Anthocyanins are absorbed in glyated forms in elderly women: a pharmacokinetic study. *Am. J. Clin. Nutr.* 73:920-926.

[296] Delgado-Vargas, F. and Paredes-López, O. (eds) (2003). *Natural colorants for food and nutraceutical uses*. CRC Press, Boca Raton. FL.

[297] Horbowicz, M., Kosson R., Grzesiuk A. and Bski H.D. (2008). Anthocyanins of fruits and vegetables – their occurrence analysis and role in human nutrition. *Vegetable Crops Res. Bull.* 68:5-22.

[298] Lila, M.A. (2004). Anthocyanins and human health: an in vitro investigative approach. *J. Biomed. Biotechnol.* 5: 306-313.

[299] Rechner, A.R. and Kroner, C. (2005). Anthocyanins and colonic metabolites of dietary polyphenols inhibit platelet function. *Thromb. Res.* 116:327-334.

[300] Noda, Y., Kaneyuki, T., Igarashi, K., Moriand, A. and Pacer, L. (1998). Antioxidant activity of nasunin, an anthocyanin in eggplant. *Res. Commun. Mol. Path. Pharmacol.* 102:175-187.

[301] Kayamori, F. and Igarashi, K. (1994). Effect of dietary nasunin on the serum cholesterol level in rats. *Biosci. Biotech. Biochem.* 58:570-571.

[302] Hou, D.X. (2003). Potential mechanisms of cancer chemoprevention by anthocyanins. *Curr. Molec. Med.* 3: 149-159.

[303] Stintzing, F.C., Stintzing, A.S., Carle, R., Frei, B. and Wrolstad, R.E. (2002). Color and antioxidant properties of cyanidin-based anthocyanin pigments. *J. Agric. Food. Chem.* 50: 6172-6181.

[304] Crozier, A., Lean, M.E., McDonaldand, M.S. and Black, C. (1997). Quantitative analysis of the flavonoid content of commercial tomatoes, onions, lettuce and celery. *J. Agr. Food Chem.* 45:590-595.

[305] Crozier, A., Burns, J., Aziz, A.A., Stewart, A.J., Rabiasz, H.S., Jenkins, G.I., Edwards, C.A., and Lean, M.E.J. (2000). Antioxidant flavonols from fruits, vegetables and beverages: measurements and bioavailability. *Biol. Res.* 33(2):79-88.

[306] Eloesser, W. and Herrmann, K. (1975). Flavanols and flavones of vegetable. V. Flavanols and flavones of root vegetables. *Z. Lebensm. Unters Forch.* 16:265-270.

[307] DuPont, M.S., Mondin, Z., Williamsonand, G. and Price, K.R. (2000). Effect of variety, processing and storage on the flavonoid glycoside content and composition of lettuce and endive. *J. Agr. Food Chem.* 48: 3957-3964.

[308] Miean, K.H. and Mohamed, S. (2001). Flavonoid (myricetin, quercetin, kaempferol, luteolin and apigenin) content of edible tropical plants. *J. Agr. Food Chem.* 49:106-112.

[309] Negre-Salvayre, A and Salvayre, R. (1992). Quercetin prevents the cytotoxicity of oxidised LDL on lymphoid cell lines. *Free Radicals Biol. Med.* 12:101-106

[310] Lio, M., Kawaguchi, H., Sakota, Y., Otonariand, J. and Nitahara, H. (1993). Effects of polyphenols, including flavonoids, on glutathione S-transferase and glutathione reductase. *Biosci. Biotech. Biochem.* 57: 1678-1680.

[311] Bohm, H., Boeing, H., Hempel, J, Raaband, B. and Kroke, A. (1998). Flavonols, flavones and anthocyanins as natural antioxidants of food and their possible role in the prevention of chronic diseases. *Z. Ernahrungswiss* 37:147-163.

[312] Gryglewski, R.J., Korbut, R. and Robak, J. (1987). On the mechanism of antithrombotic action of flavonoids. *J. Biochem. Pharmacol.* 36: 317-321.

[313] Cook, N.C. and Samman, S. (1996). Flavonoids-chemistry, metabolism, cardioprotective effects and dietary sources. *J. Nutr. Biochem.* 7:66-76.

[314] Baker, M.E., Medlockand, K.L. and Sheehan, D.M. (1998). Flavonoids inhibit estrogen binding to rat alpha-fetoprotein. *Expt. Boil. Med.* 217: 317-321.

[315] Hou, E.X., Fukuda, M., Johnson, J.A., Miyamori, K., Ushikaiand, M. and Fujii, M. (2001). Fisetin induces transcription of NADPH: quinine oxidoreductase gene through an antioxidant responsive element-involved activation. *Int. J. Oncol.* 18:1175-1179.

[316] Vinson, J.A., Jang, J., Dabbagh, Y.A., Serry, M.M. and Cai, S. (1995) Plant polyphenols exhibit lipoprotein-bond antioxidant activity using an *in vitro* oxidation model for heart disease. *J. Agric. Food Chem.* 43:2798-2799.

[317] Paganga, G., Millerand, N. and Rice-Evans, C.A. (1999). The polyphenolic content of fruit and vegetables and their antioxidant activities: What does a serving constitute? *Free Rad. Res.* 30:153-162.

[318] Cermak, R. (2008). Effect of dietary flavonoids on pathways involved in drug metabolism. *Expert Opin. Drug Metab. Toxicol.* 4(1):17–35.

[319] de Whalley, C.V., Rankin, S.M., Hoult, J.R., Hessupand, W. and Leake, D.S. (1990). Flavonoids inhibit the oxidative modification of low-density lipoproteins by macrophage. *Biochem. Pharmacol.* 39:1743-1750.

[320] Ullmannova, V. and Popescu, N.C. (2007). Inhibition of cell proliferation, induction of apoptosis, reactivation of DLC1, and modulation of other gene expression by dietary flavone in breast cancer cell lines. *Cancer Detect Prev.* 31(2):110–118.

[321] Hosny, M. and Rosazza, J.P. (2002). New isoflavone and triterpene glycosides from soybeans. *J. Nat. Prod.* 65:805-813.

[322] Xu, D.P., Xiao, K., Gu, W.Y. and Ding, X.L. (2003). Isolation of a new isoflavone from soybean germ. *Zhongcaoyao* 34:1065-1067.

[323] Halbrock, K. (1981). Flavonoids. In: Conn, E.E. (ed.). *Biochemistry of Plant.* Acad. Press, New York. Pp. 425-456.

[324] Liggins, J., Bluck, J.C., Runswick, S., Atkinson, C., Cowardand, W.A. and Bingham, S.A. (2000). Daidzein and genistein contents of vegetables. *Brit. J. Nutr.* 84:717-725.

[325] Vacek, J., Klejdus, B., Lojkova, L. and Kuban, V. (2008). Current trends in isolation, separation, determination and identification of isoflavones: a review. *J. Sep. Sci.* 31: 2054-2067.

[326] Dentith, S. and Lockwood, B. (2008). Development of techniques for the analysis of isoflavones in soy foods and nutraceuticals. *Curr. Opin. Clin. Nutr. Metab. Care* 11:242-247.

[327] Yu, O., Jung, W., Shi, J., Croes, R.A., Fader, G.M., McGonigleand, B. and Odell, J.T. (2000). Production of the isoflavones genistein and daidzein in non-legume dicot and monocot tissues. *Plant Physiol.* 124:781-794.

[328] Goldwyn, S., Lazinskyand, A. and Wei, H. (2000). Promotion of health by soy isoflavones: efficacy, benefit and safety concerns. *Drug Metabol. Drug Interact.* 17:261-289.

[329] Ren, M.Q., Kuhn, G., Wegner, J. and Chen, J. (2001). Isoflavones, substances with multi-biological and clinical properties. *Eur. J. Nutr.* 40:135-146.

[330] Messina, M., Kucuk, O., and Lampe, J.W. (2006). An overview of the health effects of isoflavones with an emphasis on prostate cancer risk and prostate-specific antigen levels. *J. AOAC Int.* 89:1121-1134.

[331] Larkin, T., Price, W.E. and Astheimer, L. (2008). The key importance of soy isoflavone bioavailability to understanding health benefits. *Crit. Rev. Food Sci. Nutr.* 48:538-552.

[332] Wiseman, H. (1999). The bioavailability of non-nutrient plant factors: dietary flavonoids and photo-estrogens. *Proc. Nutr. Soc.* 58:139-146.

[333] Sarkar, F.H. and Li, Y. (2006). Isoflavones, soybean phytoestrogens, and cancer. In: Awad, A.B. and Bradford, P.G. (eds.). *Nutrition and Cancer Prevention.* CRC Press, Boca Raton, FL. Pp 295-312.

[334] Clarkson, T.B. (2002). Soy, soy phytoestrogens and cardiovascular disease. *J. Nutr.* 132:566S-569S.

[335] Brynin, R. (2002). Soy and its isoflavones: a review of their effects on bone density. *Altern. Med. Rev.* 7:317-327.

[336] Messina, M., Ho, S. and Alekel, D.L. (2004). Skeletal benefits of soy isoflavones: a review of the clinical trial and epidemiologic data. *Curr. Opin. Clin. Nutr. Metab. Care* 7:649-658.

[337] Orgaard, A. and Jensen, L. (2008). The effects of soy isoflavones on obesity. *Exp. Biol. Med.* 233:1066-1080.

[338] Barnes, S., Boersma, B., Patel, R., Kirk, M., Darley-Usmar, V.M., Kim, H. and Xu, J. (2000). Isoflavonoids and chronic disease: mechanisms of action. *Biofactors 12:*209-215.

[339] Messina, M. (2010). A brief historical overview of the past two decades of soy and isoflavone research. *J. Nutr.* 140:1350S-1354S.

[340] Barnes, S. (2010). The biochemistry, chemistry and physiology of the isoflavones in soybeans and their food products. *Lymphat. Res. Biol.* 8:89-98.

[341] Lee, H.P., Gourley, L., Duffy, S.W., Esteve, J., Leeand, J. and Day, N.E. (1991). Dietary effects on breast cancer risk in Singapore. *Lancet* 33:1197-1200.

[342] Ziegler, R.G., Hooverand, R.N. and Hildeshein, R.N. (1993). Migration patterns and breast cancer risk in Asian-America women. *J. Nat. Cancer Inst.* 85:1819-1827.

[343] Lamartiniere, C. (2000). Protection against breast cancer with genistein: a component of soy. *Am. J. Clin. Nutr.* 71:1705S-1707S.

[344] Steiner, C., Arnould, S., Scalbert, A. and Manach, C. (2008). Isoflavones and the prevention of breast and prostate cancer: new perspectives opened by nutrigenomics. *Brit. J. Nutr.* 99(1):78-108.

[345] Messina, M.J. and Wood, C.E. (2008). Soy isoflavones, estrogen therapy, and breast cancer risk: analysis and commentary. *Nutr. J.* 7:17.

[346] Dong, J.Y. and Qin, L.Q. (2011). Soy isoflavones consumption and risk of breast cancer incidence or recurrence: a meta-analysis of prospective studies. *Breast Cancer Res. Treat.*125:315-323.

[347] Jiang, H.Y., Lv, F.J. and Tai, J.Q. (2000). Bioactive components of soybean and their function. *Soybean Sci.*19:160-164.

[348] Sarkar, F.H. and Li, Y. (2003). Soy isoflavones and cancer prevention. *Cancer Invest.* 21:744-757.

[349] Nielsen, S.E., Young, J.F., Daneshvar, B., Lauriden, S.T., Knuthsen, P., Sandrstromand, B. and Dragsted, L.O. (1999). Effect of parsley (*Petroselinum crispum*) intake on urinary apigenin excretion, blood antioxidant enzymes and biomarkers for oxidative stress in human subjects. *Brit. J. Nutr.* 81:447-455.

[350] Ching, L.S. and Mohamed, S. (2001). Alpha-tocopherol content of 62 edible tropical plants. *J. Agr. Food Chem.* 49:3101-3105.

[351] Almeida, D. (2006). *Manual de Culturas Hortícolas*. Vol. 1. Editorial Presença, Lisboa.

[352] Dhillon, N.P.S., Monforte, A.J., Pitrat, M., Pandey, S., Singh, P.K., Reitsma, K.R., Garcia-Mas, J., Sharma, A. and McCreight, J.M. (2012). Melon landraces od India: contributions and importance. *Plant Breed. Rev.* 35:85-150.

[353] Burger, Y., Yeselson, Y., Saar, U., Paris, H.S., Katzir, N., Tadmor, Y. and Schaffer, A.A. (2004). Screening of melon (*Cucumis melo*) germplasm for consistently high sucrose content and for high ascorbic acid content. In: Lebeda, A. and Paris, H.S. (eds.). *Progress in Cucurbit genetics and breeding research.* Proc. *Cucurbitaceae* 2004. Palacky University, olomouc, Czech Republic. Pp.151-155.

[354] Crosby, K.M., Lester, G.E. and Leskovar, D.I. (2006). Genetic variation for beneficial phytochemical levels in melons (*Cucumis melo*). In: Holmes, G.J. (ed.). *Cucurbitaceae 2006*. Universal Press, Raleigh, NC. Pp. 70-76.

[355] He, T., Huang, C.Y., Chen, H. and Hou, Y.H. (1999). Effects of spinach powder fat-soluble extract on proliferation of human gastric adenocarcinoma cell. *Biomed. Environ. Sci.* 12:247-252.

[356] Prakash D., Nath P. and Pal, M. (1993). Composition, variation of nutritional contents in leaves, seed protein, fat and fatty acid profile of chenopodium species. *J. Sci. Food Agric.* 62(2):203-205.

[357] Sienera, R. (2006). Oxalate contents of species of the *Polygonaceae*, *Amaranthaceae* and *Chenopodiaceae* families. *Food Chem.* 98(2):220-224.

[358] Mallillin, A.C., Trinidad, T.P., Raterta, R., Dagbay, K. and Loyola, A.S. (2008). Dietary fibre and fermentability characteristics of root crops and legumes. *Brit. J. Nutr.* 100(3):485-488.

[359] Trinidad, T.P., Mallillin, A.C., Loyola, A.S., Sagum, R.S. and Encabo, R.R. (2010). The potential health benefits of legumes as a good source of dietary fibre. *Brit. J Nutr.* 103(4):569-574.

[360] World Cancer Research Fund. (1997). *Food, nutrition and the prevention of cancer: A global perspective*. Am. Inst. Cancer Res. Washington, DC.

[361] Verhoeven, D.T.H., Goldbohm, R.A., van Poppel, G., Verhagen, H., and van den Brandt, P.A. (1996). Epidemiological studies on *Brassica* vegetables and cancer risk. *Cancer Epidem. Biomarkers Prev.* 5:733-751.

[362] Ambrosone, C.B., McCann, S.E., Freudenheim, J.L., Marshall, J.R., Zhang, Y. and Shields, P.G. (2004). Breast cancer risk in pre-menopausal women is inversely associated with consumption of broccoli, a source of isothiocyanates, but is not modified by GST genotype. *J. Nutr.* 134:1134-1138.

[363] Brennan, P., Hsu, C.C., Moullan, N., Szeszenia-Dabrowska, N., Lissowska, J., Zaridze, D., Rudnai, P., Fabianova, E., Mates, D. and Benckoet, V. (2005). Effect of cruciferous vegetables on lung cancer in patients stratified by genetic status: a mendelian randomisation approach. *Lancet* 366:1558-1560.

[364] Kirsh, V.A., Peters, U., Mayne, S.T., Subar, A.F., Chatterjee, N., Johnson, C.C., and Hayes, R.B. (2007). Prospective study of fruit and vegetable intake and risk of prostate cancer. *J. Natl. Cancer Inst.* 99:1200-1209.

[365] Traka, M. (2010). Broccoli consumption interferes with prostate cancer progression: mechanisms of action. *Acta Hort.* 867:19-25.

[366] Kohlmeier, L. and Su, L. (1997). Cruciferous vegetables consumption and colorectal cancer risk: meta-analysis of the epidemiological evidence. *FSEB J.* 11:369.

[367] Seow, A., Yuan, J.M., Sun, C.L., Van Den Berg, D., Lee, H.P., and Yu, M.C. (2002). Dietary isothiocyanates, glutathione S-transferase polymorphisms and colorectal cancer risk in the Singapore Chinese Health Study. *Carcinogenesis* 23:2055-2061.

[368] London, S.J., Yuan, J.M., Chung, F.L., Gao, Y.T., Coetzee, G.A., Ross, R.K., and Yu, M.C. (2000). Isothiocyanates, glutathione S-transferase M1 and T1 polymorphisms, and lung-cancer risk: a prospective study of men in Shanghai, China. *Lancet* 356:724-729.

[369] Fowke, J.H., Chung, F.L., Jin, F., Qi, D., Cai, Q., Conaway, C., Cheng, J.R., Shu, X.O., Gao, Y.T., and Zheng, W. (2003). Urinary isothiocyanate levels, brassica, and human breast cancer. *Cancer Res.* 63:3980-3986.

[370] Joseph, M.A., Moysich, K.B., Freudenheim, J.L., Shields, P.G., Bowman, E.D., Zhamg Y., Marshall, J.R., and Ambrosone, C.B. (2004). Cruciferous vegetables, generic polymorphism in glutathione S-transferases M1 and T1, and prostate cancer risk. *Nutr. Cancer* 50:206-213.

[371] Juge, N., Mithen, R.F., and Traka, M. (2007). Molecular basis for chemoprevention by sulforaphane: a comprehensive review. *Cell Mol. Life Sci.* 64:1105-1127.

[372] Farnham, M. W., Stephenson, K. K. and Fahey, J. (2000). Capacity of broccoli to induce mammalian chemoprotective enzyme varies among inbred lines. *J. Am. Soc. Hort. Sci.* 125:482-488.

[373] VanEtten, C.H., Dzenbichler, M.E., Williams, P. and Kwolek, W.F. (1976). Glucosinolates and derived products in cruciferous vegetables. Analysis in the edible part from twenty-two varieties of cabbage. *J. Agr. Food Chem.* 24:452-455.

[374] Carlson, D.G., Daxenbichler, M.E., Van Etten, C.H., Tookey, H.L. and Williams, P.H. (1981). Glucosinolates in crucifer vegetables: turnip and rutabagas. *J. Agr. Food Chem.* 29:1235-1239.

[375] Carlson, D.G., Daxenbichler, M.E., VanEtten, C.H., Hill, C.B. and Williams, P.H. (1985). Glucosinolates in radish cultivars. *J. Am. Soc. Hort. Sci.* 110:634-638.

[376] Carlson, D.G., Daxenbichler, M.E., VanEtten, C.H., Kwolek, W.F. and Williams, P.H. (1987a). Glucosinolates in crucifer vegetables: broccoli, Brussels sprouts, cauliflower, collards, kale, mustard green, and kohlrabi. *J. Am. Soc. Hort. Sci.* 112:173-178.

[377] Carlson, D.G., Daxenbichler, M.E., VanEtten, C.H., Kwolek, W.F., Hill, C.B. and Williams, P.H. (1987b). Glucosinolates in turnip tops and roots: cultivars grown for greens and/or roots. *J. Am. Soc. Hort. Sci.* 112: 179-183.

[378] Kushad, M.K., Brown, A.F., Kurillicn, A.C., Juvik, J.A., Klein, B.P., Wallig, M.A. and Jeffery E.H. (1999). Variation in glucosinolates in vegetable crops of *Brassica oleracea. J. Agric. Food Chem.* 47:1541-1548.

[379] Ciska, E., Martyniak-Przybyszewska, B. and H. Kozlowska, H. (2000). Content of glucosinolates in cruciferous vegetables grown at the same site for two years under different climatic condition. *J. Agr. Food Chem.* 48:2862-2867.

[380] Vallejo, F., Tomas-Barberan, F.A. and Garcia-Viguera, C. (2002). Potential bioactive compounds in health promotion from broccoli cultivars grown in Spain. *J. Sci. Food Agric.* 82:1293-1297.

[381] Vallejo, F., Tomas-Barberan, F.A., Banavent-Garcia, A.G. and Garcia-Viguera, C. (2003).Total and individual glucosinolate contents in inflorescences of eight broccoli cultivars grown under various climatic and fertilization conditions. *J. Sci. Food Agric.* 83:307-313.

[382] Nilsson, J., Olsson, K., Engqvist, G., Ekvall, J., Olsson, M., Nyman, M. and Åkesson, B. (2006). Variation in the content of glucosinolates, hydroxycinnamic acids, carotenoids, total antioxidant capacity and low-molecular-weight carbohydrates in Brassica vegetables. *J. Sci. Food and Agric.* 86:528-538.

[383] Borkowski, J., Szajdek, A., Borkowska, E.J., Ciska, E. and Zielinski, H. (2008). Content of selected bioactive components and anti oxidant properties of broccoli (*Brassica oleracea L.*). *Eur. Food Res. Tech.* 226:459-465.

[384] Cartea, M.E., Velasco, P., Obregón, S., Padilla, G. and de Haro, A. (2008). Seasonal variation in glucosinolate content in *Brassica oleracea* crops grown in northwestern Spain. *Phytochemistry* 69:403-410.

[385] Kusznierewics, B., Bartoszek, A., Wolska, L., Drzewiwcki, J., Gorinstein, S and Namiesnik, J. (2008). Partial characterization of white cabbages (*Brassica oleracea* var. *capitata* f. *alba*) from different regions by glucosinolates, bioactive compounds, total antioxidant activities and proteins. *LWT* 41:1-9.

[386] Magrath, R. and Mithen, R. (1993). Maternal effects of the expression of individual aliphatic glucosinolates in seeds and seedlings of *Brasscia napus. Plant Breed-Zeits. Pflanz.* 111:249-252.

[387] Magrath, R., Bano, F., Morgner, M., Parkin, I., Sharpe, A., Lister, C., Dean, C., Turner, J., Lydiate, D. and Mithen, R. (1994). Genetics of aliphatic glucosinolates I. Side chain elongation in *Brassica napus* and *Arabidopsis thaliana. Heredity* 72:290-299.

[388] Giamoustraris, A., and Mithen, R. (1996). Genetics of aliphatic glucosinolates. 4. Side-chain modification in *Brassica oleracea. Theor. Appl. Genet.* 93:1006-1010.

[389] Cao, G., Sofic, E. and Prior, R.L. (1996). Antioxidant capacity of tea and common vegetables. *J. Agr. Food Chem.* 44:3426-3431.

[390] Kurilich, A., Tsau, G.J., Brown, A., Howard, L., Klein, B., Jeffery, E., Kushad, M.M., Wallig, M.A. and Juvik, J.A. (1999). Carotene, tocopherol and ascorbate contents in subspecies of *Brassica oleracea. J. Agr. Food Chem.* 47:1576-1581.

[391] Konings, E.J.M., Roomans, H.H., Dorant, E., Goldbohm, R.A., Saris, W.H. and van den Brandt, P.A. (2001). Folate intake of the Dutch population according to newly established liquid chromatography data for foods. *Am. J. Clin. Nutr.* 73(4):765–776.

[392] Fermenia, A., Selvendran, R.R., Ring, S.G. and Robertson, J.A. (1999). Effects of heat treatment and dehydration on properties of cauliflower fiber. *J. Agr. Food Chem.* 47:728-732.

[393] Rahn, C.R., Bending, G.D., Lillywhite, R.D. and Turner, M.K. (1999). Chemical characterisation of vegetable and arable crops residue materials: a comparison of methods. *J. Sci. Food Agr.* 79:1715-1721.

[394] Banuelos, G. and Meek, D. (1989). Selenium accumulation in selected vegetables. *J. Plant Nutr.* 12:1255-1272.

[395] Finley, J.W., Ip, C., Davis, C.D., Hintze, K.J. and Whanger, P.D. (2001). Cancer-protective properties of high-selenium broccoli. *J. Agr. Food Chem.* 49:2679-2683.

[396] Cai, X., Block, E., Uden, P.C., Ahang, X., Quimby, B.D. and Sullivan, J.J. (1995). *Allium* chemistry: identification of selenoaminoacids in ordinary and selenium-enriched garlic, onion and broccoli using gas chromatography with atomic emission detection. *J. Food Agr. Chem.* 43:1754-1757.

[397] Finley, J.W., Davis, C.D. and Feng, Y. (2000). Selenium from high selenium broccoli protects rats from colon cancer. *J. Nutr.* 130:2384-2389.

[398] Bibak, A., Sturup, S., Knudsen, L. and Gundersen, V. (1999). Concentration of 63 elements in cabbage and sprouts in Denmark. *Commun. Soil Sci. Plant Anal.* 30:2409-2418.

[399] Hertog, M.G., Hollman, P.C. and Katan, M.B. (1992). Content of potentially anticarcinogenic flavonoids of 28 vegetables and fruits commonly consumed in The Netherlands. *J. Agr. Food Chem.* 40:2379-2383.

[400] Nielson, J.K., Olsen, C.E. and Petersen, M.K. (1993). Acylated flavonol glycosides from cabbage leaves. *Phytochemistry* 34:539-544.

[401] Kubec, R., Svobodova, M. and Velisek, J. (1999). Gas-chromatographic determination of S-alk(eny)lylcysteine sulfoxide. *J. Chromatogr.* 862: 85-94.

[402] Herrmann, K. (1976). Flavonols and flavones in food plants: a review. *J. Food Technol.* 11:433-448.

[403] Bilyk, A. and Sapers, G.M. (1985). Distribution of quercetin and kaemperol in lettuce, kale, chive, garlic chive leek, horseradish, red radish and red cabbage tissue. *J. Agr. Food Chem.* 33:226-228.

[404] Ip, C. and Lisk, D.J. (1994). Enrichment of selenium in allium vegetables for cancer prevention. *Carcinogenesis* 15:1881-1885.

[405] El-Bayoumy, K. (1991). The role of selenium in cancer prevention. In: DeVita, V.T., Hellman, S. and Rosenberg, S.A. (eds.). *Cancer principles and practices of oncology.* 4th ed. J. B. Lippincott, Philadelphia. Pp.1-15.

[406] Wang, H., Kruszewki, A. and Brautigan, D.L. (2005). Cellular chromium activation of insulin receptor kinase. *Biochemistry* 44:8167-8175.

[407] Ritsema, T. and Smeekens, S. (2003). Fructans: beneficial for plants and humans. *Curr. Opin. Plant Biol.* 6:223-230.

[408] Kruse, H.P., Kleessen, B. and Blaut, M. (1999). Effects of inulin on faecal bifidobacteria in human subjects. *Brit. J. Nutr.* 82:375-382.

[409] Kilian, S., Kritzinger, S., Rycroft, C., Gibson, G. R. and du Preez, J. (2002). The effects of the novel bifidogenic trisaccharide, neokestose, on the human colonic microbiota. *World J. Microb. and Biotech.* 18:637-644.

[410] Scholz-Ahrens, K.E., Schaafsma, G., van den Heuvel, E.G.H.M. and Schrezenmeir, J. (2001). Effects of prebiotics on mineral metabolism. *Am. J. Clin. Nutr.* 73:459S-464S.

[411] Jackson, K.G., Taylor, G.R., Clohessy, A.M. and Willieams, C.M. (1999). The effect of the daily intake of inulin on fasting lipid, inulin and glucose concentrations in middle-aged men and women. *Brit. J. Nutr.* 82:23-30.

[412] Srinivasan, K. (2005). Plant foods in the management of diabetes mellitus: Spices as beneficial anti-diabetic food adjuncts. *Int. J. Food Sci. Nutr.* 56:399-414.

[413] Sharma, K.K., Gupta, R.K., Gupta, S.C. and Samuel, K.C. (1977). Antithyperglycemic effect of onion: effect on fasting blood sugar and induced hyperglycemic in man. *Ind. J. Med. Res.* 65:422-429.

[414] Tjokroprawiro, A., Pikir, B.S., Budhiarta, A.A., Pranawa, S.H., Deonosepoetro, M., Budhianto, F.X., Wibowo, J.A., Tanuwidjaja, S.J. and Pangemanan, M. (1983). Metabolic effects of onion and green beans on diabetic patients. *Tohoku J. Exp. Med.* 141:671-676.

[415] Dorant, E., van Den Brandt, P.A., Goldbohm, R.A. and Sturnmans, F. (1996). Consumption of onions and a reduced risk of stomach carcinoma. *Gastroenterology* 110:12-20.

[416] You, W.C., Li, J.Y., Zhang, L., Jin, M.L., Chang, Y.S., Ma, J.L. and Pan, K.F. (2005). Etiology and prevention of gastric cancer: a population study in high risk area of *China. Chi. J. Dig. Dis.* 6:149-154.

[417] Galeone, C., Pelucchi, C., Levi, F., Negri, E., Franceschi, S., Talamini, R., Giacosa, A. and La Vecchia, C. (2006). Onion and garlic use and human cancer. *Am. J. Clin. Nutr.* 84(5):1027-1032.

[418] Kim, J.Y. and Kwon, O. (2009). Garlic intake and cancer risk: an analysis using the Food and Drug Administration's evidence-based review system for the scientific evaluation of health claims. *Am. J. Clin. Nutr.* 89(1):257-264.

[419] Salem, S., Salahi, M., Mohseni, M., Ahmadi, H., Mehrsai, A., Jahani, Y. and Pourmand, G. (2011). Major dietary factors and prostate cancer risk: a prospective multicenter case-control study. *Nutrition and Cancer* 63(1): 21-27.

[420] You, W.C., Zhang, L., Gail, M.H., Ma, J.L., Chang, Y.S., Blot, W.J., Li, J.Y., Zhao, C.L., Liu, W.D., Li, H.Q., Hu, Y.R., Bravo, J.C., Correa, P., Xu, G.W. and Fraumeni, J.F., Jr. (1998). *Helicobacter pylori* infection, garlic intake and precancerous lesions in a Chinese population at low risk of gastric cancer. *Int. J. Epidem.* 27(6):941-944.

[421] Takezaki, T., Gao, C.M., Ding, J.H., Liu, T.K., Li, M.S. and Tajima, K. (1999). Comparative study of lifestyles of residents in high and low risk areas for gastric cancer in Jiangsu Province, China; with special reference to allium vegetables. *J. Epidem.* 9(5):297-305.

[422] Gao, C.M., Takezaki, T., Ding, J.H., Li, M.S. and Tajima, K. (1999). Protective effect of allium vegetables against both esophageal and stomach cancer: a simultaneous casereferent study of a high-epidemic area in Jiangsu Province, China. *Japanese J. Cancer Res.* 90(6):614-621.

[423] Dorant, E., van den Brandt, P.A., Goldbohm, R.A. and Sturmans, F. (1996). Consumption of onions and a reduced risk of stomach carcinoma. *Gastroenterology* 110(1):12-20.

[424] Pelucchi, C., Bosetti, C., Rossi, M., Negri, E. and La Vecchia, C. (2009). Selected aspects of Mediterranean diet and cancer risk. *Nutrition and Cancer* 61(6):756-766.

[425] Hirohata, T. and Kono, S. (1997). Diet/nutrition and stomach cancer in Japan. *Internl. J. Cancer* 10:34-36.

[426] Munoz, N., Plummer, M., Vivas, J., Moreno, V., De Sanjose, S., Lopez, G. And Oliver, W. (2001). A case-control study of gastric cancer in Venezuela. *Int. J. Cancer* 93(3):417-423.

[427] De Stefani, E., Correa, P., Boffetta, P., Ronco, A., Brennan, P., Deneo-Pellegrini, H. and Mendilaharsu, M. (2001). Plant foods and risk of gastric cancer: a case-control study in Uruguay. *Eur. J. Cancer Prev.* 10(4): 357-364.

[428] Fleischauer, A.T., Poole, C. and Arab, L. (2000). Garlic consumption and cancer prevention: meta-analyses of colorectal and stomach cancers. *Am. J. Clin. Nutr.* 72(4):1047-1052.

[429] Hsing, A.W., Chokkalingam, A.P., Gao, Y.T., Madigan, M.P., Deng, J., Gridley, G. and Fraumeni, J.F.Jr. (2002). Allium vegetables and risk of prostate cancer: a population-based study. *J. Natl. Cancer Institute* 94(21): 1648-1651.

[430] Tanaka, S., Haruma, K., Kunihiro, M., Nagata, S., Kitadai, Y., Manabe, N., Sumii, M., Yoshihara, M., Kajiyama, G. and Chayama, K. (2004). Effects of aged garlic extract (AGE) on colorectal adenomas: a double-blinded study. *Hiroshima J. Medical Sci.* 53(3-4):39-45.

[431] Tanaka, S., Haruma, K., Yoshihara, M., Kajiyama, G., Kira, K., Amagase, H. and Chayama, K. (2006). Aged garlic extract has potential suppressive effect on colorectal adenomas in humans. *J. Nutr.* 136(3): 821S-826S.

[432] Grant, W.B. (2004). A multi-country ecologic study of risk and risk reduction factors for prostate cancer mortality. *Eur. Urol.* 45:271-279.

[433] Challier, B., Perarnau, J.M. and Viel, J.F. (1998). Garlic, onion and cereal fibre as protective factors for breast cancer: a French case-control study. *Eur. J. Epidem.* 14(8):737-747.

[434] Galeone, C., Pelucchi, C., Dal Maso, L., Negri, E., Montella, M., Zucchetto, A., Talamini, R. and La Vecchia, C. (2009). Allium vegetables intake and endometrial cancer risk. *Public Health Nutr.* 12(9):1576-1579.

[435] Satia, J.A., Littman, A., Slatore, C.G., Galanko, J.A. and White, E. (2009). Associations of herbal and specialty supplements with lung and colorectal cancer risk in the VITamins and Lifestyle study. *Cancer Epidem. Biomar. and Prev.* 18(5):1419-1428.

[436] Herman-Antosiewicz, A., Powolny, A.A. and Singh, S.V. (2007a). Molecular targets of cancer chemoprevention by garlic-derived organosulfides. *Acta Pharmac. Sinica* 28(9):1355-1364.

[437] Iciek, M., Kwiecien, I. and Wlodek, L. (2009). Biological properties of garlic and garlic-derived organosulfur compounds. *Environ. and Molecular Mutagenesis* 50(3):247-265.

[438] Melino, S., Sabelli, R. and Paci, M. (2011). Allyl sulfur compounds and cellular detoxification system: effects and perspectives in cancer therapy. *Amino acids*, 41(1):103-112.

[439] Shon, M.Y., Choi, S.D., Kahng, G.G., Nam, S.H. and Sung, N.J. (2004). Antimutagenic, antioxidant and free radical scavenging activity of ethyl acetate extracts from white, yellow and red onions. *Food Chem Toxicol.* 42:659-666.

[440] Yang, J., Meyers, K.J., vanderHeide, J. and Liu, R.H. (2004). Varietal differences in phenolic content and antioxidant and antiproliferative activities of onions. *J. Agric. Food Chem.* 52:6787-6793.

[441] Clinton, S. (1998). Lycopene: Chemistry, biology and implication for human health and disease. *Nutr. Rev.* 56:35-51.

[442] Scott, K.J. and Hart, D.J. (1995). Development and evolution of an HPLC method for the analysis of carotenoids food and the measurement of the carotenoid content vegetables and fruits commonly consumed in the UK. *Food Chem.* 54:101-111.

[443] Tonucci, L.H., Holden, J.M., Beecher G.R., Khachik, F., Davis, C.S. and Mulokozi, G. 1995. Carotenoid content of thermally processed tomato-based food products. *J. Agri. Food Chem.* 43:579-586.

[444] Gerster, H. (1997). The potential role of lycopene for human health. *J. Am. Coll. Nutr.* 16:109-126.

[445] Rao, A.V., Waseem, Z. and Agarwal, S. (1998). Lycopene contents of tomatoes and tomato products and their contribution to dietary lycopene. *Food Res. Int.* 31:737-741.

[446] Albushita, A.A., Daood, H.G. and Biacs, P.A. (2000). Change in carotenoids and antioxidant vitamins in tomato as a function of varietal and technological factors. *J. Agr. Food Chem.* 48:2075-2081.

[447] Leonardi, C., Ambrosino, P., Esposito, F. and Fogliano, V. (2000). Antioxidant activity and caroteoid and tomatine contentes in different typologies of fresh consumption tomatoes. *J. Agr. Food Chem.* 48:4723-4727.

[448] Arab, L. and Steck, S. (2000). Lycopene and cardiovascular disease. *Am. J. Clin. Nutr.* 71:1691S-1695S.

[449] Dietary Guidelines Advisory Committee (2005). Report of the Dietary Guidelines Advisory Committee. Washington, DC. http://www.health. gov/dietaryguidelines/ dga2005/report/HTML/C_ Methodology.htm.

[450] Food and Nutrition Board (1989). *Water and electrolytes. Recommended Dietary Allowances*. National Academies Press, Washington, DC. Pp. 247-261.

[451] Food and Nutrition Board (2005). Panel on Dietary Intakes for Electrolytes and Water. Potassium. *Dietary Reference Intakes for Water, Potassium, Sodium, Chloride, and Sulfate*. National Academies Press, Washington, DC. Pp.186-268.

[452] Rao, A.V. and Rao, L.G. (2007). Carotenoids and human health. *Pharmacol. Res.* 55(3):207–216.

[453] Albushita, A.A., Hebshi, E.A., Daood, H.G. and Biacs, P.A. (1997). Determination of antioxidant vitamins in tomato. *Food Chem.* 60:207-212.

[454] Stewart, A.J., Bozonnet, S., Mullen, W., Jenkins, G.I., Lean, M.E. and Crozier, A. (2000). Occurrence of flavonols in tomatoes and tomato-based products. *J. Agr. Food Chem.* 48:2663-2669.

[455] Allen, C.M., Schwartz, S.J., Craft, N.E., Giovannucci, E.L., De Groff, V.L. and Clinton, S.K. (2003). Changes in plasma and oral mucosal lycopene isomer concentrations in healthy adults consuming standard servings of processed tomato products. *Nutr. Cancer*. 47:48-56.

[456] Re, R., Mishra, G.D., Thane, C.W. and Bates, C.J. (2003). Tomato consumption and plasma lycopene concentration in people aged 65 y and over in a British national survey. *Eur. J. Clin. Nutr.* 57:1545-1554.

[457] Ganji, V. and Kafai, M.R. (2005). Population determinants of serum lycopene concentrations in the United States: data from the Third National Health and Nutrition Examination Survey, 1988-1994. *J. Nutr.* 135:567-572.

[458] Burney, P.G., Comstock, G.W. and Morris, J.S. (1989). Serologic precursors of cancer: serum micronutrients and the subsequent risk of pancreatic cancer. *Am. J. Clin. Nutr.* 49:895-900.

[459] Van Eenwyk, J., Davis, F.G. and Bowen, P.E. (1991). Dietary and serum carotenoids and cervical intraepithelial neoplasia. *Int. J. Cancer* 48:34-38.

[460] Franceschi, S., Bidoli, E., La Vecchia, C., Talamini, R., D'Avanzo, B. and Negri, E. (1994). Tomatoes and risk of digestive-tract cancers. *Int. J. Cancer* 59:181-184.

[461] Helzlsouer, K.J., Comstock, G.W. and Morris, J.S. (1989). Selenium, lycopene, alpha-tocopherol, beta-carotene, retinol, and subsequent bladder cancer. *Cancer Res.* 49:6144-6148.

[462] Gann, P.H., Ma, J. and Giovannucci, E. (1999). Lower prostate cancer risk in men with elevated plasma lycopene levels: results of a prospective analysis. *Cancer Res.* 59:1225-1230.

[463] Yuan, J.M., Ross, R.K., Gao, Y.T., Qu, Y.H., Chu, X.D. and Yu, M.C. (2004). Pre-diagnostic levels of serum micronutrients in relation to risk of gastric cancer in Shanghai, China. *Cancer Epidem. Biomarkers Prev.* 13(11): 1772-1780.

[464] Wakai, K., Ando, M. and Ozasa K. (2005). Updated information on risk factors for lung cancer: findings from the JACC Study. *J. Epidem.* 15(2): S134-S139.

[465] Foy, C.J., Passmore, A.P., Vahidassr, M.D., Young, I.S. and Lawson, J.T. (1999). Plasma chain-breaking antioxidants in Alzheimer's disease, vascular dementia and Parkinson's disease. *QJM* 92:39-45.

[466] Mecocci, P., Polidori, M.C. and Cherubini, A. (2002). Lymphocyte oxidative DNA damage and plasma antioxidants in Alzheimer disease. *Arch. Neurol.* 59:794-798.

[467] Polidori, M.C., Mattioli, P. and Aldred, S. (2004). Plasma antioxidant status, immunoglobulin G oxidation and lipid peroxidation in demented patients: relevance to Alzheimer disease and vascular dementia. *Dement Geriatr. Cogn. Disord.* 18:265-270.

[468] Sesso, H.D., Buring, J.E., Norkus, E.P. and Gaziano, J.M. (2004). Plasma lycopene, other carotenoids, and retinol and the risk of cardiovascular disease in women. *Am. J. Clin. Nutr.* 79:47-53.

[469] Maggio, D., Polidori, M.C. and Barabani M. (2006). Low levels of carotenoids and retinol in involutional osteoporosis. *Bone* 38:244-248.

[470] Anderson, L.F., Jacobs, D.R., Gross, M.D., Schreiner, P.A., Williams, D.O. and Lee, D.H. (2006). Longitudinal associations between body mass index and serum carotenoids: the CARDIA study. *Brit. J. Nutr.* 95: 358-365.

[471] Rao, L.G., Mackinnon, E.S., Josse, R.G., Murray, T.M., Strauss, A. and Rao, A.V. (2007). Lycopene consumption decreases oxidative stress and bone resorption markers in postmenopausal women. *Osteoporos Int.* 18:109-115.

[472] Yang, Z., Zhang, Z., Penniston, K.L., Binkley, N. and Tanumihardjo, S.A. (2008). Serum carotenoid concentrations in postmenopausal women from the United States with and without osteoporosis. *Int. J. Vitam. Nutr. Res.* 78:105-111.

[473] Le Marchand, L., Yoshizawa, C.N., Kolonel, L.N., Hankin, J.H. and Goodman, M.T. (1989). Vegetable consumption and lung cancer risk: a population-based case-control study in Hawaii. *J. Natl. Cancer Inst.* 81:1158-1164.

[474] De Stefani, E., Oreggia, F., Boffetta, P., Deneo-Pellegrini, H., Ronco, A. and Mendilaharsu, M. (2000). Tomatoes, tomato-rich foods, lycopene and cancer of the upper aerodigestive tract: a case-control in Uruguay. *Oral Oncol.* 36:47-53.

[475] Upritchard, J.E., Sutherland, W.H. and Mann, J.I. (2000). Effect of supplementation with tomato juice, vitamin E, and vitamin C on LDL oxidation and products of inflammatory activity in type 2 diabetes. *Diabetes Care* 23:733-738.

[476] Mucci, L.A., Tamimi, R. and Lagiou P. (2001). Are dietary influences on the risk of prostate cancer mediated through the insulin-like growth factor system? *BJU Int.* 87:814-820.

[477] Chen, L., Stacewicz-Sapuntzakis, M. and Duncan, C. (2001). Oxidative DNA damage in prostate cancer patients consuming tomato sauce-based entrees as a whole-food intervention. *J. Natl. Cancer Inst.* 93:1872-1879.

[478] Giovannucci, E., Rimm, E.B., Liu, Y., Stampfer, M.J. and Willett, W.C. (2002). A prospective study of tomato products, lycopene, and prostate cancer risk. *J. Natl. Cancer Inst.* 94:391-398.

[479] Bowen, P., Chen, L. and Stacewicz-Sapuntzakis, M. (2002). Tomato sauce supplementation and prostate cancer: lycopene accumulation and modulation of biomarkers of carcinogenesis. *Exp. Biol. Med.* 227:886-893.

[480] Hadley, C.W., Clinton, S.K. and Schwartz, S.J. (2003). The consumption of processed tomato products enhances plasma lycopene concentrations in association with a reduced lipoprotein sensitivity to oxidative damage. *J. Nutr.* 133:727-732.

[481] Kim, H.S., Bowen, P. and Chen L. (2003). Effects of tomato sauce consumption on apoptotic cell death in prostate benign hyperplasia and carcinoma. *Nutr. Cancer* 47:40-47.

[482] Bub, A., Barth, S.W., Watzl, B., Briviba, K. and Rechkemmer, G. (2005). Paraoxonase 1 Q192R (PON1-192) polymorphism is associated with reduced lipid peroxidation in healthy young men on a low-carotenoid diet supplemented with tomato juice. *Brit. J. Nutr.* 93:291-297.

[483] Jian L., Du, C.J., Lee, A.H. and Binns, C.W. (2005). Do dietary lycopene and other carotenoids protect against prostate cancer? *Int. J. Cancer.* 113:1010-1014.

[484] Stahl, W., Heinrich, U., Wiseman, S., Eichler, O., Sies, H. and Tronnier, H. (2001). Dietary tomato paste protects against ultraviolet light-induced erythema in humans. *J. Nutr.* 131:1449-1451.

[485] Stahl, W., Heinrich, U., Aust, O., Tronnier, H. and Sies, H. (2006). Lycopene-rich products and dietary photoprotection. *Photochem. Photobiol. Sci.* 5:238-242.

[486] Walfisch, S., Walfisch, Y. and Kirilov, E. (2007). Tomato lycopene extract supplementation decreases insulin-like growth factor-I levels in colon cancer patients. *Eur. J. Cancer Prev.* 16:298-303.

[487] Jacob, K., Periago, M.J., Bohm, V. and Berruezo, G.R. (2008). Influence of lycopene and vitamin C from tomato juice on biomarkers of oxidative stress and inflammation. *Brit. J. Nutr.* 99:137-146.

[488] Paran, E., Novack, V., Engelhard, Y.N. and Hazan-Halevy, I. (2009). The effects of natural antioxidants from tomato extract in treated but uncontrolled hypertensive patients. *Cardiovasc. Drugs Ther.* 23:145-151.

[489] Burton-Freeman, B. and Reimers, K. (2011). Tomato consumption and health: emerging benefits. *Am. J. Lifestyle Med.* 5(2):182-191.

[490] Edinger, M.S. and Koff, W.J. (2006). Effect of the consumption of tomato paste on plasma prostate-specific antigen levels in patients with benign prostate hyperplasia. *Braz. J. Med. Biol. Res*. 39:1115-1119.

[491] Grainger, E.M., Schwartz, S.J. and Wang, S. (2008). A combination of tomato and soy products for men with recurring prostate cancer and rising prostate specific antigen. *Nutr. Cancer* 60:145-154.

[492] Schwarz, S., Obermuller-Jevic, U.C., Hellmis, E., Koch, W., Jacobi, G. and Biesalski, H.K. (2008). Lycopene inhibits disease progression in patients with benign prostate hyperplasia. *J. Nutr*. 138:49-53.

[493] Giovannucci, E., Ashcerio, A. and Rimm, E.B. (1995). Intake of carotenoids and retinol in relation to risk of prostate cancer. *J. Natl. Cancer Inst.* 87:1767-1776.

[494] Giovannucci, E. (1999). Tomatoes, tomato based products, lycopene and cancer: review of the epidemiological literature. *J. Natl. Cancer Institute.* 91:917-331.

[495] Briviba, K., Schnabele, K., Rechkemmer, G. and Bub, A. (2004). Supplementation of a diet low in carotenoids with tomato or carrot juice does not affect lipid peroxidation in plasma and feces of healthy men. *J. Nutr*. 134:1081-1083.

[496] Steinberg D. (1997). Low-density lipoprotein oxidation and its pathobiological significance. *J. Biol. Chem*. 272:20963-20966.

[497] Bose, K.S. and Agrawal, B.K. (2006). Effect of long-term supplementation of tomatoes (cooked) on levels of antioxidant enzymes, lipid peroxidation rate, lipid profile and glycated haemoglobin in Type 2 diabetes mellitus. *West Indian Med. J*. 55:274-278.

[498] Porrini M. and Riso, P. (2000). Lymphocyte lycopene concentration and DNA protection from oxidative damage is increased in women after a short period of tomato consumption. *J. Nutr*. 130:189-192.

[499] Porrini, M., Riso, P. and Oriani, G. (2002). Spinach and tomato consumption increases lymphocyte DNA resistance to oxidative stress but this is not related to cell carotenoid concentrations. *Eur. J. Nutr*. 41: 95-100.

[500] Visioli, A., Riso, P., Grande, S., Galli, C. and Porrini, M. (2003). Protective activity of tomato products on in vivo markers of lipid oxidation. *Eur. J. Nutr*. 42:201-206.

[501] O'Kennedy, N., Crosbie, L. and Whelan, S. (2006). Effects of tomato extract on platelet function: a double-blinded crossover study in healthy humans. *Am. J. Clin. Nutr.* 84:561-569.

[502] Stahl, W. and Sies, H. (2002). Carotenoids and protection against solar UV radiation. *Skin Pharmacol. Appl. Skin Physiol*. 15:291-296.

[503] Wattanapenpaiboon, N., Lukito, W., Wahlqvist, M.L. and Strauss, B.J. (2003). Dietary carotenoid intake as a predictor of bone mineral density. *Asia Pac. J. Clin. Nutr.* 12:467-473.

[504] Sahni, S., Hannan, M.T., Blumberg, J., Cupples, L.A., Kiel, D.P. and Tucker, K.L. (2009a). Protective effect of total carotenoid and lycopene intake on the risk of hip fracture: a 17-year follow-up from the Framingham Osteoporosis Study. *J. Bone Miner. Res*. 24:1086-1094.

[505] Sahni, S., Hannan, M.T., Blumberg, J., Cupples, L.A., Kiel, D.P. and Tucker, K.L. (2009b). Inverse association of carotenoid intakes with 4-y change in bone mineral density in elderly men and women: the Framingham Osteoporosis Study. *Am. J. Clin. Nutr.* 89:416-424.

[506] Rinaldi, P., Polidori, M.C. and Metastasio A. (2003). Plasma antioxidants are similarly depleted in mild cognitive impairment and in Alzheimer's disease. *Neurobiol. Aging* 24:915-919.

[507] Polidori, M.C. and Mecocci, P. (2002) Plasma susceptibility to free radical-induced antioxidant consumption and lipid peroxidation is increased in very old subjects with Alzheimer disease. *J. Alzheimer's Dis.* 4:517-522.

[508] Raben, A., Tagliabue, A., Christensen, N.J., Madsen, J., Holst, J.J. and Astrup, A. (1994). Resistant starch: the effect on postprandial glycemia, hormonal response, and satiety. *Am. J. Clin. Nutr.* 60(4):544–551.

[509] Cummings, J.H., Beatty, E.R., Kingman, S.M., Bingham, S.A. and Englyst, H.N. (1996). Digestion and physiological properties of resistant starch in the human large bowel. *Brit. J. Nutr.* 75(5):733–747.

[510] Hylla, S., Gostner, A. and Dusel, G. (1998). Effects of resistant starch on the colon in healthy volunteers: possible implications for cancer prevention. *Am. J. Clin. Nutr.* 67(1):136–142.

[511] Englyst, H.N., Kingman, S.M. and Cummings J.H. (1992). Classification and measurement of nutritionally important starch fractions. *Eur. J. Clin. Nutr.* 46:S33–S50.

[512] University of California and National Center for Food Safety and Technology. (2010). Lose the Weight, NOT the Potatoes. University of California, Davis and the National Center for Food Safety and Technology, Illinois Institute of Technology. http://www.healthypotato. com/MediaCenter/LosetheWeightNOTthePotatoes.php.

[513] Fernandes, G., Velangi, A. and Wolever, T.M.S. (2005). Glycemic index of potatoes commonly consumed in North America. *J. Am. Dietetic Assoc.* 105(4):557–562.

[514] McCay, C.M., McCay, J.B. and Smith, O. (1987). The nutritive value of potato. In: Talburt, W.F. and Smith, O. (eds.). *Potato processing*. AVI, Westport, CT. Pp. 287-331.

[515] Okeyo, J. and Kushad, M. (1995). Composition of four potato cultivars in relation to cold storage and reconditioning. *Hort. Technology* 5:250-253.

[516] Friedman, M. (1996). The nutritional value of proteins from different food sources: a review. *J. Agr. Food Chem.* 44:6-29.

[517] Friedman, M. (1997). Chemistry, biochemistry and dietary role of potato polyphenols: a review. *J. Agr. Food Chem.* 45:1523-1540.

[518] Jung, C.S., Griffiths, H.M., De Jong, D.M., Cheng, S., Bodis, M. and De Jong, W.S. (2005). The potato P locus codes for flavonoid 3',5'-hydroxylase. *Theor. Apl. Genet.* 110(2): 269–275.

[519] Reeve, R.M., Hautala, E. and Weaver, M.L. (1969). Anatomy and compositional variation within potatoes II. Phenolics, enzymes and other minor components. *Am. Potato. J.* 46:374-386.

[520] Thomas, P. and Joshi, M.R. (1977). Prevention of after-cooking darkening of irradiated potatoes. *Potato Res.* 20:77-84.

[521] Al-Saikhan, M.S., Howard, L.R. and Miller, J.C. (1995). Antioxidant activity and total phenolics in different genotypes of potato (*Solanum tuberosum* L.). *J. Food Sci.* 60:341-344.

[522] Hasegawa, D., Johnson, R.M. and Gould, W.A. (1966). Effect of cold storage on chlorogenic acid content of potatoes. *J. Agr. Food Chem.* 14:165-169.

[523] Cieslik, E. (1994). The effect of naturally occurring vitamin C in potato tubers on the levels of nitrates and nitriles. *Food Chem.* 49:233-235.

[524] Hagg, M., Hakkinen, R., Kumpulainen, J., Ahvenainen, R. and Hurme, E. (1998). Effect of preparation procedures, packaging and storage on nutrient retention of peeled potato. *J. Sci. Food Agr.* 77:519-526.

[525] Ong, A.S. and Tee, E.S. (1992). Natural sources of carotenoids from plants and oils. *Meth. Enzymol.* 213:142.

[526] Packer, L. (1994). Vitamin E is natures' master antioxidant. *Sci. Med.* 11: 54-63.

[527] Lachman, J., Hamouz, J.K., Orsak, M. and Pivec, V. (2000). Potato tubers as a significant source of antioxidants in human nutrition. *Rostl. Vyroba* 46:231-236.

[528] Djujic, I., Djujic, B. and Trajkovic, L. (1995). Dietary intake of selenium in Serbia: results for 1991. Conf. Selenium. *Nauc. Skup. Srp. Akad. Nauk. Umet* 6:81-87.

[529] Bosland, P.W. (1996). Capsicums: Innovative uses of an ancient crop. In: Janick, J. (ed.). *Progress in new crops.* ASHS Press, Arlington, VA. Pp. 479-487.

[530] Howard, L.R., Talcott, S.T., Brenes, C.H. and Villalon, B. (2000). Changes in phytochemical and antioxidant activity of selected pepper cultivars (*Capsicum* species) as influenced by maturity. *J. Agr. Food Chem.* 48:1713-1720.

[531] Howard, L.R., Smith, R.T., Wagner, A.B., Villalon, B. and Burns, E.E. (1994). Provitamin A and ascorbic acid content of fresh pepper cultivars (*Capsicum annum*) and processed jalapenos. *J. Food Sci.* 59:362-365.

[532] Lee, Y., Howard, R. and Villalon, B. (1995). Flavonoids and antioxidant activity of fresh pepper (*Capsicum annum*) cultivars. *J. Food Sci.* 60:473-476.

[533] Nelson, E.K. (1919). The constitution of capsaicin – the pungent principle of capsicum. *J. Agr. Chem. Soc.* 41:1115-1117.

[534] Suzuki, T. and Iwai, K. (1984). Vol. XXIII. In: Brossi, A. (ed.). *The Alkaloids.* Academic Press, San Diego, CA. Pp. 227-229.

[535] Krajewska, A.M. and Powers, J.J. (1988). Sensory properties of naturally occurring capsaicinoids. *J. Agr. Food Sci.* 53:902-905.

[536] Thomas, B.V., Schreilber, A.A. and Weisskopf, C.P. (1998). Simple method for quantitation of capsaicinoids in pepper using capillary gas chromatography. *J. Agr. Food Chem.* 46:2655-2663.

[537] Szallasi, A. and Blumberg, P.M. (1999). Vanilloid (capsaicin) receptors and mechanisms. *Pharmacol. Rev.* 51:159-211.

[538] Noda, Y., Kneyuki, T. and Igarashi, K. (2000). Antioxidant activity of nasunin, an anthocyanin in eggplant peels. *Toxicology* 148(2-3):119-123.

[539] Matsuzoe, N., Yamaguchi, M., Kawanobu, S., Watanabe, Y., Higashi, H. and Sakata, Y. (1999). Effect of dark treatment of the eggplant on fruit skin color and its anthocyanin components. *J. Japan. Soc. Hort. Sci.* 68:138-145.

[540] Ben-Amos, A. and Fishler, R. (1998). Analysis of carotenoids with emphasis on 9-*cis* β-carotene in vegetables and fruits commonly consumed in Israel. *Food Chem.* 62:515-520.

[541] Ensminger, A.H., Esminger M.E., Kondale, J.E. and Robson, J.R.K. (1986). *Food for Health: A Nutrition Encyclopedia.* Pegus Press, Clovis, California.

[542] Wood, R. (1988). *The Whole Foods Encyclopedia.* Prentice-Hall Press; New York, NY.

[543] Jorge, P.A., Neyra, L.C. and Osaki, R.M. (1998). Effect of eggplant on plasma lipid levels, lipidic peroxidation and reversion of endothelial dysfunction in experimental hypercholesterolemia. *Arq. Bras. Cardiol.* 70(2):87-91.

[544] Guimarães, P.R., Galvão, A.M., Batista, C.M., Azevedo, G.S., Oliveira, R.D., Lamounier, R., Freire, N., Barros, A., Sakurai, E., Oliveira, J., Vieira, E. and Alvarez-Leite, J. (2000). Eggplant (*Solanum melongena*) infusion has a modest and transitory effect on hypercholesterolemic subjects. *Brazil J. Med. Biol. Res.* 33:1027-1036.

[545] Praça, J.M., Thomaz, A. and Caramelli, B. (2004). Eggplant (*Solanum melongena*) extract does not alter serum lipid levels. *Arq. Bras. Cardiol.* 82(3):273–276.

[546] Nagase, H., Sasaki, K., Kito, H., Haga, A. and Sato, T. (1998). Inhibitory effect of delphinidin from *Solanum melongena* on human fibrosarcoma HT-1080 invasiveness *in vitro*. *Planta Med.* 64:216-219.

[547] Kwon, Y.I., Apostolidis, E. and Shetty, K. (2008). *In vitro* studies of eggplant (*Solanum melongena*) phenolics as inhibitors of key enzymes relevant for type 2 diabetes and hypertension. *Bioresour. Technol.* 99(8): 2981-2988.

[548] Domino, E.F., Hornbach, E. and Demana, T. (1993). The nicotine. *New Engl. J. Med.* 32986):437.

[549] Parivar, F., Low, R.K. and Stoller, M.L. (1996). The influence of diet on urinary stone disease. *J. Urol.* 155(2):432-440.

[550] Assimos, D.G. and Holmes, R.P. (2000). Role of diet in the therapy of urolithiasis. *Urol. Clin. North Am.* 27(2):255-268.

[551] Kikunaga, S., Arimori, M. and Takahashi, M. (1988). The bioavailability of calcium in spinach and calcium-oxalate to calcium-deficient rats. *J. Nutr. Sci. Vitaminol* 34(2):195-207.

[552] McCollum, E.V. and Davis, M. (1913). The necessity of certain lipids in the diet during growth. *J. Biol. Chem.* 15:167-175.

[553] Steenbock, H. (1919). White corn vs. yellow corn and a probable relation between the fat-soluble vitamin and yellow plant pigments. *Science* 50:352-353.

[554] Troyer, A.F. (1999). Background of U.S. hybrid corn. *Crop Sci.* 39:601-626.

[555] Hayes, H.K., and Garber, R.J. (1927). *Breeding Crop Plants*. McGraw Hill, New York.

[556] Allard, R.W. (1960). *Principles of Plant Breeding*. Wiley, New York.

[557] Fehr, W.R. (1987). *Principles of Cultivar Development*. Vol.1 Theory and Technique. Macmillan Publishing Company, New York.

[558] Lincoln, R.E., Zscheile, F.P., Porter, J.W., Kohler, G.W. and Caldwell, R.M. (1943). Provitamin A and vitamin C in the genus *Lycopersicon*. *Bot. Gaz.* 105:113-115.

[559] Lincoln, R.E. and Porter, J.W. (1950). Inheritance of beta-carotene in tomatoes. *Genetics* 35:206-211.

[560] Tomes, M.L., Quackenbush, F.W., Nelson, O.E., Jr. and North, B. (1953). The inheritance of carotenoid pigment systems in the tomato. *Genetics* 38:117-127.

[561] Gabelman, W.H., and Peters, S. (1979). Genetical and plant breeding possibilities for improving the quality of vegetables. *Acta Hort.* 93:243-270.

[562] Simon, P.W. (2000). Domestication, historical development, and modern breeding of carrot. *Plant Breed. Rev.* 19:157-190.

[563] Sudhakar, P., Singh, J., Upadhyay, A.K. and Ram, D. (2002). Genetic variability for antioxidants and yield components in pumpkin (*Cucurbita moschata* Duch. Ex Poir.). *Veg. Sci.* 29:123-126.

[564] Frossard, E., Brucher, M., Mozafar, A. and Hurrell, R. (2000). Potential for increasing the content and bioavailability of Fe, Zn and Ca in plants for human nutrition. *J. Sci. Food Agr.* 80:861-879.

[565] Brown, M.J., Ferruzzi, M.G., Nguyen, N.L., Cooper, D.A., Eldridge, A.L., Schwartz, S.J. and White, W.S. (2004). Carotenoid bioavailability is higher from salads ingested with full-fat than with fat-reduced salad dressings as measured with electrochemical detection. *Am. J. Clin. Nutr.* 80:396-403.

[566] Diwadkar-Navsariwala, V., Novotny, J.A., Gustin, D.M., Sosman, J.A., Rodvold, K.A., Crowell, J.A., Stacewicz-Sapuntzakis, M. and Bowen, P.E. (2003). A physiological pharma-cokinetic model describing the disposition of lycopene in healthy men. *J. Lipid Res.* 44:1927-1939.

[567] Kurilich, A.C., Clevidence, B.A., Britz, S.J., Simon, P.W. and Novotny, J.A. (2005). Plasma and urine responses at lower for acylated vs. nonacylated anthocyanins from raw and cooked purple carrots. *J. Agr. Food Chem.* 53:6537-6542.

[568] Khush, G.S. (2002). The promise of biotechnology in addressing current nutritional problems in developing countries. *Food Nutr. Bul.* 23:354-357.

[569] King, J.C. (2002). Evaluating the impact of plant biofortification on human nutrition. *J. Nutr.* 132:511S-513S.

[570] Carvalho, P.G.B., Machado, C.M.M., Moretti, C.L. and Fonseca, M.E.N. (2006). Vegetable crops as functional food. *Hort. Brasileira* 24:367-404.

[571] Graham, R.D., Welch, R.M., Saunders, D.A., Ortiz-Monasterio, I., Bouis, H.E., Bonierbale, M., Haan, S., Burgos, G., Thiele, G., Dominguez, M.R.L., Meisner, C.A., Beebe, S.E., Potts, M.J., Kadian, M., Hobbs, P.R., Grupta, R.K. and Twomlow. S. (2007). Nutritious subsistence food systems. *Adv. Agron.* 92:1-74.

[572] Hotz, C. and McClafferty, B. (2007). From harvest to health: Challenges for developing biofortified staple foods and determining their impact on micronutrient status. *Food Nutr. Bul.* 28:S271-S279.

[573] AVRDC (World Vegetable Center) 2009. AVRDC – The World Vegetable Center in South Asia. Vegetable Research and Development in South Asia. ICRISAT, Andhra Pradesh, India. https://www.avrdc.org/ publications/AVRDC-India-2009.pdf.

[574] Fraser, P.D., Romer, S., Shipton, C.A., Mills, P.B., Kiano, K.W., Misawa, N., Drake, R.G., Schuch, W. and Bramley, P.M. (2002). Evaluation of transgenic tomato plants expressing an additional phytoene synthase in a fruit specific manner. *Proc. Nat. Acad. Sci. (USA)* 99: 1092–1097.

[575] Mehta, R.A., Cassol, T., Li, N., Ali, N., Handa, A.K. and Mattoo, A.K. (2002). Engineered polyamine accumulation in tomato enhances phytonutrient content, juice quality and vine life. *Nature Biotech.* 20: 613–618.

[576] Premanchandra, B.R. (1986). Genetic regulation of carotene biosynthesis in selected tomato strains: aspects of beta-carotene biosynthesis and B gene specificity. *Intl. J. Vitam. Nutr. Res.* 56:35-43.

[577] Markovic, Z., Zdravkovic, J., Mijatovic, M. and Damjanovic, M. (2002). Breeding potential of local tomato populations for beta-carotene and vitamin C. *Acta Hort.* 579:157-161.

[578] Stommel, J.R., and Haynes, K.G. (1994). Inheritance of beta carotene content in the wild tomato species *Lycopersicon cheesmanii*. *J. Hered.* 85:401-404.

[579] Andre, C.M., Ghislain, M., Berlin, P., Oufir, M., Herrera, M.R., Hoffman, L., Hausman, J.F., Larondelle, Y. and Evers. D. (2007). Andean potato cultivars (*Solanum tuberosum* L.) as a source of antioxidant and mineral micronutrients. *J. Agr. Food Chem.* 55:366-378.

[580] Brown, C.R., Culley, D., Yang, C.P., Durst, R. and Wrolstad. R. (2005). Variation of anthocyanin and carotenoid contents and associated antioxidant values in potato breeding lines. *J. Am. Soc. Hort. Sci.* 130: 174-180.

[581] Reyes, L.F., Miller Jr., J.C. and Cisneros-Zevallos, L. (2005). Antioxidant capacity, anthocianins and total phenolics in purple-and red-fleshed potato (*Solanum tuberosum* L.) genotypes. *Am. J. Pot. Res.* 82:271-277.

[582] Stevens, M.A. (1986). Inheritance of tomato fruit quality components. *Plant Breed. Rev.* 4:273-311.

[583] Biacs, P.A., Daood, H.G., Huszka, T.T. and Biacs, P.K. (1993). Carotenoids and carotenoide esters from new cross-cultivars of paprika. *J. Agr. Food Chem.* 41:1864-1867.

[584] Hanson, P.M., Yu, Y.R., Lin, S., Tsou, S.C.S., Lee, T.C., Wu, J., Jin, S., Gniffke, P. and Ledesma, D. (2004). Variation for antioxidant activity and antioxidants in subset of AVRDC – the World Vegetable Center Capsicum core Collection. *Plant Gen. Res.: Characterization and Utilization* 2:153.

[585] Dickson, M.H., Lee, C.Y. and Blamble, A.E. (1988). Orange-curd high carotene cauliflower inbreds, NY 156, NY 163, and 165. *Hort. Science* 23:778-779.

[586] Simon, P.W., and Navazio, J.P. (1997). Early orange mass 400, early orange mass 402, and late orange mass 404: high-carotene cucumber germplasm. *Hort. Science* 32:144-145.

[587] Simon, P.W., Pollak, L.M., Clevidence, B.A., Holden, J.M. and Haytowitz, D.B. (2009). Plant breeding for human nutritional quality. *Plant Breed. Rev.* 31: 325-392,

[588] Romer, S., Fraser, P.D., Kiano, J.W., Shipton, C.A., Misawa, N., Schuch, W. and Bramley, P.M. (2000). Elevation of the provitamin A content of transgenic tomato plants. *Nat. Biotech.* 18:666–669.

[589] Fraser, P.D., Romer, S., Shipton, C.A., Mills, P.B., Kiano, K.W., Misawa, N., Drake, R.G., Schuch W. and Bramley, P.M. (2002). Evaluation of transgenic tomato plants expressing an additional phytoene synthase in a fruit specific manner. *Proc. Natl. Acad. Sci. (USA)* 99:1092–1097.

[590] Rosati, C., Aquilani, R., Dharmapuri S., Pallara, P., Marusic C., Tavazza R., Bouvier, F., Camara, B. and Giuliano, G. (2000). Metabolic engineering of beta-carotene and lycopene content in tomato fruit. *Plant J.* 24:413–419.

[591] Lu, S., van Eck, J., Zhou, X., Lopez, A.B., O'Halloran D.M., Cosman K.M., Conlin B.J., Paolillo, D.J., Garvin, D.F., Vrebalov, J., Kochian, L.V., Kupper, H., Earle, E.D., Cao, J. and Li, L. (2006). The cauliflower Or gene encodes a DnaJ cysteine-rich domain-containing protein that mediates high levels of b-carotene accumulation. *Plant Cell* 18:3594–3605.

[592] Wahlroos, T., Susi, P., Solovyev, A., Dorokhov, Y., Morozov, S, Atabekov, J. and Korpela, T. (2004). Increase of histidine content in *Brassica rapa* subsp. *oleifera* by over-expression of histidine-rich fusion proteins. *Mol. Breed.* 14:455–462.

[593] Cho, E.A., Lee, C.A., Kim, Y.S., Baek, S.H., Reyes, B.G. and Yun, S.J. (2005). Expression of *g-tocopherol methyltransferase* transgene improves tocopherol composition in lettuce (*Lactuca sativa* L.). *Molecules and Cells* 19:16–22.

[594] Diaz de la Garza, R.I., Quinlivan, E.P., Klaus, S.M.J., Basset, G.J.C., Gregory, J.F. and Hanson, A.D. (2004). Folate biofortification in tomatoes by engineering the pteridine branch of folate synthesis. *Proc. Natl. Acad. Sci. (USA)* 101:13720–13725.

[595] Diaz de la Garza, R.I., Gregory, J.F. and Hanson, A.D. (2007). Folate biofortification of tomato fruit. *Proc. Natl. Acad. Sci. (USA)* 104:4218–4222.

[596] Park, S., Kim, C.K., Pike, L.M., Smith, R.H., Hirschi, K.D. (2004). Increased calcium in carrots by expression of an *Arabidopsis* H^+/Ca^{2+} transporter. *Mol. Breed.* 14:275–282.

[597] Morris J., Hawthorne, K.M., Hotze, T., Abrams, S.A. and Hirschi, K.D. (2008). Nutritional impact of elevated calcium transport activity in carrots. *Proc. Natl. Acad. Sci. (USA)* 105:1431–1435.

[598] Zuo, X., Zhang, Y., Wu, B., Chang, X. and Ru, B. (2002). Expression of the mouse metallothionein mutant b-cDNA in the lettuces (*Lactuca sativa* L.). *Chinese Sci. Bull.* 47:558–562.

[599] Schijlen, E., Ric de Vos, C.H., Jonker, H., van den Broeck, H., Molthoff, J., van Tunen, A., Martens, S. and Bovy, A. (2006). Pathway engineering for healthy phytochemicals leading to the production of novel flavonoids in tomato fruit. *Plant Biotech. J.* 4:433–444.

[600] Liu, S., Hu, Y., Wang, X., Zhong, J. and Lin, Z. (2006). High content of resveratrol in lettuce transformed with a stilbene synthase gene of *Partenocissus henryana*. *J. Agr. Food Chem.* 54:8082–8085.

[601] Sparrow, P.A.C., Dale, P.J. and Irwin, J.A. (2004). The use of phenotypic markers to identify *Brassica oleracea* genotypes for routine high-throughput *Agrobacterium* mediated transformation. *Plant Cell Rep.* 23:64–70.

[602] Braun, R.H., Morrison, S.C., Schwinn, K.E. and Christey, M.C. (2006). *Agrobacterium* mediated transformation of *Brassica oleracea* with the Lc locus. International Association for Plant Tissue Culture and Biotechnology, Beijing, 13–18 August. p. 115.

[603] Lancaster, J.E. and Collin, H.A. (1981). Presence of alliinase in isolated vacuoles and of alkyl cysteine sulphoxides in the cytoplasm of bulbs of onion (*Allium cepa* L.). *Plant Sci. Lett.* 22:169–176.

[604] Eady, C.C., Davis, S., Farrant, J., Reader, J. and Kenel, F. (2003). *Agrobacterium tumefaciens* mediated transformation and regeneration of herbicide resistant onion (*Allium cepa*) plants. *Ann. Appl. Biol.* 142:213–217.

[605] Sun, H.J., Cui, M.L., Ma, B. and Ezura, H. (2006). Functional expression of the taste modifying protein, miraculin, in transgenic lettuce. *FEBS Letters* 580:620–626.

[606] Bartoszewski, G., Niedziela, A., Szwacka, M. and Niemirowicz-Szczytt. K. (2003). Modification of tomato taste in transgenic plants carrying a thaumatin gene from *Thaumatococcus daniellii* Bent. *Plant Breed.* 122:347–351.

[607] Li, H.Q., Sautter, C., Potrykus, I. and Pounti-Kaerlas, J. (1996). Genetic transformation of cassava (*Manihot esculenta* Crantz). *Nat. Biotech.* 14:736–740.

[608] Raemakers, C.J.J.M., Sofiari, E., Taylor, N.J., Henshaw, G.G., Jacobsen, E. and Visser. R.G.F. (1996). Production of transgenic cassava plants by particle bombardment using luciferase activity as the selection marker. *Mol. Breed.* 2:339–349.

[609] Schopke, C., Taylor, N.J., Carcamo, R., Konan, N.K., Marmey, P., Henshaw, G.G., Beachy, R.N. and Fauquet, C.M. (1996). Regeneration of transgenic cassava plants (*Manihot esculenta* Crantz) from microbombarded embryogenic suspension cultures. *Nat. Biotech.* 14: 731–735.

[610] Siritunga, D. and Sayre, R.T. (2003). Generation of cyanogen-free transgenic cassava. *Planta* 217:367–373.

[611] Siritunga, D. and Sayre. R.T. (2004). Engineering cyanogen synthesis and turnover in cassava (*Manihot esculenta*). *Plant Mol. Biol.* 56:661–669.

[612] Siritunga, D., Arias-Garzon, D., White, W. and Sayre, R.T. (2004). Over-expression of hydroxynitrile lyase in transgenic cassava roots accelerates cyanogenesis and food detoxification. *Plant Biotech. J.* 2:37–43.

[613] Langridge, W.H.R. 2000. Edible vaccines. Scientific Am. Sept. p. 66–71.

[614] Pascual, D.W. (2007). Vaccines are for dinner. *Proc. Natl. Acad. Sci. (USA)* 104:10757–10758.

[615] SunilKumar, G.B., Gananpathi, T.R. and Bapat, V.A. (2007). Production of hepatitis B surface antigen in recombinant plant systems. *Biotech. Prog.* 23:523–529.

[616] McGarvey, P.B., Hammond, J., Dienelt, M.M., Hooper, D.C., Fu, Z.F., Dietzschold, B., Koprowski, H. and Michaels, F.H. (1995). Expression of the rabies virus glycoprotein in transgenic tomatoes. *Bio/Technology* 13:1484–1487.

[617] Ma, Y., Zhang, J., Lin, S.Q. and Xia. N.S. (2001). Genetic engineering vaccines produced by transgenic plants. *Xiamen Daxue Xuebao* 40: 71–77.

[618] Chen, H.F., Chang, M.H., Chiang, B.L. and Jeng. S.T. (2006). Oral immunization of mice using transgenic tomato fruit expressing VP1 protein from enterovirus 71. *Vaccine* 24:2944–2951.

[619] Benner, G., Andrews, G., Byrne, W., Strachan, S., Sample, A. and Heath, D. (1999). Immune response to *Yersinia* outer proteins and other *Yersinia pestis* antigens after experimental plague infection in mice. *Infection and Immunity* 67:1922–1928.

[620] Alvarez, M.L., Pinyerd, H.L., Crisantes, J.D., Rigano, M.M., Pinkhasov, J., Amanda, M., Walmsley, A.M., Masona, H.S. and Cardineau. G.A. (2006). Plant-made subunit vaccine against pneumonic and bubonic plague is orally immunogenic in mice. *Vaccine* 24:2477–2490.

[621] WHO (World Health Organization) (2002). 20 preguntas sobre los alimentos modificados genéticamente. WHO, Geneva, Switzerland. www.who.int/foodsafety/publications/biotech/en/20questions_es.pdf.

[622] FAO (2000). *FAO statement on biotechnology*. Food and Agriculture Organization of the United Nations, Rome, Italy. www.fao.org/biotech/ stat.asp?lang=en.

[623] Royal Society of London, the US National Academy of Sciences, the Brazilian Academy of Sciences, the Chinese Academy of Sciences, the Indian National Science Academy, the Mexican Academy of Sciences, and the Third World Academy of Sciences (2000). Transgenic plants and world agriculture. National Academies Press, Washington, DC.

[624] Chassy, B.M. (2002). Food safety evaluation of crops produced through biotechnology. *J. Am. Coll. Nut.* 21:S166–S173.

[625] Society of Toxicology (2002). *The safety of genetically modified foods produced through biotechnology. Society of Toxicology*, Virginia. www.toxicology.org/ai/gm/GM_Food.asp.

[626] British Medical Association (2004). *Genetically modified foods and health: A second interim statement.* British Medical Association, London, UK. http://www.bma.org.uk/images/GM_tcm41-20804.pdf.

[627] Union of the German Academies of Science and Humanities (2006). Are there health hazards for the consumers from eating genetically modified foods? *Interacademy Panel Initiative on Genetically Modified Organisms,* Berlin, Germany. www.akademienunion.de/_files/ memorandum_gentechnik/GMGeneFood.pdf.

[628] Federal Office of Consumer Protection and Food Safety (Germany) and partners (2009). Long-term effects of genetically modified (GM) crops on health and the environment (including biodiversity): *Prioritisation of potential risks and delimitation of uncertainties.* Federal Office of Consumer Protection of Food Safety, Berlin, Germany. http://bch.cbd. int/database/record-v4-shtml?documentid=101007.

In: Cultivars
Editor: Katya Carbone
ISBN: 978-1-62417-101-7

Chapter 2

CULTIVAR IDENTIFICATION AND VARIETAL TRACEABILITY IN PROCESSED FOODS: A MOLECULAR APPROACH

Antonella Pasqualone *
University of Bari "Aldo Moro", DIBCA Department,
Food Science and Technology Unit, Bari, Italy

ABSTRACT

Aside from the influence of production processes, the quality of food is related to the quality characteristics of raw materials which, in turn, depend on the interaction between intrinsic, genetically regulated traits, and the environment. The effort of breeders aimed to select superior genotypes in the main crops leads to the release of a number of cultivars. Prior to marketing, there is a need to assess their distinctness, uniformity and stability according to the principles established by the Union pour la Protection des Obtentions Végétales (UPOV). Moreover, in some cases, such as Protected Designation of Origin (PDO) products, the use of specific cultivars is required, and is essential to guarantee consumers the authenticity of products. Historically, varietal characterization has been based on profiling of morphological, biochemical and agronomical descriptors. However, these methods are affected by environmental conditions, and are not applicable to foodstuffs. An accurate and unambiguous cultivar identification, independent from environmental influence, cannot be performed without the use of molecular markers. Fingerprinting methods are able to track high-value cultivars, and DNA analysis is particularly suitable to this purpose.

This chapter gives an insight into the relevant research results regarding cultivar identification in wheat by molecular techniques, and describes the main methods set up to trace cultivars from raw material to the main end-products, such as bread and pasta. The new frontiers of genotyping by Single Nucleotide Polymorphisms (SNPs) are focused, with applicative examples in wheat and olive cultivar identification, and extra-virgin olive oil varietal traceability.

* E-mail address: antonella.pasqualone@agr.uniba.it

Introduction

Aside from the production process, the quality of food is related to the quality characteristics of raw materials which, in turn, depend on the interaction between intrinsic, genetically regulated characters and the environment. The effort of the breeders aimed to select superior genotypes in the main crops lead to the release of a number of cultivars. Prior to marketing, it is needed to assess their distinctness, uniformity and stability (DUS) according to the principles established by the Union pour la Protection des Obtentions Végétales (UPOV). A reliable cultivar identification method is needed for raw materials, such as seed stocks, and for manufactured products to protect from deliberate or accidental mislabeling. Moreover, to preserve the features and typicality of food products, the European Commission (EC) enforces two certification labels: Protected Designation of Origin (PDO) and Protected Geographical Indication (PGI), according to the EC regulation No. 510/2006 [1]. These labels are aimed to guarantee, preserve and promote both the peculiar quality characteristics of raw materials from certain areas and the specific processing technology adopted to transform them into high quality end-products. The official production protocol of PDO and PGI products must meet specific varietal requirements. Therefore, is essential to trace cultivars in raw and processed materials in order to guarantee the consumer knowledge of the authenticity of these specialties. Finally, besides PDO and PGI labels (which imply a specific geographical origin), processed foods made from a certain variety are being increasingly introduced in the markets, irrespective of their provenience. This tendency is well-established in oenology and, more recently, also some pasta types and many virgin olive oils are labeled claiming the use of specific cultivars. Historically, varietal characterization has been based on profiling of morphological, biochemical and agronomical descriptors, as discussed along the chapter. However, these methods are affected by environmental factors, and generally are not applicable to foodstuffs. An accurate and unambiguous cultivar identification, independent from environmental influence, can be performed with the use of DNA molecular markers. Their analysis allows to fingerprint and track high-value cultivars and presents high analytical specificity and sensitivity. Another advantage is the higher thermal stability of DNA molecules, compared to other compounds such as proteins, which allows to analyze it also in thermally processed foodstuffs. This chapter gives an insight into the relevant research results regarding cultivar identification in wheat (both *Triticum aestivum* L. Thell. subsp. *vulgare* Vill. MK., or bread wheat, and *Triticum turgidum* L. Thell. subsp. *turgidum* convar. *durum* Desf. MK., or durum wheat) by molecular techniques, and describes the main methods set up to trace cultivars from raw material to the main end-products, such as bread and pasta. The new frontiers of genotyping by Single Nucleotide Polymorphisms (SNPs) are focused, with applicative examples in wheat and olive (*Olea europaea* L.) cultivar identification, and in extra-virgin olive oil varietal traceability.

Categories of Molecular Markers

Many classes of Polymerase Chain Reaction (PCR) based molecular markers are able to allow the detection of DNA polymorphism among different cultivars, so as to achieve varietal fingerprinting and identification. According to their ability to amplify genomic DNA without needing prior knowledge of its sequence for primer design, or to amplify target nucleotide sequences present only in the species of interest, they can be classified in i) non-species-specific and ii) species-specific molecular marker systems.

Non-specific markers include Random Amplified Polymorphic DNA (RAPD) [2, 3], Inter-Simple Sequence Repeats (ISSR) [4], and Amplified Fragment Length Polymorphisms (AFLP) [5], while specific markers are microsatellites or Simple Sequence Repeats (SSR) [6, 7], Selective Amplification of Microsatellite Polymorphic Loci (SAMPL) [8], Sequence Characterized Amplified Regions (SCAR) [9], and SNP [10, 11].

RAPD AND SCAR MARKERS

RAPD markers are generated by using relatively short (10-mer) arbitrary "universal" primers, which randomly anneal to several loci and amplify discrete random sequences.

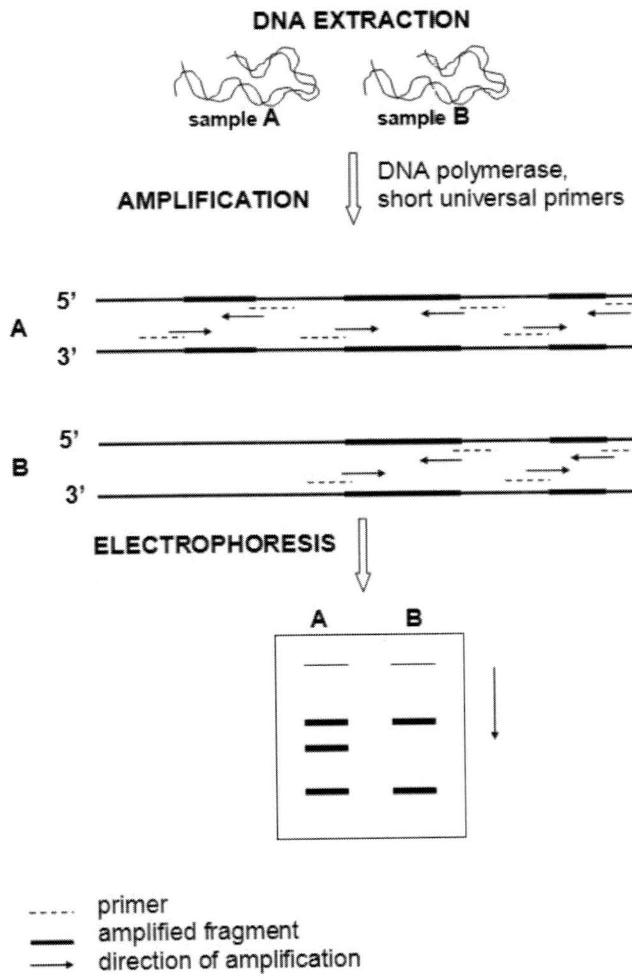

Figure 1. Steps of the RAPD technique.

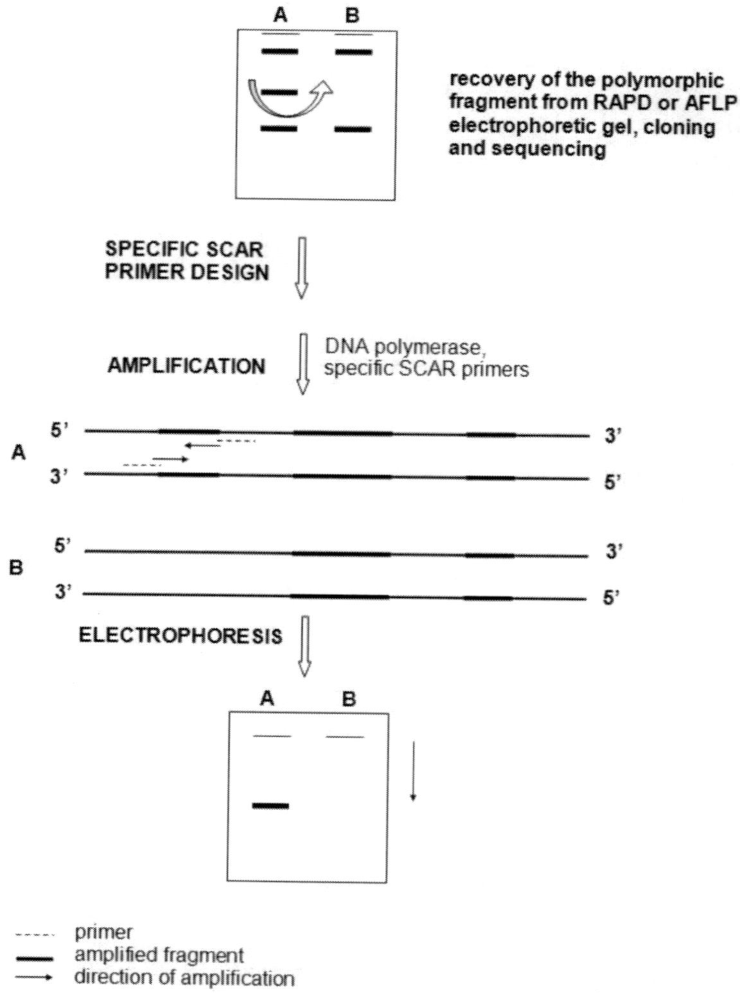

Figure 2. Conversion of RAPD markers into SCAR markers.

The amplification leads to multi-fragment PCR patterns and allows to screen polymorphism in several regions contemporarily (Figure 1), resulting from DNA sequence variation at, and between, primer-binding sites.

The use of universal (non species-specific) primers is suitable in absence of information about the genome of the species under study. However, some concerns about the reproducibility of RAPD analysis within and among laboratories may remain unresolved, mainly due to low annealing temperature related to the limited length of primers [12]. RAPD markers can be converted into SCARs. SCAR markers are short unique sequences that identify a specific DNA locus amplified by PCR.

The procedure, schematized in Figure 2, involves identification of polymorphic band(s) generated in a RAPD assay, cloning the fragment(s), sequencing the clones, and designing the PCR primers for the locus-specific amplification product. This strategy implies the use of stringent PCR conditions, hence reduces the unwanted background signals produced in RAPD gels. The SCARs developed in such a way may be dominant or co-dominant, while RAPD markers show only a dominant behavior.

ISSR MARKERS

Similarly to RAPD markers, also ISSRs do not need preliminary sequencing of genome to be applied. They involve amplification of DNA regions situated between adjacent repetitive sequences by using a long single primer composed of a repeated sequence of 16-18 base pairs (bp) anchored, at the 3' or 5' end, to 2-4 arbitrary nucleotides (Figure 3).

Since the repeated sequences are abundant in plant genome, the primer anneals to several regions, hence allowing simultaneous screening of a large number of loci. The result is a rich amplification pattern showing fragments ranging from 100 bp to 1500 bp in size, with a higher frequency around 700 bp [4].

Coupled with the separation of the amplified products on polyacrylamide gel electrophoresis, ISSR amplification can reveal a much larger number of fragments than RAPD analysis. Moreover, ISSR are more reproducible than RAPD markers, due to the use of longer primers and higher annealing temperatures.

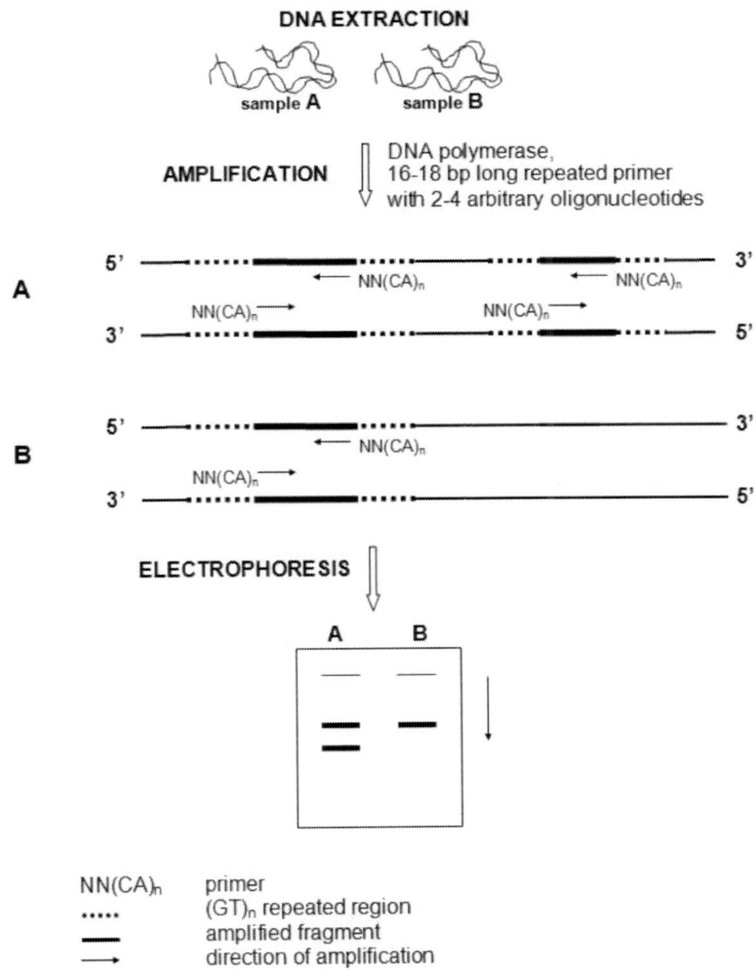

Figure 3. Steps of the ISSR technique.

AFLP MARKERS

The AFLP technique is based on selective PCR amplification of restriction fragments and requires the following steps (Figure 4): (i) restriction digestion of total genomic DNA with two endonucleases (one rare cutter and the other frequent cutter); (ii) ligation of adapters of known sequences to the two ends of each fragment, so that the sequences of adapters and the adjacent restriction sites serve as primer-binding sites; (iii) selective amplification of restriction fragments by using primers designed to be complementary to the adapters and the restriction sites, along with 1-3 selective bases added at their 3' ends.

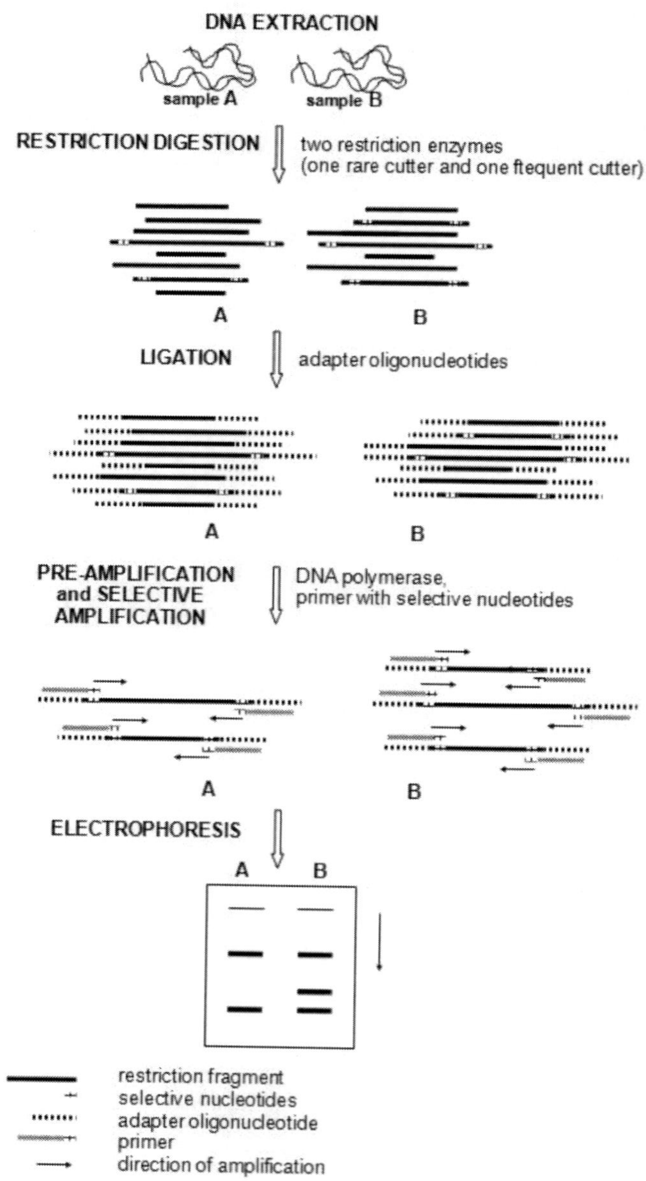

Figure 4. Steps of the AFLP technique.

The use of selective bases allows amplification of only a subset of the restriction fragments, which still generate a large number of bands facilitating the detection of polymorphism. Amplification may require two steps, the first (pre-amplification) using primers with fewer (0-1) selective nucleotides and the second (selective amplification) using primers with 3 selective nucleotides. Also the AFLP approach [5] needs no cloning and sequencing, and gives a very large number of fragments, which enhance its power to detect DNA polymorphism. These multi-locus markers allow reliable analyses due to highly stringent PCR primer annealing conditions. Such in case of RAPD markers, also from AFLP profiles it is possible to develop SCAR markers (Figure 2).

Microsatellites or SSR Markers

Microsatellites are repetitive DNA sequences ubiquitously interspersed in eukaryotic genomes. They show a high frequency of variation in the number of repeats, probably due to slippage during DNA replication. The polymorphisms are easily detectable by DNA amplification with specifically designed primers flanking the repeated sequences (Figure 5) [6].

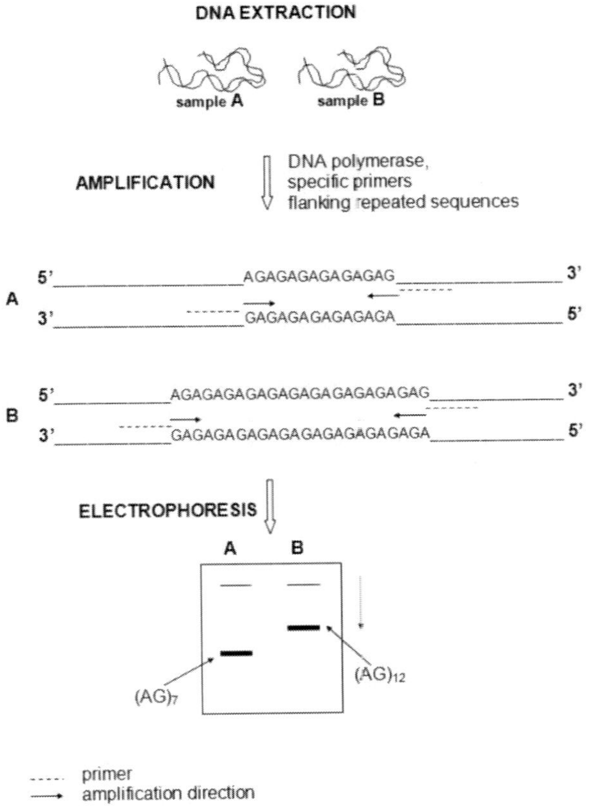

Figure 5. Steps of the SSR technique. The example considers a microsatellite with a di-nucleotide base unit (AG)n.

While RAPD, ISSR and AFLP markers can be analyzed in absence of preliminary sequence knowledge, microsatellites, or SSRs, need a longer procedure. Genomic libraries from the target species must be screened for di-, tri-, or tetra-nucleotide repeat motifs in tandem arrays. Clones containing the repeats have to be isolated and the regions flanking each tandem array have to be sequenced, then primers are designed for the amplification of the repeated sequence.

The hypervariability observed at SSR loci is due to the unique mechanism by which this variation is generated: replication slippage is thought to occur more frequently than single nucleotide mutations and insertion/deletion events, which generate the polymorphisms detectable by AFLP and RAPD analyses [13, 14]. Besides, the codominant nature of microsatellite markers permits the detection of a high number of alleles per locus and contributes to higher levels of expected heterozygosity (also called diversity index) than would be possible with RAPDs and AFLPs [14].

Hence, when sequences of genomic DNA are available, microsatellites are the markers of choice for varietal identification studies [15]. In fact, microsatellites are highly polymorphic, multi-allelic, locus-specific markers with relatively simple banding patterns and a co-dominant inheritance.

They result cheaper and less laborious than AFLPs and do not require special equipment and manipulation of radio-labeled substances, making possible to transfer SSR protocols for routine analyses.

SAMPL Markers

Some of the merits of SSR and AFLP markers have been combined into a single assay, called SAMPL [8], a microsatellite-based marker system consisting in a modification of the AFLP methodology [16]. Adapter-ligated restriction fragments of a conventional AFLP procedure are used, but the final PCR selective amplification is achieved using primers which differ from those employed in AFLP. One of the primers is a 5′ self-anchoring compound SSR (the "SAMPL primer"), and the other is an AFLP primer that is designed on the basis of the sequences of the synthetic adapter plus the restriction site, and carries 2-3 selective nucleotides [17, 18]. The use of SAMPL technique is particularly suitable when low genetic variation is expected, since the primers target the hyper-variable microsatellite loci [17].

SNP Markers

SNP is a ubiquitous type of genetic variation in eukaryotic genomes. It occurs when different nucleotide alternatives (alleles) for a single base pair position in the genomic DNA exist. The principle of SNP is to distinguish a perfect match from a single base mismatch, at the SNP site, between a probe of known sequence and the target DNA containing the SNP site (hence they need only a plus/minus assay, permitting easy automation). While microsatellites are due to length variation and are multi-allelic, SNPs are due to sequence variation and are bi-allelic [10, 11].

SNPs are the most abundant marker systems known so far. Both in plant and animal genomes their extraordinary abundance has been estimated in one SNP every 100-300 bp [10]. They may be found both in the non-repetitive coding or regulatory sequences and in the repetitive non-coding sequences.

SNP technology does not require the time-consuming and expensive gel-based assays that are needed for most molecular markers (RAPD, ISSR, AFLP, and often also SSR). Several approaches have been set up for SNP identification, suitable for automation and high throughput [10, 11]. The high density oligonucleotide arrays on DNA chips and the matrix-assisted laser desorption/ionization-time of flight mass spectrometry (MALDI-TOF/MS) allow analysis of large numbers of SNP markers. Hence, SNP markers are increasingly becoming more popular respect to other marker classes.

Varietal Characterization of Wheat by Classical Methods

Besides the analysis of plant morpho-physiological descriptors, the first methods for varietal characterization in wheat grains have been based on the polymorphism of endosperm storage proteins. Electrophoresis [19-21], chromatographic techniques [22, 23], and near infrared reflectance (NIR) spectroscopy [24] have been used to set up databases of protein patterns of wheat collections [25, 26]. The ISO standard No. 8981 of 1993 [27], and the ICC standard No. 143 of 1995 [28], both applicable to lots of grains, as well as to flour or semolina, rely on electrophoresis of gliadins. More recently, other techniques such MALDI-TOF/MS, combined with an artificial neural network (ANN), have been proposed [29, 30], as well as NMR spectroscopy and isotopic ratios evaluation [31].

However, although the environment does not influence protein patterns qualitatively, there is a significant effect on the results of the quantitative analysis, which can produce errors in the automated varietal identification procedures [32]. Moreover, dubious classifications occur in case of high genetic similarity of some cultivars, which is frequent due to the fact that crosses are often performed among elite lines with similar agronomic and end-use characteristics. To take in account these drawbacks, and to search for higher discrimination power, so as to achieve cultivar identification more effectively, various studies have explored the possibility of analyzing wheat directly at DNA level.

Varietal Characterization of Wheat by Molecular Markers

Valuable markers have been generated in wheat from RAPD, ISSR and AFLP technologies, especially by turning them into SCARs. Microsatellites, both genomic and functional, have been developed successively and have been proven to be more effective for cultivar identification. More recently, the increasing amount of sequence information coupled to high throughput genotyping methods, including DNA chips, induced to prefer new marker types, such as SNPs. The majority of reports regarded bread wheat, with a few studies devoted to durum and diploid wheat. The number of cultivars examined, their geographical origin and the type and number of markers tested in molecular studies reported up to now is summarized in Table 1.

Table 1. Molecular studies about cultivar identification in wheat, ordered chronologically

Cultivars examined (Number)	Origin of cultivars	Markers tested		Reference
		Number	Type	
12[a]	UK	10	RAPD	Devos and Gale, 1992
15[a]	USA	40	RAPD	Joshi and Nguyen, 1993a
30[b]	Israel, Turkey, Jordan, Syria, US	40	RAPD	Joshi and Nguyen, 1993b
18[a]	Germany	15	SSR	Röder et al., 1995
40[a,b]	Europe	23	SSR	Plaschke et al., 1995
18[a]	-	32	SSR	Ma et al., 1996
10[a]	France, UK	153	SSR	Bryan et al., 1997
10[a]	Russia, South Africa	29	RAPD	Myburg et al., 1997
6[a,b,d]	-	100	ISSR	Nagaoka and Ogihara, 1997
55[a]	UK	6	AFLP	Law et al., 1998
56[a,d]	Pacific Northwest	16	AFLP	Barrett and Kidwell, 1998
55[a]	29 countries in 6 continents	20	SSR	Prasad et al., 2000
52[b]	Italy	9	ISSR	Pasqualone et al., 2000
16[b]	Turkey	7	SSR	Dograr et al., 2000
500[a]	Europe	30-40	SSR	Vosman et al., 2001
55[a]	29 countries in 6 continents	2	SAMPL	Roy et al., 2002
64[b]	Algeria, Canada, Ethiopia, France, Greece	42	20 SSR 22 EST-SSR	Ejuayl et al., 2002
38[a]	-	16	SNP	Kirkpatrick et al., 2002
502[a]	Europe	20	SSR	Röder et al., 2002
8[a]	France	688	EST-SSR	Nicot et al., 2004
18[b]	Canada	7	SSR	Perry, 2004
480[a]	Europe	39	SSR	Roussel et al., 2005
58[b]	Italy, France, USA, Mexico, Tunisia	86	16 AFLP 70 SSR	Maccaferri et al., 2007
69[b]	Italy	105	7 AFLP 98 SSR	Noli et al., 2008
34[a,b]	Turkey	8	5 AFLP 3 SAMPL	Altintaş et al., 2008
58[a]	Japan, USA, Australia, Canada	10	EST-SSR	Fujita et al., 2009
347[a]	Europe, Asia	38	SSR	Mitrofanova et al., 2009
91[a,b]	USA	96	SNP	Akhunov et al., 2009
20[b]	USA	359	SNP	Chao et al., 2009
80[b]	Italy	11	SSR	Mangini et al., 2010
23[a,b,d]	UK	1114	SNP	Allen et al., 2011
8[a]	China	3	1 ISSR 2 SSR	Zhu et al., 2011

[a]Bread wheat; [b]durum wheat; [c]einkorn; [d]diploid wheat.

RAPDs were evaluated as a DNA marker system in wheat since the beginning of the Nineties. Reproducible RAPD products were obtained, but reaction conditions had to be rigorously optimized in terms of annealing temperature as well as concentration of DNA, Mg^{2+}, and polymerase, all influencing the RAPD patterns. Taq polymerase concentrations of 2.5 U in 50 μL reaction volume resulted in non-specific DNA amplifications and concentrations of 0.8 U in 50 μL reaction volume were found to be optimal [33]. RAPD primers were used to assess the extent of genetic diversity among varieties [34] and for cultivar fingerprinting, provided that only reproducible and unambiguously scored amplification products were considered [35, 36]. The development of SCAR markers by direct sequencing of RAPD products was also reported, essentially for marker assisted selection aims [35, 37-39].

The applicability of ISSR markers in wheat was evaluated some years later than RAPDs. The extent of polymorphism was found to be greater than RAPD markers [40]. High efficiency for cultivar identification was observed [41], especially by using a fluorescence-labeling detection system that allowed to appreciate very small differences in length (1-2 bp) among amplified fragments [42].

AFLP markers were first used at the end of the Nineties [43]. The optimal number of polymorphic bands to discriminate among varieties was assessed between v and 2v, where v is the number of varieties under test. Discrimination levels were adversely affected if the number of bands was below v/2 [44]. AFLPs were applied alone [45] and with microsatellite markers [44, 46] to assess distinctness in panels of varieties representative of different international breeding programs, and were able to discriminate all the genotypes, even when tightly related.

SAMPL technology was used for the first time in 2002 for bread wheat genotype identification [16]. SAMPL markers were also used with AFLPs, and several cultivar-specific bands, useful for cultivar identification, were observed both in bread and durum wheat [47].

Development of Wheat Microsatellites

The development of informative microsatellite markers in bread wheat has been difficult and time-consuming due to its large genome size (16×10^9 bp), polyploidy, and high level of repetitive sequences in the genome (>80%) [16]. In spite of this, the reliability and discrimination power of these markers induced several authors to devote their studies to the isolation and characterization of microsatellites in wheat genome, and a great number of primers has been rendered available for various applications. Hence, respect to other markers, the number of studies which considered microsatellites is definitely greater.

First studies began in 1995, when Röder et al. [48] isolated 70 microsatellites of bread wheat (generating the primer series 'WMS', Wheat Micro-Satellites), and investigated the abundance and variability of 15 of them in a panel of accessions. The microsatellite frequency was observed to be of the same order of magnitude in wheat as those reported for other plants. Besides, about 20-30% of microsatellites amplified more than one locus. Ma et al. [49] set up 32 pairs of primers for microsatellite sequences and, by surveying 18 bread wheat cultivars, they also observed that many primers gave additional fragments respect to the expected one. Moreover, some of these were polymorphic and had a dominant behavior. Seven primers gave polymorphic products among the cultivars examined. Bryan et al. [50] isolated over 200 clones containing microsatellites and tested 153 primer pairs (named ‘PSP’) in wheat varieties. They found a high level of polymorphism, and observed that a high proportion of primers, approximately two-thirds, detected additional loci generating additional amplification products, generally smaller in size than the expected. Song et al. [51] developed 540 microsatellite-flanking primer pairs (the ‘BARC’ series, from the acronym for the USDA-ARS Beltsville Agricultural Centre).

The unexpected multi-locus amplification, observed as smears or stuttered patterns, implied the need of a careful selection of the ‘best’ primers, coupled to more efficient strategies in clone purification which might reduce the proportion of non-useful amplification products. These additional products might be eventually due to the isolation of clones which contain, in the flanking region of the microsatellite sequence, short repetitive motifs embedded in the flanking region itself, allowing annealing to more than one locus [50]. Another point was the possible occurrence of homoeologous amplifications, when homoeologous SSR sequences are located more than one of the genomes of wheat (A, B or D), eventually different in length [50]. Moreover, besides the number of repeats, polymorphism of microsatellite can affect the priming site, giving rise to null alleles, which detection is quite common [50, 52].

These observations were taken into account in successive works. In fact, Pestsova et al. [53] isolated 65 microsatellite markers specific for the D genome of bread wheat (the series ‘GDM’, Gatersleben D-genome Microsatellites), and selected only primers that amplified fragments of the expected size, discarding those giving either a smear, fragments of wrong size, or nothing. Stephenson [54] mapped 53 microsatellites of the PSP series and scored them on a 1-5 scale according to the complexity of their amplification profile.

Due to recognized usefulness of these markers, a 39-member international consortium (Wheat Microsatellite Consortium, WMC) was established to develop microsatellites in bread wheat. The results of this collaborative study were 396 microsatellite primer pairs, the ‘WMC series’, available in the ‘Graingenes’ database [55, 56]. Some of them amplified more than one locus (microsatellites having multiple loci), in the same proportion found by Röder et al. (20-30%) [57].

All the microsatellites developed in these studies were sited in non-coding DNA regions. More recently, the increasing availability of expressed sequence tags (EST) in bread wheat, which represents only expressed genes, provided a valuable resource of ‘non-anonymous’ microsatellite markers. EST-microsatellites are generally characterized by robust and high-quality markers with strong bands, while some of genomic microsatellites may show faint bands or stuttering. However, EST-microsatellite markers have been reported to be less polymorphic compared with genomic SSR in crop plants because of greater DNA sequence conservation in transcribed regions. EST-SSR sequences containing dinucleotide motifs were found to be significantly more polymorphic (74%) than those containing trinucleotides (56%), and multiple-loci were detected by 39% of the primer pairs [58].

Nicot et al. [59] screened the ESTs from the public databases and identified 3,530 microsatellite sequences. Seventy-four percent of the sequences allowed primer pairs to be designed, 70% led to an amplification product, mainly of high quality, and 53% detected polymorphism among the 8 cultivars examined. In the framework of mapping studies, Gao et al. [60] developed 478 EST-microsatellites and mapped 101 of them in bread wheat. EST-derived microsatellites were used to identify 41 Japanese and 17 US, Australian and Canadian bread wheat cultivars [61].

Up to now, almost 2000 microsatellite primer pairs have been rendered available in bread wheat, accessible by the scientific community. Hence, in addition to the cited studies, aimed to sequencing, primer development and first evaluation of primer sets for microsatellite sequences, several researches apply the already developed primers to cultivar identification, further evidencing the potential of these markers [62], and remarking that a relatively small number of them can be used for cultivar identification in wheat. Besides, microsatellites are useful to assess genetic diversity and varietal uniformity, evidencing their potential use in DUS testing [63-65].

Transferability of Microsatellites to Durum Wheat

Durum wheat attracted minor attention than bread wheat, and no microsatellites were specifically developed in durum wheat. However, it was demonstrated that it is possible to apply to durum wheat the microsatellites developed in bread wheat, apart those located in genome D, absent in durum wheat [63, 66]. Also EST-microsatellites developed in bread wheat were found to be useful for fingerprinting durum wheat accessions [67]. A high transferability of bread wheat EST-microsatellites to other cereals was evidenced, with the highest percentage of successful amplifications in durum wheat (90.4%) [68].

Bread wheat microsatellite markers were effectively used to identify winter type durum wheat varieties from Anatolia [69], durum wheat varieties registered in Canada for commercial production [70], and wheat cultivars included in the Italian Register of Varieties [71].

Other studies applied the joint analysis of microsatellites and AFLP markers to durum wheat genotype identification and assessment of varietal distinctness [44, 46]. A high correlation between similarity estimates obtained with microsatellite markers and AFLPs was observed. Besides, while molecular data obtained from both SSR and AFLP markers were highly effective in accurately determining the pattern of relatedness among durum accessions, this was not the case with the phenotypic data. It was pointed out that a set of 28 informative SSR markers (one per chromosome) is a useful tool to assess varietal distinctness in durum wheat [46].

Set up of Large Databases of Microsatellite Profiles

All the above mentioned studies demonstrated the feasibility of microsatellite analysis in wheat for cultivar identification purposes. However, they focused on relatively limited number of samples, respect to the high number of registered cultivars available on the market. Hence, at the beginning of the 2000s, a EU project was set up to demonstrate the technical viability of the use of microsatellite markers for identification of large cultivar sets. The aim was to compare efficiently each new cultivar, candidate to registration in national lists, against all varieties of common knowledge. The approach to the problem was to set up consensus databases of microsatellite profiles of cultivars, generated according to an agreed protocol. Wheat was one of the two crops chosen as case study, the other being tomato. Initially, about 30-40 markers were used to identify a smaller set of 20 good markers and to standardize methodology and interpretation of the results in different laboratories [72]. For example, to take in account different fragment-analysis procedures, including analysis on automated sequencers, it was found that the size of the amplified fragments (alleles) may vary by a few base pairs in different experiments. Hence, this study pointed out that for a correct allele-recognition it is necessary to include reference cultivars, which amplify representative alleles and can be used for direct comparisons.

The selected markers were then tested on 502 European bread wheat cultivars (plus 52 controls), to build a large database of their patterns [73]. Nineteen microsatellites allowed discrimination among all of the samples, except duplicates and cultivars derived from identical parents. A similar observation was described by Plaschke et al. [63] that could not distinguish two sister lines with 23 microsatellites. The differentiation of lines originating from identical parents may pose a general difficulty for the molecular identification of varieties or accessions. In such cases, only limited regions of the genome are different between samples, so that a higher number of markers is required for detecting differences.

Comparisons of data generated by different laboratories revealed a high degree of agreement [72]. Only 34 discrepancies between duplicate samples analyzed in different laboratories were observed, out of 11,080 data-points, indicating an accuracy of >99.5% for the database. These discrepancies may either be due to experimental error or to heterogeneity of the sample [72]. Although wheat is a self-pollinating plant, evidence of internal heterogeneity, defined as the identification of more than one allele for a given marker in a single variety, was found. Some wheat varieties were heterogeneous with respect to several markers, perhaps indicating the presence of mixtures in the seed samples. For other varieties heterogeneity was found for a single marker only. In total, more than 25% of all of the varieties analyzed were non-uniform with at least one of the markers, with the highest level of heterogeneity in south-eastern European samples. In most cases, it was possible to resolve the problem by investigating pooled seed samples; however, in a few cases heterogeneity between duplicate samples of the same variety persisted.

This may cause a problem for the unequivocal identification of varieties by molecular markers, if a perfect match criterion (100%) is applied stringently. Therefore, the threshold of genetic similarity for considering two varieties as identical was lowered to 95%, which led to correct identification results in nearly all cases by ensuring that most of the heterogeneities observed within duplicate samples were accepted as identical varieties.

Large cultivar sets were considered also by Roussel et al. [74] who analyzed 480 bread wheat varieties from 15 European geographical areas, released from 1840 to 2000, with a set of 39 microsatellite markers. In another study, Mitrofanova et al. [75] examined 38 microsatellites in 347 bread wheat genotypes, including landraces and cultivars.

Cultivar Traceability in Pasta and Bread

Molecular markers represent a great support to traceability since they allow monitoring DNA at each stage in the food chain. In spite of the large number of studies regarding cultivar identification in wheat, very few researches applied DNA analysis to processed wheat-based end-products with cultivar traceability purposes.

The most important drawback for such studies is the difficult extraction of 'good quality' DNA from food matrix, that is a critical point for PCR. In particular, the main factors negatively affecting DNA quality are the presence of PCR inhibitors, abundant in the food matrices, and DNA damaging (depurination and fragmentation). These factors, in turn, are related to the intrinsic properties of food matrix (i.e. pH and intensity of the treatments needed for food production) and to the method utilized to extract DNA.

Physical and chemical treatments cause random breaks in DNA strands, thus reducing the average DNA fragment size [76]. DNA is very sensitive to acid and alkaline agents because of the mechanism of its hydrolytic degradation. For example, at acid pH, purines are removed from the nucleic acid backbone as a result of the cleavage of N-glycosidic bonds between deoxyribose residues and bases. Subsequently, adjacent 3',5'-phosphodiester linkages are hydrolyzed, leading to the shortening of DNA strands [76]. Besides, thermal treatments induce the Maillard's reaction in several foodstuffs, that may result in cross-linking of macromolecules producing "tangles" of proteins and nucleic acids. This may prevent DNA extraction or successive amplification [76].

DNA may be extracted from foods either using commercial kits or by classic extraction protocols based on in-house reagents (Cetyltrimethylammonium bromide buffer (CTAB), and phenol-chloroform or chloroform-octanol solutions) [77-79]. After cell lysis, and elimination of solid residues, DNA is selectively bound to membranes or adsorption resins packed into micro-columns, and washed to remove contaminating molecules (salts, peptides, and polysaccharides) that may act as PCR inhibitors. Post-extraction procedures have to be carried out for further DNA purification. They include solvent precipitation, enzymatic treatments, and chromatographic separation. These additional purifications result in a better removal of PCR inhibitors, and ensure higher amounts of amplification products, provided that the initial amount of target DNA is sufficiently high.

DNA extractions carried out in various cereal-based food matrices (flour, polenta, crackers, bread, toasted bread) were reported to give almost intact high molecular weight genomic DNA only in the case of flour. Due to more intense treatments of mechanical, biochemical, and thermal nature respect to flour, processed foodstuffs lead to fragments sized 200-300 bp [76]. In fact, it has been reported that both the quality and quantity of DNA recovered from food products tend to decrease with the extent to which the product is processed [80-81].

Tilley [82] evaluated the effect that bread-making steps have on wheat DNA size and PCR-based detection of sequences. DNA was extracted by a commercial kit from wheat kernels, milling fractions (bran and shorts), flour, and from samples taken at various steps during and after the baking process. High molecular weight DNA (>12,000 bp) was extracted from kernels, whereas flour DNA exhibited a broad range of fragments, from >12,000 bp to <300 bp. A marked reduction in DNA yield and size occurred after the first 5 min. of baking, resulting in DNA with an average molecular size of about 300 bp. These findings agreed with those of other authors [83-84]. The successful amplification of a 238 bp fragment was obtained from all samples, including those collected at various stages of bread-making, fully baked bread and even bread loaves 5 days after baking. However, PCR conditions required the maximum product size be no more than the average molecular weight of the DNA recovered from the source. In fact, an amplicon of 900 bp was obtained only from DNA extracted from flour and unbaked dough.

Several researches set up methods to extract DNA from pasta with the aim of subsequent detection of the adulteration with bread wheat flour, a topic of great interest in the pasta chain [85-90]. All the reports demonstrated the feasibility of DNA extraction and amplification from pasta. DNA appeared similarly degraded as bread, with fragments sized less than 500 bp [89].

Table 2 summarizes the molecular studies about cultivar traceability in wheat-based foods. Terzi et al. [91] applied AFLP markers to set up a method to test the authenticity of monovarietal pasta, prepared under controlled conditions from eight different certified cultivars. Only AFLP markers sized 79-400 bp were scored in pasta, due to DNA fragmentation during pasta-making process. In this range, AFLP profiles corresponded to those of starting durum wheat cultivars. Hence, it was possible to verify the identity and correspondence with labeling of monovarietal pasta.

The small dimensions of microsatellites induced to expect better results than AFLPs in case of degraded DNA, such as that extractable from pasta, or bread. Microsatellite analysis was effective in checking the authenticity of the Protected Designation of Origin (PDO) bread ‘Pane di Altamura’, obtained from the durum wheat cultivars Appulo, Duilio, Arcangelo and Simeto [92].

The method required the preparation of reference breads starting from semolina obtained from certified grains of the above cited cultivars. The four types of semolina were blended to obtain all the possible combinations and bread was produced from the various combinations and from monovarietal semolina. Ten microsatellite markers were tested, but four were sufficient to distinguish all the possible blends.

The microsatellite amplicon sizes were comprised in the molecular weight range between 115 and 272 bp. In the amplification pattern, analyzed by capillary electrophoresis, it was possible to ascertain the contribution of the single cultivars required for PDO mark and to verify their presence, single or in combination, also when other cultivars were added up to 20%.

Fujita et al. [61] successfully applied 10 selected EST-derived microsatellite markers to varietal identification of commercial flour, cookies, bread and boiled traditional ‘udon’ noodles (thick wheat flour noodles typical of Japanese gastronomy), detecting the contribute of single cultivars in ternary mixtures. Besides checking the end-products, it is essential also to examine seed stocks or flours to avoid the possibility of seed mixture or substitutions.

Table 2. Molecular studies about varietal traceability in wheat-based foods

Samples examined		Markers tested		Reference
Type	Number	Type	Number	
Monovarietal semolina	20	SSR	15	Pasqualone et al., 1999
Seed stocks, monovarietal semolina	52	ISSR	9	Pasqualone et al., 2000
Monovarietal pasta	8	AFLP	3	Terzi et al., 2004
PDO bread ‘Pane di Altamura’	15	SSR	10	Pasqualone et al., 2009
Flour, cookies, bread, ‘udon’ noodles	4	EST-SSR	10	Fujita et al., 2009

Microsatellites were used to set up an identification key of monovarietal semolina from the 20 most diffused Italian durum wheat cultivars [93]. Out of 15 primers used, the combined analysis of only 5 microsatellite electrophoretic profiles enabled the identification of all the monovarietal semolina samples.

The varietal identification of seed stocks and semolina from 52 durum wheat certified grains was successfully carried out also by ISSR markers, provided a preliminary selection of most repeatable markers was made [41]. DNA extracted from durum wheat seeds and from semolina appeared to be amplifiable by ISSR markers due to its integrity (in seeds) or low degradation level (in semolina). Two primers out of nine were able to distinguish all the cultivars, evidencing the effectiveness of these markers to characterize high number of samples with few amplifications.

The New Frontier: Development and Application of SNP Markers

SNP markers have recently become the most widely used markers in many applications. In wheat, an average of one SNP per 540 bp was detected in coding regions [94]. A number of SNP-based platforms have been developed and successfully employed to genotype large numbers of SNPs in wheat [95-99]. A total of 53 durum wheat and 38 bread wheat lines were genotyped at 96 SNP loci by a platform (Illumina BeadArray technology coupled with GoldenGate assay) that did not need prior PCR amplification step. The approach was proved to be cost- and time-effective, able to genotype thousands of wheat lines at thousands of polymorphic sites [96]. Allen et al. [100] mined elite wheat germplasm for intervarietal SNPs and developed and validated 1114 markers on a panel of hexaploid, diploid and tetraploid wheat varieties. Seventy-six percent of putative varietal SNPs were polymorphic between the different varieties, 4% were monomorphic in the hexaploid varieties, and 29% failed to generate a useful amplification signal. Kirkpatrick [101] distinguished 38 wheat genotypes by using 16 SNPs. Chao et al. [99] developed 359 primers for SNP markers and used them to evaluate SNP polymorphism in wheat. These SNPs showed an average polymorphism information content (PIC) of 0.18 among 20 US elite wheat cultivars representing seven market classes. This value increased to 0.23 when SNPs were pre-selected for polymorphisms among a diverse set of 13 wheat accessions. A lower polymorphism degree was observed for SNP markers in the D genome, probably imputable to the selection for developing modern commercial cultivars. Additional SNP discovery efforts targeted to the D genome in elite wheat germplasm will likely be required to offset the lower diversity of this genome.

SNP Markers in Other Crops: Application to Olive Cultivar Identification and Olive Oil Traceability

Several studies were aimed to develop SNP markers in olive (*Olea europaea* L.). A set of 8 SNP markers, identified in coding sequences of different olive genes, was developed to genotype 65 cultivars collected in Europe and Australia. Analyses of the sequences of 3 genes from 11 cultivars revealed a SNP frequency of one per 190 bp in exons and one per 149 bp in introns. The markers were able to discriminate 77% of the samples [102]. Moreover, the utility of SNPs as tool for resolving nomenclature issues was demonstrated. In fact, samples, putatively of the same cultivar but derived from different source, were confirmed as identical (Leccino from Italy and Australia; Spanish and Australian samples of Picual, Arbequina and Hojiblanca). On the other hand, the two cultivars Ottobratica and Mirtolia, considered to be synonyms, were clearly distinguished. In few cases, the markers were not able to distinguish among cultivars (i.e. the couples Sevillano and Lechìn de Sevilla, Manzanilla Cacereña and Manzanilla de Jaén), probably due to very near or coincident geographical origin that genuinely caused a very close genetic relationship. For the identification of such closely related cultivars the utilization of bi-allelic markers such as SNPs may not be sufficient unless a large number is used [102].

Recently, 9 new SNPs were developed in olive by Hakim et al. [103] and were then used to genotype 24 Tunisian olive cultivars to assess their potential in cultivar identification and to compare their performance with 10 previously characterized microsatellites. SNPs showed lower genotyping costs but also lower discrimination power than microsatellites, hence the authors selected a combination of 6 more informative SNPs and one microsatellite marker that can be used for cultivar authentication.

A technique which coupled DNA analysis with a microarray platform was applied for identifying SNP markers in olive, with the aim of discriminating among a panel of 49 varieties selected among the most widely cultivated for olive oil production in the Mediterranean area [104].

The assay distinguished alleles in a ligation reaction, with subsequent fluorescent detection by hybridization on a universal array. This approach was used to detect 17 SNPs in olive genomic sequences previously amplified by PCR of DNA extracted from olive leaves. This semi-automated SNP genotyping assay enabled to unequivocally discriminate all the examined cultivars with the investigated markers.

Recently, high-resolution melting (HRM) analysis has been used to detect SNPs in olive, allowing rapid and cost-effective olive cultivar genotyping [105].

The very small size of SNP amplified fragments renders their analysis suitable in case of degraded DNA, such as olive oil DNA. Hence, with the aim of reliably determine and trace the genotype of monovarietal oils verifying their authenticity, a topic of great impact on the oil industry, Consolandi et al. [106] applied the developed multiplex amplification to DNA extracted from 8 monovarietal olive oils. Preliminary set up of a reliable method of DNA extraction from olive oil was needed, specifically designed for disrupting neutral micelles, able to incorporate DNA.

It implied the addition of a surfactant, after oil mixing with exane, and subsequent recovering of DNA from the oily, rather than from the aqueous, phase. Two sequences containing loci for 4 SNPs were not amplified. Hence, respect to leaves, the multiplex PCR procedure on oil-extracted DNA permitted the analysis of only 11 DNA fragments containing 13 SNPs, and successive application of the ligation detection reaction/universal array platform [106].

The analysis enabled to unequivocally identify the different 8 monovarietal oils, but it was estimated that it was potentially able to distinguish all the 49 cultivars object of study and, consequently, the corresponding oils.

Conclusion

Many classes of molecular markers allowed the detection of DNA polymorphisms among different cultivars in agricultural species. It has been shown that is possible to build large data-bases of DNA profiles of varieties and to populate them with data from different laboratories, if certain precautions are taken with regard to marker selection and general methodology.

DNA analysis is a powerful tool for reliable cultivar identification, with the aim of protecting the efforts of the breeders from deliberate or accidental mislabeling. Moreover, good levels of amplification of molecular markers can be obtained on food DNA, provided that the amount of DNA extracted is sufficiently great and its purity sufficiently high. The step of enzymatic digestion in the AFLP technique can represent a critical point when processing degraded DNA. Microsatellite markers are superior in traceability of processed foods due to their ability to amplify degraded DNA. The availability of methods to extract adequately pure DNA from food, coupled with amplification of the smallest extracted fragments with selected molecular markers, enable to identify and trace the varieties present in PDO and PGI food products, essential in validating their conformity.

In recent years, high expectations are being put into SNP markers, which offered great promise for rapid and highly automated genotyping. They are very abundant and relatively simple to manage. Multiplexed and cost-efficient technologies with high throughput are available for their analysis. The increasing number of genomic sequence data that are accumulating in the databases will be extremely useful in future for discovering new SNP markers.

Acknowledgments

The author gratefully acknowledges Bob Pasqualone for linguistic revision.

References

[1] EC Council Regulation No 510/2006 of 20 March 2006, *Off. J. Eur. Union*, L93/12 (2006).

[2] J. G. K. Williams, A. R. Kubelik, K. J. Livak, J. A. Rafaski and S. V. Tingey, *Nucleic Acids Res.* 18, 6531 (1990).

[3] J. Welsh and M. McClelland, *Nucl. Acids Res.* 18, 7213 (1990).

[4] E. Zietkiewicz, A. Rafalski and D. Labuda, *Genomics* 20, 176 (1994).

[5] P. Vos, R. Hogers, M. Bleeker, M. Reijans, T. van de Lee, M. Hornes, A. Frijters, J. Pot, J. Peleman, M. Kuiper and M. Zabeau, *Nucleic Acids Res.* 23, 4407 (1995).

[6] D. Tautz, *Nucleic Acids Res.* 17, 6463 (1989).

[7] P. K. Gupta, H. S. Balyan, P. C. Sharma and B. Ramesh, *Curr. Sci.* 70, 45 (1996).

[8] M. Morgante and J. Vogel, *United States Patent* No. 08/326456, (1994).

[9] Paran and R. Michelmore, *Theor. Appl. Genet.* 85, 985 (1993).

[10] P. K. Gupta, J. K. Roy and M. Prasad, *Curr. Sci.* 80, 524 (2001).

[11] J. Brookes, *Gene* 234, 177 (1999).
[12] G. A. Penner, A. Bush, R. Wise, W. Kim, L. Domier, K. Kasha, A. Laroche, G. Scoles, S. J. Molnar and G. Fedak, *PCR Methods and Applications* 2, 341 (1993).
[13] W. Powell, M. Morgante, C. Andre, M. Hanafey, J. Vogel, S. Tingey and A. Rafalski, *Mol. Breed.* 2, 225 (1996).
[14] Belaj, Z. Satovic, G. Cipriani, L. Baldoni, R. Testolin, L. Rallo and I. Trujillo, *Theor. Appl. Genet.* 107, 736 (2003).
[15] M. Morgante and A. M. Olivieri, *Plant J.* 3, 175 (1993).
[16] J. K. Roy, H. S. Balyan, M. Prasad and P. K. Gupta, *Theor. Appl. Genet.* 104, 465 (2002).
[17] H. Witsenboer, J. Vogel, and R. W. Michelmore, *Genome* 40, 923 (1997).
[18] M. Rakoczy-Trojanowska and H. Bolibok, *Cell. Mol. Biol. Lett.* 9, 221 (2004).
[19] W. Bushuk and R. R. Zillman, *Can. J. Plant Sci.* 58, 505 (1978).
[20] C. W. Wrigley, J. C. Autran and W. Bushuk, *Adv. Cereal Sci. Technol.* 5, 211 (1982).
[21] W. E. Werner, J. E. Wiktorowicz and D. D. Kasarda, *Cereal Chem.* 71, 397 (1994).
[22] L. Batey, *J. Cereal Sci.* 2, 241 (1984).
[23] J. A. Bietz, T. Burnouf, L. A. Cobb and J. S. Wall, *Cereal Chem.* 61, 129 (1984).
[24] D. Bertrand, P. Robert and W. Loisel, *J. Sci. Food Agric.* 36, 1120 (1985).
[25] E. V. Metakovsky, N. E. Pogna, A. M. Biancardi and R. Redaelli, *J. Genet. Breed.* 48, 55 (1994).
[26] E. Aguiriano, M. Ruiz, R. Fité and J. Carrillo, *Genet. Resour. Crop Evol.* 53, 1543 (2006).
[27] ISO, *Standard 8981: 1993. Wheat. Identification of varieties by electrophoresis*, International Standardization Organization, Geneva (1993).
[28] ICC, *Standard 143: 1995. Wheat. Identification of varieties by electrophoresis*, International Association for Cereal Science and Technology, Vienna (1995).
[29] H. A. Sørensen, M. M. Sperotto, M. Petersen, C. Keşmir, L. Radzikowski, S. Jacobsen and I. Søndergaard, *Rapid Comm. Mass Spectr.* 16, 1232 (2002).
[30] W. H. Vensel, L. Harden, C. H. Tanaka, W. J. Hurkman and W. F. Haddon, *J. Biomol. Tech.* 13, 95 (2002).
[31] M. A. Brescia, G. Di Martino, C. Fares, N. Di Fonzo, C. Platani, S. Ghelli, F. Remiero and A. Sacco, *Cereal Chem.* 79, 238 (2002).
[32] B. A. Marchylo, J. E. Kruger and D. W. Hatcher, *Cereal Chem.* 67, 372 (1990).
[33] K. M. Devos and M. D. Gale, *Theor. Appl. Genet.* 84, 567 (1992).
[34] C. P. Joshi and H. T. Nguyen, *Plant Sci.* 93, 95 (1993a).
[35] A. Myburg, A.-M. Botha, B. D. Wingfield and W. J. M. Wilding, *Cereal Res. Comm.* 25, 875 (1997).
[36] P. Joshi and H. T. Nguyen, *Genome* 36, 602 (1993b).
[37] P. Hernández, A. Martín and G. Dorado, *Mol. Breeding* 5, 245 (1999).
[38] Z. Liu, Q. Sun, Z. Ni, T. Yang and R. A. McIntosh, *Plant Breeding* 118, 215 (1999).
[39] O. Robert, C. Abelard and F. Dedryver, *Mol. Breeding* 5, 167 (1999).
[40] T. Nagaoka and Y. Ogihara, *Theor. Appl. Genet.* 94, 597 (1997).
[41] Pasqualone, C. Lotti, A. Bruno, P. De Vita, N. Di Fonzo and A. Blanco, *Options Méditerranéennes A* 40, 157 (2000).
[42] Y. Zhu, J. Hu, R. Han, Y. Wang and S. Zhu, *Australian J. Crop Sci.* 5, 846 (2011).

[43] J. R. Law, P. Donini, R. M. D. Koebner, J. C. Reeves and R. J. Cooke, *Euphytica* 102, 335 (1998).

[44] M. Maccaferri, S. Stefanelli, F. Rotondo, R. Tuberosa and M. C. Sanguineti, *Genome* 50, 373 (2007).

[45] B. A. Barrett and K. K. Kidwell, *Crop Sci.* 38, 1261 (1998).

[46] E. Noli, M. S. Teriaca, M. C. Sanguineti and S. Conti, *Mol. Breeding*, 22, 301 (2008).

[47] S. Altıntaş¸ F. Toklu, S. Kafkas, B. Kilian, A. Brandolini and H. Özkan, *Plant Breeding* 127, 9 (2008).

[48] M. S. Röder, J. Plaschke, S. U. Konig, A. Borner, M. E. Sorrels, S. D. Tanksley and M. W. Ganal, *Mol. Gen. Genet.* 246, 327 (1995).

[49] Z. Q. Ma, M. S. Röder and M. E. Sorrels, *Genome* 39, 123 (1996).

[50] G. J. Bryan, A. J. Collins, P. Stephenson, A. Orry, J. B. Smith and M. D. Gale, *Theor. Appl. Genet.* 94, 557. (1997).

[51] Q. J. Song, J. R. Shi, S. Singh, E. W. Fickus, J. M. Costa, J. Lewis, B. S. Gill, R. Ward and P. B. Cregan, *Theor. Appl. Genet.* 110, 550 (2005).

[52] P. Rallo, G. Dorado and A. Martín, *Theor. Appl. Genet.* 101, 984 (2000).

[53] E. Pestsova, M. W. Ganal and M. S. Röder, *Genome*, 43, 689 (2000).

[54] P. Stephenson, G. J. Bryan, J. Kirby, A. J. Collins, K. Devos, C. Busso and M. D. Gale, *Theor. Appl. Genet.* 97, 946 (1998).

[55] Agricultural Research Service of the US Department of Agriculture, *Graingenes 2.0: a Database for Triticaeae and Avena*, at http://wheat.pw.usda.gov/GG2/ (2009).

[56] P. Gupta, H. Balyan, K. Edwards, P. Isaac, V. Korzun, M. Röder, M F. Gautier, P. Joudrier, A. Schlatter, J. Dubcovsky, R. De la Pena, M. Khairallah, G. Penner, M. Hayden, P. Sharp, B. Keller, R. Wang, J. Hardouin, P. Jack and P. Leroy, *Theor. Appl. Genet.* 105, 413 (2002).

[57] M. S. Röder, V. Korzun, K. Wendehake, J. Plaschke, M.-H. Tixier, P. Leroy and M. W. Ganal, *Genetics* 149, 2007 (1998).

[58] J.-K. Yu, T. M. Dake, S. Singh, D. Benscher, W. Li, B. Gill and M. E. Sorrells, *Genome* 47, 805 (2004).

[59] N. Nicot, V. Chiquet, B. Gandon, L. Amilhat, F. Legeai, P. Leroy, M. Bernard and P. Sourdille, *Theor. Appl. Genet.* 109, 800 (2004).

[60] L. F. Gao, R. L. Jing, N. X. Huo, Y. Li, X. P. Li, R. H. Zhou, X. P. Chang, J. F. Tang, Z. Y. Ma and J. Z. Jia, *Theor. Appl. Genet.* 108, 1392 (2004).

[61] Y. Fujita, H. Fukuoka and H. Yano, *Breeding Sci.* 59, 159 (2009).

[62] P. Donini, P. Stephenson, G. J. Bryan and R. M. D. Koebner, *Genet. Resour. Crop Evol.* 45, 415 (1998).

[63] J. Plaschke, M. W. Ganal and M. S. Röder, *Theor. Appl. Genet.* 91, 1001 (1995).

[64] R. J. Cooke, G. M. M. Bredemeijer, M. W. Ganal, R. Peeters, P. Isaac, S. Rendell, J. Jackson, M. S. Röder, V. Korzun, K. Wendehake, T. Areshchenkova, M. Dijcks, D. Laborie, L. Bertrand and B. Vosman, *Euphytica* 132, 331 (2003).

[65] M. Prasad, R. K. Varshney, J. K. Roy, H. S. Balyan and P. K. Gupta, *Theor. Appl. Genet.* 100, 584 (2000).

[66] V. Korzun, M. S. Röder, K. Wendehake, A. Pasqualone, C. Lotti, M. W. Ganal and A. Blanco, *Theor. Appl. Genet.* 98, 1202 (1999).

[67] Eujayl, M. E. Sorrells, M. Baum, P. Wolters and W. Powell, *Theor. Appl. Genet.* 104, 399 (2002).

[68] L. Y. Zhang, M. Bernard, P. Leroy, C. Feuillet and P. Sourdille, *Theor. Appl. Genet.* 111, 677 (2005).
[69] N. Dograr, S. Akin-Yalin and M. S. Akkaya, *Plant Breed.* 119, 360 (2000).
[70] D. J. Perry, *Theor. Appl. Genet.* 109, 55 (2004).
[71] G. Mangini, F. Taranto, S. L. Giove, A. Gadaleta and A. Blanco, *Cereal Res. Comm.* 38, 155 (2010).
[72] B. Vosman, R. Cooke, M. Ganal, R. Peeters, P. Isaac and G. Bredemeijer, *Acta Hort.* 546, 307 (2001).
[73] M. S. Röder, K. Wendehake, V. Korzun, G. Bredemeijer, D. Laborie, L. Bertrand, P. Isaac, S. Rendell, J. Jackson, R. J. Cooke, B. Vosman and M. W. Ganal, *Theor. Appl. Genet.* 106, 67 (2002).
[74] V. Roussel, L. Leisova, F. Exbrayat, Z. Stehno and F. Balfourier, *Theor. Appl. Genet.* 111, 162 (2005).
[75] O. P. Mitrofanova, P. P. Strelchenko, A. V. Konarev and F. Balfourier, *Genetika* 11, 1530 (2009).
[76] G. Di Bernardo, S. Del Gaudio, U. Galderisi, A. Cascino and M. Cipollato, *Biotechnol. Progress* 23, 297 (2007).
[77] J. J. Doyle and J. L. Doyle, *Focus* 12, 13 (1990).
[78] M. G. Murray and W. F. Thompson, *Nucleic Acids Res.* 8, 4321 (1980).
[79] J. Sambrook, E. F. Fritsch and T. Maniatis, *Molecular cloning: a laboratory manual*, 2nd edn., Cold Spring Harbor Laboratory Press, Cold Spring Harbor, New York (1989).
[80] C. Peano, M. C. Samson, L. Palmieri, M. Gulli and N. Marmiroli, *J. Agric. Food Chem.* 52, 6962 (2004).
[81] D. S. Smith, P. W. Maxwell and S. H. De Boer, *J. Agric. Food Chem.* 53, 9848 (2005).
[82] M. Tilley, *Cereal Chem.* 81, 44 (2004).
[83] M. Allman, U. Candrian and J. Lüthy, *Lancet* 339, 309 (1992).
[84] J. A. Straub, C. Hertel and W. P. Hammes, *Z. Lebensm. Unters. Forsch A* 208, 77 (1999).
[85] R. Alary, A. Serin, M. P. Duviau, P. Joudrier and M. F. Gautier, *Cereal Chem.* 79, 553 (2002).
[86] M. Arlorio, J. Coïsson, E. Cereti, F. Travaglia, M. Papasso and A. Martelli, *Eur. Food Res. Technol.* 216, 253 (2003).
[87] G. J. Bryan, A. Dixon, M. D. Gale and G. Wiseman, *J. Cereal Sci.* 28, 135 (1998).
[88] V. Terzi, M. Malnati, M. Barbanera, A. M. Stanca and P. Faccioli, *J. Cereal Sci.* 38, 87 (2003).
[89] Pasqualone, C. Montemurro, A. Grinn-Gofron, G. Sonnante and A. Blanco, *J. Agric. Food Chem.* 55, 3312 (2007).
[90] G. Sonnante, C. Montemurro, A. Morgese, W. Sabetta, A. Blanco and A. Pasqualone, *J. Agric. Food Chem.* 57, 10199 (2009).
[91] V. Terzi, C. Morcia, D. Giovanardi, M. G. D'Egidio, A. M. Stanca and P. Faccioli, *Eur. Food Res. Technol.* 219, 428 (2004).
[92] Pasqualone, V. Alba, G. Mangini, A. Blanco and C. Montemurro, *Eur. Food Res. Technol.* 230, 723 (2010).
[93] Pasqualone, C. Lotti and A. Blanco, *Eur. Food Res. Technol.* 210, 144 (1999).
[94] D. J. Somers, R. Kirkpatrick, M. Moniwa and A. Walsh, *Genome* 49, 431 (2003).

[95] M. Akbari, P. Wenzl, V. Caig, J. Carling, L. Xia, S. Yang, G. Uszynski, V. Mohler, A. Lehmensiek, H. Kuchel, M. J. Hayden, N. Howes, P. Sharp, P. Vaughan, B. Rathmell, E. Huttner and A. Kilian, *Theor. Appl. Genet.* 113, 1409 (2006).
[96] E. Akhunov, C. Nicolet and J. Dvorak, *Theor. Appl. Genet.* 119, 507 (2009).
[97] Bérard, M. C. Le Paslier, M. Dardevet, F. Exbrayat-Vinson, I. Bonnin, A. Cenci, A. Haudry, D. Brunel and C. Ravel, *Plant Biotechnol. J.* 7, 364 (2009).
[98] N. Bernardo, P. J. Bradbury, H. Ma, S. Hu, R. L. Bowden, E. S. Buckler and G. Bia, *BMC Genomics* 10, 251 (2009).
[99] S. Chao, W. Zhang, E. Akhunov, J. Sherman, Y. Ma, M. C. Luo and J. Dubcovsky, *Mol. Breed.* 23, 23 (2009).
[100] M. Allen, G. L. A. Barker, S. T. Berry, J. A. Coghill, R. Gwilliam, S. Kirby, P. Robinson, R. C. Brenchley, R. D'Amore, N. McKenzie, D. Waite, A. Hall, M. Bevan, N. Hall and K. J. Edwards, *Plant Biotechnol. J.* 9, 1086 (2011).
[101] R. Kirkpatrick, D. J. Somers, M. Moniwa, A. Walsh and E. Riemer, *Variety identification using single nucleotide polymorphisms in hexaploid wheat.* The plant, animal and microbe genomes X Conference. January 12-16, 2002. San Diego, CA (2002).
[102] S. Reale, S. Doveri, A. Díaz, A. Angiolillo, L. Lucentini, F. Pilla, A. Martín, P. Donini and D. Lee, *Genome* 49, 1193 (2006).
[103] R. Hakim, N. Grati Kammoun, E. Makhloufi and A. Rebaï, *Diversity* 2, 17 (2010).
[104] Consolandi, L. Palmieri, S. Doveri, E. Maestri, N. Marmiroli, S. Reale, D. Lee, L. Baldoni, N. Tosti, M. Severgnini, G. De Bellis and B. Castiglioni. *J. Biotechnol.* 129, 565 (2007).
[105] R. Muleo, M. C. Colao, D. Miano, M. Cirilli, M. C. Intrieri, L. Baldoni and E. Rugini, *Genome* 52, 252 (2009).
[106] Consolandi, L. Palmieri, M. Severgnini, E. Maestri, N. Marmiroli, C. Agrimonti, L. Baldoni, P. Donini, G. De Bellis and B. Castiglioni, *Eur. Food Res. Technol.* 227, 1429 (2008).

In: Cultivars
Editor: Katya Carbone

ISBN: 978-1-62417-101-7

Chapter 3

ANTIOXIDANT COMPOUNDS AND NUTRACEUTICAL BENEFITS OF MEDITERRANEAN RED FRUIT

***Teresa Casacchia*[1,2] *and Adriano Sofo*[3]**
[1] Freelance Nutritionist, Presidente Associazione Nutrizione Umana, Cosenza, Italy
[2] Agricultural Research Council of Italy Olive Growing and Olive Oil Industry Research Centre, Rende, Italy
[3] Department of Crop Systems, Forestry and Environmental Sciences, University of Basilicata, Potenza, Italy

ABSTRACT

Today, the responsibility of reactive oxygen species (ROS) on the onset of non-transmissible chronic degenerative diseases has been clearly recognized by the scientific community. Epidemiological studies show that eating habits can significantly influence the incidence of neurodegenerative diseases. A conspicuous part of the protective effects of the Mediterranean diet is due to the presence of antioxidants in foods and beverages obtained from red fruits. The protective effect of dietary antioxidants from red fruits is supported by data obtained from animals, showing that the supplementation of the diet with red fruits can have beneficial effects against different types of cancer, neurodegenerative diseases and cardiovascular damage. This chapter aims to contribute to a better understanding of the abilities of the antioxidant compounds of red fruits to counteract the damaging effects of ROS and will focus on the relevance of their antioxidant action.

1. INTRODUCTION

Today, the responsibility of reactive oxygen species (ROS) on the onset of non-transmissible chronic degenerative diseases has been clearly recognized by the scientific community. Indeed, ROS react with amino acid residues (in particular cysteine, proline, arginine and histidine), lipids and DNA, so inducing the structural and functional alterations capable of inhibiting the biological activity of cells [1]. The chemical species ROS are very unstable and highly reactive intermediates that are generated as a result of external stimuli (ionizing radiation, chemicals, drugs, stress) and of normal cellular metabolic reactions [2]. It is well known that during cellular respiration, which occurs at the level of the inner membrane of mitochondria, cells perform the complete oxidation of organic substrates at the expense of oxygen to obtain carbon dioxide (CO_2) and water. In the mitochondrial respiratory processes, the main substrate is pyruvate, which comes from glycolysis, but fatty acids and organic acids are also used.

Within the mytochondrial respiratory electron transport chain, the reduction of molecular oxygen (O_2) in superoxide ($O_2{\bullet}^-$) occurs. This latter is the precursor of many other ROS, such as H_2O_2 (hydrogen peroxide), HO• (hydroxyl radical), $HO_2\bullet$ (hydroperoxide radical), $^1O_2$ (singlet oxygen), and nitric oxide NO•. The presence of the unpaired electron in the last orbital make these chemical species highly unstable and able to bind with biological macromolecules, in turn converted into free radicals. Approximately 3 to 10% of the oxygen used by the tissues is converted to its reactive intermediates, which damages the functioning of cells and tissues [3]. These indirect negative effects of ROS can cause lipid oxidation, that inevitably leads to inefficiencies in trans-membrane processes, protein oxidation (with particular regard to thiol groups) and consequent change in the kinetics and inactivation of many enzymes [4], alteration of the mechanisms involved in cellular regulation, signal transduction and gene transcription [5, 6], oxidative damage at the level of DNA and telomeres with consequent cell death apoptosis, cell degeneration and neoplastic transformation [7]. These negative effects on cell metabolism are further increased when an imbalance between the amount of ROS produced and the control systems for their scavenging occurs.

It is certain and scientifically proved the involvement of ROS in the processes of the pathogenesis of many forms of disease, ranging from inflammations to immunosuppression, from liver cirrhosis to atherosclerosis, from diabetes to Alzheimer's disease (AD) and various types of cancer [1, 8]. However, epidemiological studies have clearly shown that the adoption of the Mediterranean diet, based on fruits and vegetables rich in specific micronutrients, vitamins and plant pigments (phytochemicals) helps to restore a balance between the production of free radicals and control ROS-scavenging systems, impacting positively on the reduction of the oxidative damage [8 - 16].

2. Oxidative Stress in Humans

The term oxidative stress refers to a condition, which is characterized by the accumulation of non-enzymatic oxidative damage to molecules that threatening the normal function of the cell or the organism [17]. Several authors have shown that there is a close relationship between oxidative damage, cell aging and the onset of many diseases [18 - 20], ranging from various forms of cancer, neurodegenerative and cardiovascular damage. Observational studies on vegetarians, regular consumers of natural, fresh and simple food, rich in bioactive compounds, highlighted the low incidence rate of the onset of cancer and dementia [1 - 3, 21]. Likewise, women with breast cancer, have shown reduced recurrence rates, as well as less risks of mortality, following a antioxidant-based diet [4, 22]. However, it should be recalled that the production ROS is a condition absolutely physiological and in normal cases not related to cytotoxic effects. In fact, they are molecules that play essential roles for cell metabolism acting, in some cases, as second messengers [23]. The mechanisms of action of ROS have been highlighted not only in eukaryotic cells, but also in bacteria and in plants [22, 24]. It seems that ROS are mainly involved in the regulation of some transcription factors [24 - 28], cellular signal transduction [29 - 37], and cell growth, death by apoptosis, differentiation and immune response.

Indeed, high concentrations of ROS are a source of damage to all biological macromolecules. The imbalance between the production of molecules and the control systems can cause modifications of proteins, lipids and DNA. Proteins undergo fragmentation of the aminoacidic chains, especially in correspondence with arginine, proline and histidine residues. The oxidative damage appears also to induce the formation of crosslinks between two or more proteins by the oxidation of sulfhydryl groups of cysteine [38]. Overall, these alterations result in a premature degradation of damaged proteins and the consequent reduction of their biological function. In this regard, some enzymes are very sensitive to the oxidative damage. These include mitochondrial aconitase [39], all metalloenzymes [40], the enzymes involved in the Calvin cycle (in particular, fructose bisphosphatase and glyceraldehyde-3-phophate dehydrogenase) [41]. Another indication of protein oxidation is the production of nitrotyrosine from peroxynitrite, which could be a useful clinical parameter of the presence of oxidative stress in neurodegenerative diseases, as high levels of nitrotyrosine have been found, after autopsy, in patients with Parkinson's disease (PD) and Amyotrophic Lateral Sclerosis (ALS) [42 - 47].

Lipid peroxidation is considered the main source of damage by ROS. This induces a series of chain reactions (initiation, propagation and termination), which destabilize the structure of the phospholipid bilayer of cellular membranes, with oxidation of polyunsaturated fatty acids (PUFA), formation of toxic species, such as isoprostanes and aldehydes. These include propanal, butanal, pentanal, hexanal and 4-hydroxy-2-trans-nonenal (4-HNE). This last product appears to be involved in neuronal death, and in particular in the pathogenesis of PD [48, 49] and in the consequent complete destruction of the lipid bilayer of cellular membranes.

Oxidative damage to DNA consists in the rupture or rearrangement of the double helix following the formation of hydantoins, glycols, deaminated products and rearrangements in the pyrimidine ring [50].

2.1. Damage Caused by Oxidative Stress and Apoptosis

Any damage to the cells, which have been originated by chemical agents or toxic stimuli, can trigger a complex series of events culminating in cell death. This process can follow two distinct pathways, called the accidental necrosis and the programmed cell death. The former generally appears as a result of pathological processes, such as loss of membrane integrity, cell lysis and consequent release of cytoplasmic components in the extracellular space, inflammatory state, and damage amplification. The latter does not take place only after to pathological events, but is a physiological mechanism occurring during normal body development and growth.

From a morphological point of view, the cells destined to apoptosis lose contact with the surrounding cells and undergo condensation and fragmentation. During apoptosis, cell cytoplasmic organelles assume a compact form and chromatin condenses, resulting in DNA destruction. The resulting DNA fragments are surrounded by the plasma membrane, looking like bubbles (blebbing), which tend to weld together and give rise to "apoptotic bodies", without any damage to neighboring cells. However, particular attention should be placed on the nerve cells, highly sensitive to this type of mechanism. Indeed, a vast scientific literature reports that there is a relationship between ROS production, apoptosis and pathogenesis of neurodegenerative disorders [51 - 53].

2.1.1. Neurodegenerative Damage

If we consider that the human brain is only 2% of body weight, but it consumes an amount of oxygen significantly higher than that of all other organs (about 20% of the quantity available), and it has a high metabolic activity combined with a reduced capacity for cell regeneration and a lower level of antioxidant defenses, it is easy to understand why the structures of the central nervous system (CNS) are particularly susceptible to oxidative damage. In particular, the dopaminergic neurons are strongly susceptible, as the metabolism of dopamine (DA) generates hydrogen peroxide and peroxide radicals, and dopamine oxidation generates dopamine-quinone [54], a protein capable of damaging other proteins and with an essential role in the onset of a Parkinsonian symptomatology.

Also, excessive exposure to ROS seem to be implicated, as demonstrated in experiments of neurotoxicity *in vitro* and *in vivo*, in the production of amyloid peptides, responsible for the onset of AD [55 - 58].

It seems that ROS are eliminated from the cellular antioxidant systems, that includes enzymes as glutathione peroxidase (GPX), ascorbate peroxidase (APX), catalase (CAT), and superoxide dismutase (SOD). Indeed, it has been shown that brain tissues of PD-affected patients, the concentration and activity of these enzymes is significantly lower if compared to normal patients. This suggests that oxidative stress, which is generated as a result of the reduction of cellular antioxidant defenses, not only may be an incidental damage to the cells, but could also trigger intracellular signaling pathways that lead to their death.

In PD, the oxidative damage of neurons, which are highly sensitive to ROS [59], also determines the oxidation of proteins and their subsequent precipitation. It has been demonstrated that α-synuclein, a small presynaptic protein whose physiological role is not yet known, shows increased ability to aggregation and precipitation when oxidized, and it also appears to exist a relation between the oxidative damage to this protein and exposure to herbicides [60 - 62].

Several studies have proved that alterations in the functionality of mytochondrial complex I can expose the cells not only to an energy deficit, but also to oxidative damage [63]. The increase of ROS could amplify the presence of structurally abnormal proteins, which require an increased functionality of the ubiquitin-proteasome system to be eliminated. The alterations of this complex, reducing the availability of energy, may limit the vesicular storage of dopamine, whose level could increases in the intercellular spaces and interact with other cellular proteins. This would explain why dopaminergic neurons are so sensitive to oxidative damage. While the alterations of the mitochondria have been deeply documented, it is impossible to verify with autopsy, as dopaminergic neurons die before the analysis.

However, at present, it is unclear whether there is a final common pathway or if different stimuli lead to cell death. The most important type of programmed cell death is apoptosis. In PD, has not yet been established whether the process of apoptosis is responsible for the primitive cell death, or whether it simply performs its original function, namely to remove irreversibly damaged cells.

2.1.2. Cardiovascular Diseases

Atherosclerosis is a multifactorial disease that represents the leading cause of death in the world. High levels of LDL (low density lipoprotein) in plasma and low levels of HDL (High density lipoprotein) are associated with a high risk of atherosclerosis, following oxidation of LDL. Specifically, LDL is a spherical particle composed of fats and proteins, whose central nucleus contains triglycerides and cholesterol esters. It has been seen that these particles also contain antioxidants, such as vitamin E. The LDL represents a serious danger to the arterial walls, especially when an oxidative process modifies them. The oxidation of LDL, or peroxidation, is a chain reaction caused by free radicals. By means of *in vitro* experiments, it has been shown that the oxidation of these lipoproteins occurs less frequently in the presence of antioxidants. Therefore, flavonoids appear to have the ability to block LDL oxidation and platelet aggregation, so reducing the risk of atherosclerosis.

Many epidemiological studies correlate a moderate consumption of red wine and green tea, particularly rich in flavonoids, with a lower risk of cardiovascular diseases. In addition, olive oil appears to strongly reduce LDL oxidation, and its antioxidant effect seems to be due to the presence of polyphenolic compounds such as hydroxytyrosol and its precursor oleuropein.

The high content of monounsaturated fatty acids has an important protective role on human health. In fact, it was tested that substitution of saturated fats with monounsaturated or polyunsaturated fats in the diet reduces the levels of LDL, which are therefore less available for oxidative processes. Experimental studies have also shown that the LDL rich in oleic acid, the most abundant fatty acid of olive oil, are much more resistant to oxidation.

2.1.3. Endothelial Damage

Several authors have shown that oxidative damage greatly influence vascular function by modulating the activity of contraction and expansion, as well as cell growth, apoptosis, migration, and inflammation [64 - 67]. The mechanism of action by which ROS act at the level of the vascular endothelium is strongly related to their ability to modulate the expression and function of some important genes, including the vascular endothelial growth factor (VEGF), fibroblast growth factor (FGF) and platelet-derived growth factor (PDGF), which appears to play an important atherogenic role in the regulation of cell growth and differentiation [68].

2.1.4. Metabolic Disorders

According to Reitman et al. [69], obesity, a disease with a complex etiopathogenesis, seems to be related to a decrease in the concentration of plasma antioxidants. Furthermore, fat increase in adipocytes appears to be closely related to the markers of oxidative stress [70] and is widely correlated with body mass index (BMI) [71]. On this basis, it is possible to define obesity as a situation of chronic inflammation [72].

Generally, obesity is frequently associated with metabolic diseases like diabetes, and it has been found that the increase of free fatty acids (FFA) in the pre-diabetic phase induces oxidative stress due to increased mitochondrial activity (for example, in the case of increased beta oxidation and mitochondrial uncoupling), with a significant increase in ROS production. In knockout mice, it has also been demonstrated that ROS alter the metabolism of methionine, involved in the enhancement of insulin resistance, and that hyperglycemia is itself an activator of oxidative stress.

These phenomena increase the risk of the onset of a common disease called NAFLD (non-alcoholic fatty liver disease), which in turn can lead to hepatocarcinome (HCC). Probably, the mechanisms responsible of this disease are multiple and include mitochondrial dysfunction, oxidative stress, lipid peroxidation, inflammation and alterations in methionine metabolism [73 - 75].

2.1.5. Oxidative Damage and Cancer

If the concentration of ROS is controlled, their action on cells can be protective, for example inducing phenomena as stasis of the cell cycle, senescence, apoptosis or necrosis of the altered cells. On the other hand, it has been shown that during a condition of ROS imbalance, the action of ROS promotes the proliferation, invasiveness, and metastasis formation of tumor cells.

In addition to directly cause oxidative damage, ROS appear to influence the behavior of cells in many ways. Depending on the types and levels of ROS, ROS exposure time, ROS-removing cellular antioxidant defenses, and activities of cell repair systems, the responses or cells exposed to ROS can include increased proliferation, dysfunctions in cell cycle, senescence, apoptosis or necrosis. The responsibility of ROS in promoting initiation and progression of the tumor may be due to their effects on cell cycle, gene expression, direct or indirect damage to DNA, and other forms of apoptosis and cell death.

It was observed that knockout mice lacking Cu/Zn-SOD, a strong antioxidant enzyme defined as the main scavenger of superoxide radical ($O_2{\bullet}^-$), have a high chance of developing liver cancer. However, at mitochondrial level, Mn-SOD plays a key role, and its deficiency in mice is lethal. Heterozygous animals (50% of normal mitochondrial Mn-SOD) survive, but show increased risk of developing lymphomas, adenocarcinomas and pituitary adenomas with age.

3. Main Types of Antioxidants

An antioxidant is defined as “any substance that, present in very low concentration compared to that of an oxidisable substrate, is able to significantly retard or inhibit the oxidation of that substrate” [76]. An oxygen atom has eight external electrons. The metabolic reactions can deprive the atom of an electron, turning it into a free radical. When the free radical subtracts an electron from a molecule located in the cell membrane, a new free radical is formed and a chain reaction starts. The chain of "kidnapped electrons" damages the cell membrane, leading to the disintegration of the cell and opening the way for cancer and other diseases.

On the basis of their origin, antioxidants can be divided into endogenous or exogenous, depending if they are produced by the body or are obtained from the diet, respectively. In relation to their mechanism of action, antioxidants are distinguished in:

a) preventive or primary antioxidants that, through the chelation of transition metals or the "quenching" of the oxygen singlet, prevent the upstream ROS generation. In this way, the sequence of chain reactions of radical production is not just triggered.
b) the secondary antioxidants or the chain-breaking type antioxidants (i.e. polyphenols), that are agents able to block free radicals, transforming them in less reactive compounds from the starting ones. In this way, these antioxidants are able to act in the phases of initiation and propagation of radical reactions thereby slowing down their speed.

However, in nature, the limits between these classes of antioxidants are not so clear either because some mechanisms of action of certain antioxidants have not been well defined or because there are substances, such as phenolic compounds, which can act simultaneously as first and second type antioxidants.

3.1. Endogenous Antioxidants

The main endogenous enzymes arrayed against the excess of oxygen free radicals include SOD, GPX, APX, CAT, other peroxidases using different substrates (PDX), and a series of water-soluble or liposoluble agents distributed both inside the cells and in extracellular fluids, such as, among others, ascorbate, glutathione, tocopherols, lactoferrin and transferrin.

The antioxidant enzymes cited above act synergically to reduce the $O_2\bullet^-$ to H_2O_2, a less reactive chemical species. Therefore, this transformation allows limiting the concentration of ROS within the cell.

Among the non-enzymatic endogenous antioxidants, we have referred to glutathione, able not only to neutralize free radicals, but also to regenerate in their active reduced forms other antioxidants, such as vitamins C (ascorbate) and E (α-tocopherol). Lactoferrin and transferrin play a preventive antioxidant function, expressed in terms of iron chelation. In this way, iron is prevented to react with peroxides and these latter are not transformed into strongly harmful free radicals, such as alkoxy and peroxyl, by Fenton reaction.

3.2. Exogenous Antioxidants

Exogenous antioxidants, widely present in fruits and vegetables, can be divided into three main classes:

a) Vitamins
b) Carotenoids
c) Phenolic compounds

3.2.1. Vitamins

Among vitamins, vitamin C (ascorbic acid) and vitamin E (α-tocopherol) play an important role. Indeed, vitamin C is the most powerful water-soluble antioxidant present in blood plasma and also acts as a regenerator of vitamin E [77] and cationic species radical of carotenoids [78]. Block et al. [79] observed that vitamin C may reduce the levels of C-reactive protein (CRP), a marker used to monitor the inflammation of the cardiovascular diseases. It is well known that vitamin C is capable of removing $O_2{\bullet}^-$ and HO•, and regenerating tocopherols [80].

Vitamin E, a generic term to define the family of derivatives of α-tocopherol (tocopherols and tocotrienols), is a fat-soluble vitamin that extrinsic its function through the donation of a hydrogen atom with the following formation of tocopheroxyl radical [81]. Vitamin E has strong protective effects against cardiovascular and coronary diseases due to its inhibiting action toward the oxidation of low-density lipoprotein-cholesterol (LDL-cholesterol) [82].

3.2.2. Carotenoids

Carotenoids are a family of yellow, orange and red pigments produced in plant tissues. They are compounds resulting from the head-to-tail condensation (with 1,4 bounds) of isoprenic units (C_5H_8). All the double bonds present in a carotenoid may be in *cis* or *trans* form.

The hydrocarbon carotenoids, formed only by carbon and hydrogen, are called carotenes (tetraterpenes), while those containing also oxygen are called xanthophylls. Many carotenoids have been isolated and the most important ones are lycopene, β-carotene, α-carotene, phytoene, phytofluene and lutein. While carrots, papaya, and mango are rich in β-carotene, tomatoes contain particularly high levels of lycopene.

The carotenoids scavenge singlet oxygen ($^1O_2$) by a "quenching" effect, acting on the transfer of energy between molecules. In fact, carotenoids are able to receive the molecular energy of $^1O_2$ by producing stabilized oxygen (O_2) and excited carotenes at the triplet state. Instead of generating further chemical reactions, carotenoids dissipate the excess energy in the medium in which they are located. Furthermore, as in the course of these reactions carotenoids do not change, they are again available for other cycles of $^1O_2$-quenching [83]. The effectiveness of carotenoids as quenchers is related to the number of double bonds, so the N-carotene, zeaxanthin, cryptoxanthin, W-carotene and, especially lycopene molecules are particularly active from this point of view.

Generally, all carotenoids, and in particular β-carotene, being liposoluble are localized in the lipids of the cell membrane, so providing a protective barrier against ROS. Several studies have shown a lower incidence of lung, prostate and stomach cancer, as well as cardiovascular diseases in individuals subjected to a carotenoid-rich diet. In particular, the consumption of foods rich in lycopene was associated with a lower incidence of prostate cancer [84] and infarct [85].

It was observed that carotenoids and in particular lycopene and β-carotene, are associated with various classes of lipoprotein; specifically, 75% of carotenoids are associated to low density lipoprotein (LDL) and 25% to high-density lipoprotein (HDL) [86]. Carotenoids also have a high antioxidant capacity, due to the ability of their conjugated double bonds to delocalize unpaired electrons. Therefore, their antioxidant action is of chain-breaking type. Human body enacts a series of reactions, through which the correct balance of oxidants/antioxidants is maintained: on the one hand, $O_2{\bullet}^-$ gives rise to secondary ROS that can interact with proteins, lipids and DNA, and on the other hand, the permanent of antioxidants tend to limit such damage, detoxifying and blocking ROS, so bringing lipids and vitamins in their original un-damaged conditions.

3.2.3. Polyphenols

Epidemiological studies have clearly demonstrated that consumption of fruits and vegetables, rich in phytochemicals with a high phenolic content, is strongly correlated with a reduction in cardiovascular and cerebrovascular diseases, as well as with a decrease in cancer mortality [87].

Polyphenols are the most abundant antioxidants of Mediterranean diet and are found in fruits, fresh and dried vegetables, cereals, olives, chocolate and beverages, such as tea, coffee and wine. Polyphenols are products of the secondary metabolism of plants, whose function is to protect the plant from the attacks of pathogenic parasites, and also to give color to some plant organs (e.g., fruits, flowers and leaves), making them attractive to insects and birds (pollination and seed dispersal).

Despite their wide distribution, studies on the effects of polyphenols have begun only in recent years. Several hydroxyl groups placed on aromatic rings characterize the basic structure polyphenol, but molecules with a single phenolic ring can also be considered polyphenols (e.g., phenolic acids and phenolic alcohols). The natural polyphenols are divided into several classes, depending on the number of phenolic rings and to the structural elements that bind each ring with one another.

3.2.3.1. Phenolic Acids

Both grapes and wines contain simple phenolic acids like benzoic acid and cinnamic acid, whose concentration varies from 100 to 200 mg L^{-1} in red wines and from 10 to 20 mg mL^{-1} in white wines. There are seven known types of benzoic acids (C6-C1). Two of them, salicylic acid (*o*-hydroxybenzoic acid) and gentisic acid (2,5-hydroxybenzoic acid) are present only in traces. Benzoic acids differ for the types of substituents of the benzene ring and are found in grapes, especially in the form of glycosides, which are liberated by acid hydrolysis, or in the form of esters (gallic and ellagic tannins), from which they are liberated by alkaline hydrolysis. The free forms are mostly present in red wines, because of the hydrolysis of their combined forms and of degradation reactions of more complex molecules, as it happens for anthocyanins under the action of heat. In grapes and wines, cinnamic-type phenolic acids, with a C6-C3 structure, are also present. They are located essentially as esters of tartaric acid or as glycosides of glucose, while the free forms are poorly represented.

3.2.3.2. Flavonoids

Flavonoids belong to a group of natural substances with variable phenolic structures and particularly abundant in fruits, vegetables, cereals, flowers, tea and wine. Over 4000 varieties of flavonoids have been identified, and many of them are responsible for the attractive colors of flowers, fruit and leaves. Structurally, they are formed by three phenolic rings (two of them, called A and B, are aromatic,) joined by a heterocyclic one (ring C) formed by three carbon atoms and one atom of oxygen [87, 88].

They are divided according to the structure of the ring C in: flavonols (e.g., quercetin, myricetin), flavones (e.g., luteolin and apigenin), flavanols (e.g., catechin and epicatechin), flavans, anthocyanidins and isoflavones [89, 90]. The functional properties of the flavonoids are influenced by a number of structural characteristics, here reported:

a) the presence of a hydroxyl group in the position 3 of the ring C: flavonoids aglycons having a 3-OH (e.g., quercetin, myricetin) are antioxidants more powerful than those having the same group substituted (e.g., diosmentin, apigenin, hesperetin and naringenin);
b) the presence of a double bond between C2 and C3 increases their antioxidant capacity;
c) their antioxidant activity increases with increasing the number of hydroxyl groups (-OH) on the ring B;
d) the presence of glycosidic groups causes a steric hindrance that reduces their antioxidant capacity.

Among flavonoids, anthocyanins are glycosides having a sugar, usually glucose, bonded to the C-3 position (anthocyanin = anthocyanidin aglycon + sugar). The presence of this sugar determines an increase of water solubility and allows anthocyanins to diffuse into the must and wine during maceration, contributing to the color intensity and properties of red wines. Based on the position of hydroxyl and methyl groups on the second phenyl ring, anthocyanin pigments are classified into five groups: cyanins, petunins, peonins, malvins, and delphinins.

Generally, flavonoids inhibit the activity of many enzymes, including lipoxygenases, cyclooxygenase, monooxygenase, xanthine oxidase, NADH oxidase, phospholipase A2, some protein kinases, and transcription factors such as NF-kB [91]. Because of the acidic character of the hydroxyl groups and of the nucleophilic properties of the phenolic rings, flavonoids are highly reactive. They appear to have antiviral, antibacterial, immunostimulant, anti-ischemic, anti-neoplastic, anti-inflammatory, anti-allergic, anti-lipoperoxidative and gastroprotective properties [92-94].

3.2.3.3. Resveratrol

Resveratrol (3,5,4'-trihydroxystilbene) belongs to the family of stilbenes. These are phenolic compounds, with low molecular weight, characterized by the presence of two aromatic rings joined by an ethane or by an ethenic bridge (C6-C2-C6). They are widely distributed in some Bryophytes and in higher plants, where they usually act as phytoalexins and as growth regulators. For example, lunularic acid is a dihydroxystilbene, which acts as a growth inhibitor in a manner analogous to abscisic acid (ABA). Resveratrol is a natural phytoalexin synthesized by plant in response to fungal attacks, or to abiotic agents, such as exposure to ultraviolet rays and drought. It is mainly synthesized in the berry skin of red grapes, and is transferred into wine [95 - 99].

4. Antioxidants in Mediterranean Red Fruits

4.1. Red Grape and Wine

4.1.1. Bioactive Compounds in Red Grapes

The main bioactive compounds extracted from the processes of grape transformation are essentially resveratrol, anthocyanins, proanthocyanidins (tannins), flavonols, and catechins [100]. Wine marc contains significant amounts of substances that have beneficial effects on human health, such as fibers (17-21%), tannins (16-27%) and other polyphenolic compounds (2-6.5%), fats (7-12%), sugars (3%), and salts of tartaric acid. In particular, polyphenols (mainly ellagic acid and quercetin) and resveratrol have relevance for their antioxidant properties and their ability to scavenge ROS [101]. Therefore, the use of residues of grape processing as a raw material for the production of bioactive compounds may constitute an interesting option.

4.1.2. Metabolic Effects of Grape Antioxidants

Grape polyphenols have antioxidant characteristics, with potential beneficial effects on human health, such as the reduction in the risk of cardiovascular diseases and cancer. Polyphenols have also been studied as an additional source of benefits of organic products, even though no conclusive evidence has come yet. They can bind the non-heme iron, a less absorbable form of iron (approximately 60% of the iron in meat and 100% of the iron in vegetables, fruits, milk and dairy products are in this form), so increasing iron absorption in the body. The recommended daily dose of polyphenols from red wine is 100 to 300 mg, which corresponds to about two glasses of red wine day^{-1}.

The wide range of health-promoting antioxidant compounds in grape and wine suggests that several different and interrelated mechanisms of action are involved. The study of the components present in grapes and wine, as in the case of flavonols and anthocyanins, may also contribute to winegrape taxonomic characterization [102 -104] and to certificate wine quality and origin.

4.1.2.1. Resveratrol

Red wines contain high levels of resveratrol (Figure 1), as do grapes, raspberries, peanuts, and other cultivated plant species [105]. This compound is located in the skin of red grapes and is a constituent of red wine.

On the basis of the results obtained with tests on animals, the average level of resveratrol in wine seems to be not sufficient to justify the "French paradox". Several plants produce Resveratrol, perhaps due to its antifungal properties. It is one of the most important grape polyphenols. Natural Resveratrol can be extracted from red grapes and wine but also by their roots.

Numerous beneficial effects of resveratrol on health, such as anti-cancer, anti-viral, neuroprotective, anti-aging, anti-inflammatory and "stretch-life" activity were found in animals (e.g., rats). Resveratrol was effective against dysfunction and death early detection of nerve cells, and it could also act against diseases such as Huntington's and Alzheimer's diseases. Resveratrol has been shown to reduce tumor incidence in animals by affecting one or more stages of cancer development. For all these reasons, the strong antioxidant and radical-scavenging properties of resveratrol has been intensively studied in both grapes and wines. Interesting studies supported Sinclair's hypothesis that the effects of resveratrol are indeed due to the activation of the Sirtuin 1 gene which is involved in life extension [106, 107]. Resveratrol has not yet been tested in human trials for any disease.

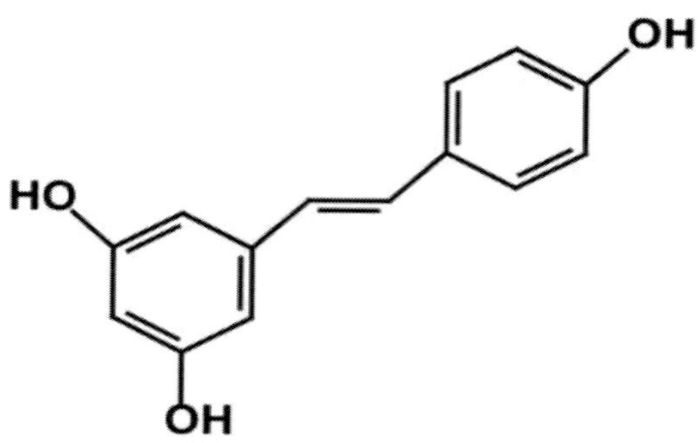

trans-resveratrol

Figure 1. Molecular structure of trans-resveratrol.

Resveratrol has a wide range of pharmacological properties and it has been invoked to explain the "French paradox". This paradox dates back to the epidemiological studies conducted in the '70s that revealed an inverse correlation between red wine consumption and cardiovascular disease in France, a country where the consumption of saturated fats is particularly high. Other clinical studies have demonstrated that a moderate consumption of red wine determines a reduction of the risk factors for atherosclerosis. In red wine consumers, a reduction of platelet aggregation, an increase in plasma levels of HDL-cholesterol (responsible for the disposal of the excess cholesterol in peripheral tissues) and a lower oxidation of LDL occur. These events are associated with a lower formation of atherosclerotic plaques in blood vessels [95] and thus to a reduction of cardiovascular events, making resveratrol also a strong cardioprotective agent. Resveratrol, like many phytoalexins, possesses many activities, such as inhibition of platelet aggregation and lipid peroxidation, anti-inflammatory and vessel-relaxing properties, inhibition of the damage induced by ROS, and agonist properties for estrogen receptors. In the last ten years, it has been shown that resveratrol, flavonoids (e.g., genistein, quercetin) and other polyphenols are able to induce apoptosis of cancer cells [96, 97, 98]. In 1997, Jang et al. [99] reported that resveratrol possesses a strong antitumoral capacity, reducing tumor in rats. These authors demonstrated that resveratrol blocks the in vivo the three stages of carcinogenesis: initiation, promotion and progression. Other studies on cell lines deriving from human tumors confirmed that resveratrol is a chemopreventive agent.

However, the average amount of resveratrol present in any wine is negligible if compared to the amount necessary to obtain beneficial effects on health: 1-10 mg L^{-1} of wine against hundreds of thousands of mg day^{-1}. The amount of resveratrol in food varies greatly. The skins of fresh grapes contain from 50 to 100 μg g^{-1} of resveratrol. European red wines contain from 0.2 to 30.88 mg L^{-1} of total resveratrol, depending on the variety of grape, while white wines contain much less amounts. This is due to the fact that red wine is obtained by the fermentation of must deriving from grape skins rich in resveratrol. On the contrary, white wine is obtained by fermenting the must after berry skin removal, and thus the resulting wine is particularly poor of resveratrol.

4.1.2.2. Anthocyanins and Flavonols

Anthocyanins are a very large group of red-blue polyphenol plant pigments, located within vacuoles of the skin cells of red grape berries, characterized by a positive charge on the molecule [101, 108, 109] (molecular formula in Figure 2). They are also synthesized by some bacteria. In plants, they are responsible for the coloration of leaves, buds, roots, flowers and fruits.

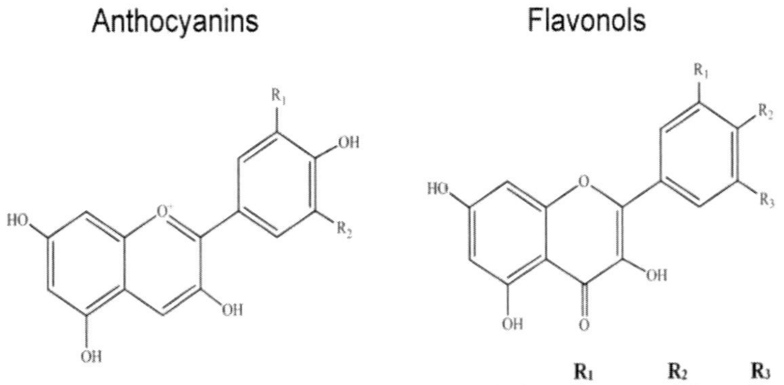

	R_1	R_2
Cyanidin	OH	H
Peonidin	OCH_3	H
Delphinidin	OH	OH
Petunidin	OCH_3	OH
Malvidin	OCH_3	OCH_3

	R_1	R_2	R_3
Kaempferol	H	OH	H
Quercetin	OH	OH	H
Isorhamnetin	OCH_3	OH	H
Myricetin	OH	OH	OH
Laricitrin	OCH_3	OH	OH
Syringetin	OCH_3	OH	OCH_3

Figure 2. The anthocyanins and flavonols commonly found in red grape berry skin and wine.

Blueness and redness levels of these compounds are directly related to the number of free hydroxyl groups of methoxy groups. An example of this phenomenon is malvidine, the main anthocyanin of red grapes, that has the greatest degree of methylation and thus a reddest color.

The polyhydroxylated anthocyanins are able to react with oxidants, such as ROS, thus reducing the damage that these molecules may lead to cells and tissues. On the basis to this ROS-scavenging activity, such substances can be very important for their uses in medicine. For instance, anthocyanins appear to protect against capillary fragility and various processes of aging due to cellular changes caused by ROS, such as inflammatory processes and cancerous changes. Some of these activities have been also demonstrated for red wine consumption.

Anthocyanins have strong anti-inflammatory properties [110]. These compounds are used to fight coronary heart disease [111] and they are a dietary supplement for the prevention and treatment of metabolic disorders, in particular obesity, weight gain, insulin resistance syndromes, diabetes, fasting hyperlipidemia and osteoarthritis [112, 113].

In particular, cyanidin-3-glucoside, delphinidin-3-glucoside, pelargonidin-3-galactoside are the major anthocyanins that promote insulin secretion (1.4-fold at 4 mM glucose concentration) [114, 115]. In red wines, the free anthocyanin fraction is the main contributor to its total antioxidant capacity [116, 117]. Glycosilated and methoxy-derivatives of anthocyanins, like malvidin-3-glucoside, are the most effective as inhibitors at both the initiation and promotion-propagation stages of tumor promotion-carcinogenesis, and protect cells against oxidative damage. These antioxidant properties are due to the flavylium ion (Figure 3).

The presence and concentration of each of the five groups of anthocyanins strongly depend on the environmental conditions and viticultural practices [118 - 121]. Moreover, the pattern of anthocyanin accumulation appears to be closely related to genetic characteristics of the grape itself, and these pigments can be considered as chemical markers to differentiate grape cultivars [122 - 123].

Anthocyanins are also used as food additives and in particular constitute the die anthocyanin red (E163), used in jams and other normally acid food, such as yogurt. Anthocyanins are extracted from red grape skin, as by-product of wine production and processing. The extraction takes place with dilute acids and the product is a liquid containing sugars, acids, salts and pigments originally present in the berry skins. With drying, a water soluble powder relatively rich in these pigments is obtained.

In plants, anthocyanins act as powerful anti-oxidant, helping to protect the plant from ROS produced by the action of UV radiation and during metabolic processes (e.g., electron transport chain during mitochondrial respiration and chloroplastic photosynthesis). The antioxidant power of anthocyanins is maintained even after their consumption by other organisms, which provides a clear explanation of why fruits and vegetables with a red peel are an excellent source of nutrition. It is possible to extract from 30 to 75 mg of anthocyanins from 100 g of red grapes, and 24 to 35 mg from 100 mL of red wine.

On the other hand, flavonols (molecular formula in Figure 1) are pigments that vary in colour from white to yellow distributed in the cellular vacuoles of the epidermis and the first sub epidermal cell layers of berry skins and leaves of white and red grapes.

Noteworthy, flavonols play a protective role against UV radiations and their biosynthesis has found to be light induced in many plant species [101, 104]. In combination with anthocyanins, flavonols are involved in the long-term color stability of red wines by forming intermolecular co-pigments and pigmented polymers with tannins, also precipitating in colloidal coloring matter during wine aging, so positively affecting the aging of red wines [124, 125]. Furthermore, flavonols are considered beneficial because of its antioxidant protective role in the pathogenesis of multiple diseases associated to oxidative stress such as cancer, CHD and arteriosclerosis.

Quercetin, myricetin, laricitrin, kaempherol, isorhamnetin, syringetin and their derivatives are the most commonly flavonols found in grape and wine. Particularly, quercetin resulted to be the most active flavonoid, and high quercetin content is responsible for most of antioxidant activity of medicinal plants. Quercetin has demonstrated significant anti-inflammatory activity by inhibiting the beginning of inflammatory processes. For example, inhibits both the production and the release of histamine and of other allergic/inflammatory mediators.

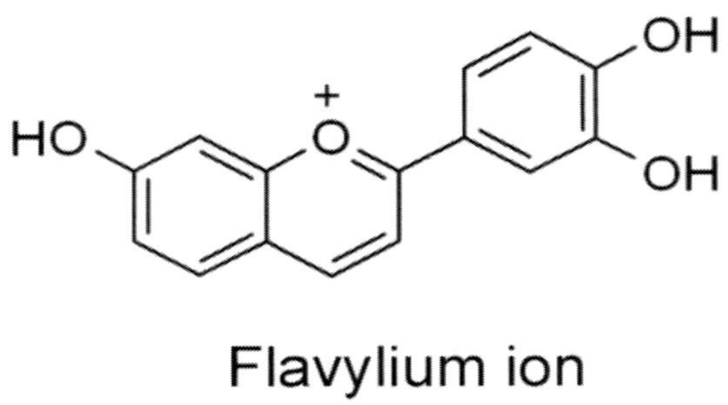

Figure 3. Molecular structure of the flavylium ion.

In addition, it exerts a potent antioxidant and protective action against vitamin C. Quercetin may have positive effects in combating or helping to prevent cancer, prostatitis, heart disease, cataracts, allergies/inflammations, and respiratory diseases, such as bronchitis or asthma. Kaempherol is able to prevent arteriosclerosis by inhibiting the oxidation of low-density lipoprotein and the formation of platelets in the blood. Current evidence indicates that kaempherol not only protects LDL from oxidation but also prevents atherogenesis through suppressing macrophage uptake of oxLDL. Numerous studies showed kaempferol might have health benefits for people at risk of cancer [126, 127, 128]. Finally, isorhamnetin inhibits adipogenesis through down-regulation the genes PPAR-gamma and C/EBP-alpha and represses adipogenesis in 3T3-L1 cells [129].

Although relevant advances in the knowledge of the structures of wine compounds, no systematic research on the qualitative and quantitative determination of the anthocyanins and flavonols has been carried out and the chemistry of wine phenolics subjected to different irradiance levels is still largely a matter of speculation [119]. It is known that the level of light can alter the chemical composition of the fruit both directly, through the activation of specific metabolic pathways or indirectly through control of the berry size [130, 131] or by the control of gas exchange or modification of the source-sink relationships [121]. At the same time, the qualitative and quantitative profiles of grapevine anthocyanins and flavonols also depend on berry skin characteristics, such as thickness and total surface per unit of volume, that in turn are affected by berry size [130].

4.1.2.3. Proanthocyanidins (Tannins)

Proanthocyanidins, another class of grape flavonoids, are polymers of catechin derivatives (Figure 4). They are also known as oligomeric proanthocyanidins (OPCs), tannins or condensed flavan-3-ols.

Jacques Masquelier, who was the first to develop techniques for their extraction from certain plant species, discovered proanthocyanidins in 1936. Because of incorrect food habits, most of the people do not assume sufficient amount of these compounds.

Proanthocyanidins can be found in many plants, particularly in the cortex of the pine, in grape skins and seeds, and in red wines. Since high concentrations of OPCs are present only in some parts of plants such as bark, seeds and skin, is practically impossible to assume the necessary amount of OPCs from our daily diet.

In particular, seeds, containing high amounts of OPCs, are rarely present in food, as sold table grapes are usually seedless. A conspicuous part of the beneficial effects of red wine

(French paradox) is due to its OPC content. A daily consumption of wine, however, is not sufficient to provide the necessary amount of OPCs, as it would be necessary to consume daily two liters of wine per person.

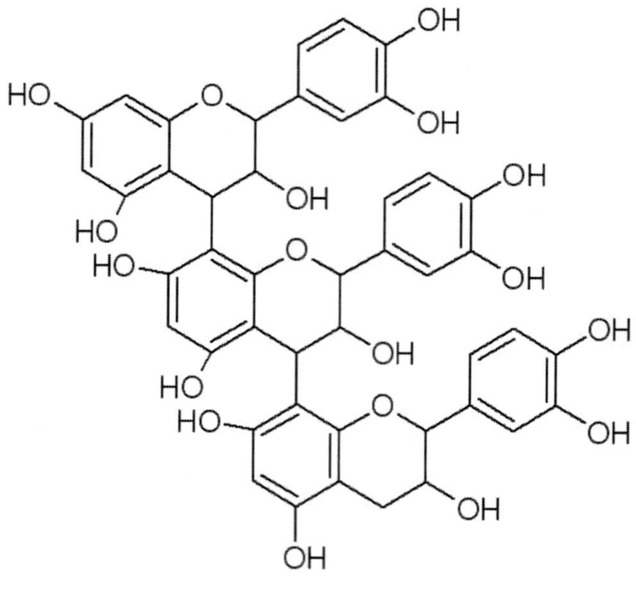

Figure 4. Building blocks of a proanthocyanidin.

The OPCs are powerful antioxidants, able to reduce ROS within the organism. These free radicals are produced by the normal oxygen metabolism, as well as by exposure to sunlight, chemicals, nicotine, alcohol, bacteria, parasites, some fatty acids, and many other substances and conditions. ROS can destroy cell membranes, damage connective tissues, disrupt important physiological processes, and create mutations within the DNA. In particular, collagen and elastin are the main proteins of the so-called connective tissue. With age, these tissues are oxidized, becoming wrinkled and stiff and losing their elasticity. In combination with the vitamin C, OPCs create a favorable environment for the biosynthesis of new collagen and elastin. The antioxidant effects of OPCs and their affinity for connective tissues are the reasons that justify their use, as they stabilize collagen and elastin, making skin young and smooth.

The oxidation of LDL-cholesterol is one of the main responsible for the hardening of the arteries and heart disease. The OPCs have proved very effective antioxidant properties and for this reason different doctors prescribe OPC-based products for the treatment of vascular diseases. Several studies have shown that OPCs are able to reduce tumor activity and propagation. However, these antioxidants cannot be considered effectively a cure for cancer, but rather have a preventive effect and strengthen the immune system by balancing the daily diet. It is believed that the ideal dosage of OPC is from 1 to 3 mg kg^{-1} of body weight day^{-1}.

4.2. Other Mediterranean Red Fruits

The consumption of other Mediterranean red fruits and their juices (e.g., pomegranate, watermelon strawberry, orange, small fruits) have received insufficient attention in relation to their antioxidant activity.

For example, pomegranate juice has become very popular because of the attribution of important biological actions [132]. The strong antioxidant and anti-tumor activity of tannins in the bark of pomegranate (punicacortein) [133, 134] and the antioxidant activity of pomegranate-fermented juice [135] were reported. However, detailed investigations of phenolic compounds and the antioxidant activity of pomegranate juice has not yet been fully clarified. In the next paragraphs, the antioxidant properties of these Mediterranean red fruits will be discussed in detail.

4.2.1. Pomegranate

It is not easy to explain in details all the antioxidant and health-promoting properties of pomegranate [136], so we decided to summarize the main benefits caused by their consumption (as fruits or juices), listed as follows:

- reduction of the risk of breast cancer [137 - 140];
- slowing of prostate cancer: studies on mice show that when fed with red fruits or juices, a slowing of prostate cancer occurs [137, 138, 141]. In a study on 50 men who regularly consumed a glass of pomegranate juice a day, PSA (Prostate Specific Antigen) levels remained stable, reducing the need for further treatments, such as chemotherapy or hormonal therapy. The protective activity against cancer pathologies is due to the high content of flavonoids, powerful ROS-scavenging antioxidants;
- protection of neonatal brain: studies show that their consumption is capable of protecting a child from the disorder that affects the neural tube [142];
- prevention of osteoarthritis: several researches show that they may prevent cartilage deterioration [143];
- protective action for arteries and heart. They combat hardening of the arteries and related diseases, like heart attacks and strokes [144];
- reduction in the blood levels of LDL-cholesterol (bad cholesterol) and increase in HDL-cholesterol (good cholesterol) [145];
- reduction of blood pressure: for instance, a study showed that drinking 50 mL of pomegranate juice day^{-1} can lower systolic blood pressure by 5% [146];
- prevention of AD. After a diet with red fruits or their juice, mice reduce or delay the manifestation of this disease, compared to the control that had instead followed a normal diet [147];
- reduction of the problems of erectile dysfunction, a disorder that affects 1/10 men worldwide and 10 million to 30 million of men in the U.S. Alone. Erectile dysfunction can be caused by several factors: high blood pressure, heart disease, diabetes, neurological damage, endocrine imbalance or depression [148]

The recommended dose of pomegranate juice is 250 mL day^{-1}, capable of supplying about 50% of the Recommended Daily Allowance (RDA) of vitamins A, C and E, and 13% RDA of potassium for adults. It is absolutely recommended the consumption for all women who want to prepare for pregnancy. For instance, pomegranate juice adds up to 100% RDA of folic acid, whose role is crucial for preventing serious diseases in children, such as the disorder of the neural tube.

4.2.2. Watermelon

In watermelon, an important protective role is played by citrulline (Figure 5), a non-essential α-amino acid synthesized by the human body but whose levels are significantly higher in this fruit. It is not encoded by mRNA codons, although it is part of some proteins, such as histones. Thus, the citrulline contained in proteins derives from a post-translational process called "citrullination", or deimination of arginine.

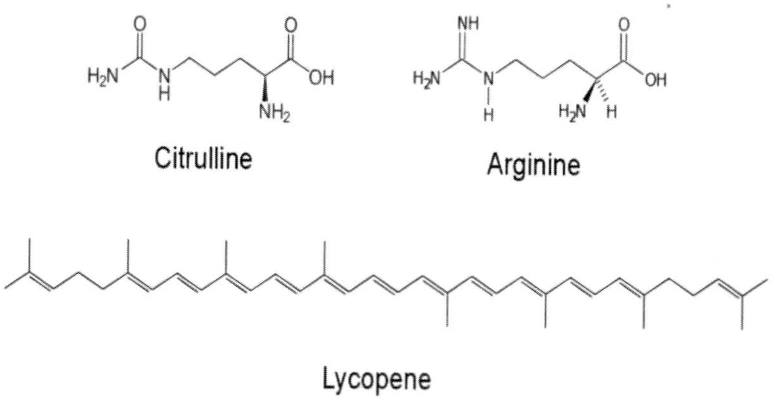

Figure 5. Main antioxidant compounds of watermelon.

This latter falls into the category of "conditionally essential amino acids" together with glycine, glutamine, proline and taurine. Citrulline is used in the human body to produce arginine (Figure 5), an amino acid essential for cell multiplication and the removal of ammonia from the body. Some authors suggest that a diet rich in watermelon can be a real remedy for the prevention of atherosclerosis and of other vascular complications, and for the control of high blood pressure and high levels of glucose in the blood [149 - 152].

Furthermore, according to a study conducted at the University of Texas, the arginine which originates from citrulline increases the concentration of nitric oxide, which prevents the hardening of blood vessels with a Viagra-like effect [153]. For sure, it has not the same effect of drugs, but may bring some benefits without side effects. Finally, it has been discovered the antibody against citrulline, which is a highly specific index for rheumatoid arthritis [154].

Watermelon is also a good dietary source of lycopene (Figure 5), an interesting antioxidant which seems to be implicated, due to its high number of conjugated dienes, in reducing the risk of various chronic degenerative diseases [155, 156]. In particular, research suggests that high blood levels of lycopene, are associated with a considerable reduction of risk of cardiovascular diseases [157].

These benefits appears to be related to the concomitant presence of other interesting phytonutrients, such as the glycoside esculeoside A, the flavonoids chalconaringenin, naringenin, rutin, and kaempferol, and the fatty acid 9-oxo-octadecadienoic acid. These phytonutrients seem to be implicated in the regulation of lipid metabolism and thus contribute to the decrease in total cholesterol, LDL-cholesterol and triglycerides.

4.2.3. Strawberry

Significant levels of antioxidants in the blood caused by consumption of strawberries [158, 159] have been observed in many studies, and their beneficial effects include the decrease of atherogenic risk [160 - 162], diabetes [163], multiple sclerosis [164], and different forms of cancer [165 -167], as well as a higher neuroprotective capacity [168]. The antioxidant capacity of strawberries appears to be the highest among other fruits, such as apples, peaches, pears, grapes, tomatoes, oranges and kiwifruit.

The benefits of the consumption of strawberries can be attributed to the very high content of vitamin C (Figure 6). For instance, in the summer, 5-6 strawberries can substitute one orange fruit, as 100 of strawberries contain 54 mg of vitamin C vs 50 mg of vitamin C in 100 g of oranges.

Furthermore, the antioxidant activity of strawberries is associated with the presence of ellagic acid (Figure 7), which is the most prevalent polyphenol and whose levels strongly depend on storage conditions, and of *p*-coumaric acid (Figure 5) [169 - 172].

The former is found in mean concentrations of 39.6 and 52.2 mg 100 g^{-1} of strawberries and decreases during the process of fruit ripening; the latter, whose concentration depends on climatic conditions, has average values ranging around 0.9 to and 4.1 mg 100 g fruits [169, 170].

Among flavonoids, ripe strawberries is almost absent in kaempferol (Figure 2). This likely depends on the extraction methods and times, and on storage temperatures. According to D'Antuono et al. [173], organic strawberries have higher antioxidant power, total phenolic content and ellagic acid levels if compared to those deriving from integrated cultivation. The anthocyanin content greatly affects the antioxidant capacity of strawberries, as these compounds represent between 52 and 92% of the total content of flavonoids [169].

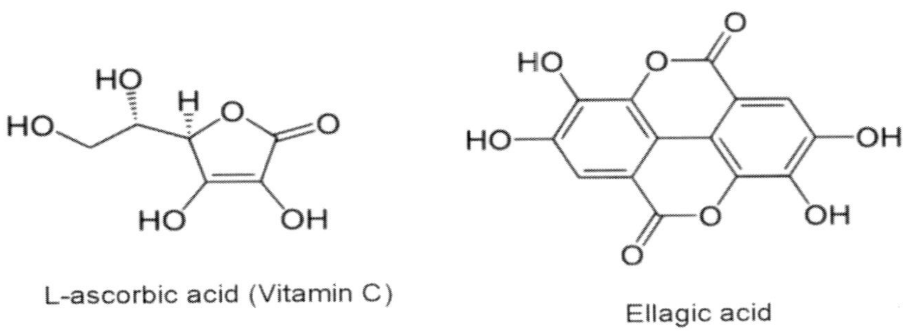

Figure 6. Main antioxidant compounds of strawberry.

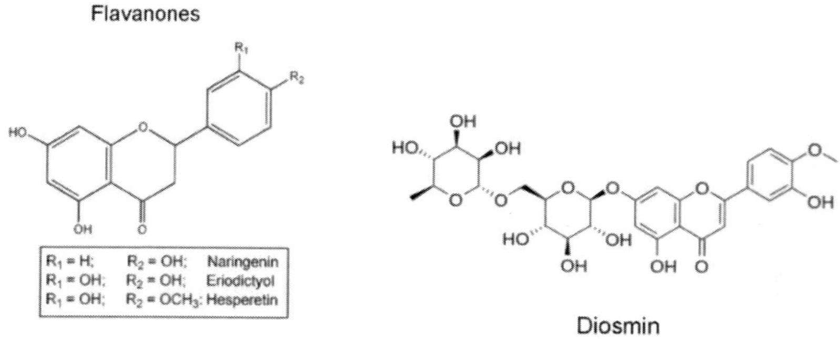

Figure 7. Main antioxidant compounds of orange.

4.2.4. Orange

Oranges are a rich source of antioxidants and nutraceutical activities if consumed regularly during the winter season. They have higher levels of vitamin C (Figure 7), whose concentration in the freshly pressed orange juice ranges between 35-56 mg 100 mL^{-1} [174]. Oranges also contain important ROS-scavenging anthocyanins, responsible for the red color of the varieties Sanguinella, Moro and Tarocco, such as cyanidin-3-glucoside, peonidin-5-glucoside, delphinidin-3-glucoside and petunidin-3-glucoside (Figure 2). In particular, cyanidin-3-glucoside has antioxidant properties higher than those of resveratrol and vitamin C, which results in a greater inhibition of LDL oxidative process [175] with consequent demonstrated protection of the myocardium [176]. Some studies [177, 178] examined the uptake of physiological concentrations of cyanidin-3-glucoside into human vascular endothelial cells. The experimental observations were carried out in isolated plasma membrane vesicles and intact endothelial cells from human endothelial cells and on an ischemia–reperfusion model in isolated rat hearts. Cyanidin-3-glucoside was transported via bilitranslocase into endothelial cells, where it acted as a powerful intracellular antioxidant and a cardioprotective agent in the reperfusion phase after ischemia [179]. Furthermore, recent studies demonstrated the anticancer activity of cyanidin-3-glucoside [180].

In oranges, phenolic acids (ferulic, coumaric, synapic, and ceffeic), flavanones (hesperidin, neohesperdin, hesperetin, poncirin, eriocitrin, neoeriocitrin, naringenin, eriodictyol, and narirutin; Figure 8) and flavones (diosmin; Figure 7) play significant roles in the prevention of cardiovascular diseases [181 - 185]. These molecules seem to have powerful anticancer and neuroprotective actions [186, 187], and are able to induce apoptosis [188 - 190].

4.2.5. Small Fruits (Blackberry, Blueberry, Raspberry)

Small fruits contain significant concentrations of several antioxidants [191 - 194], including vitamin C (approximately 19.0 mg in 100 g of edible portion), procyanidins, ellagic acid derivatives, and the very complex molecules ellagitannins (Figure 8). For instance, recent studies suggest that a diet rich in raw blackberry seems to have anticancer [195, 196] and anti-inflammatory actions [197].

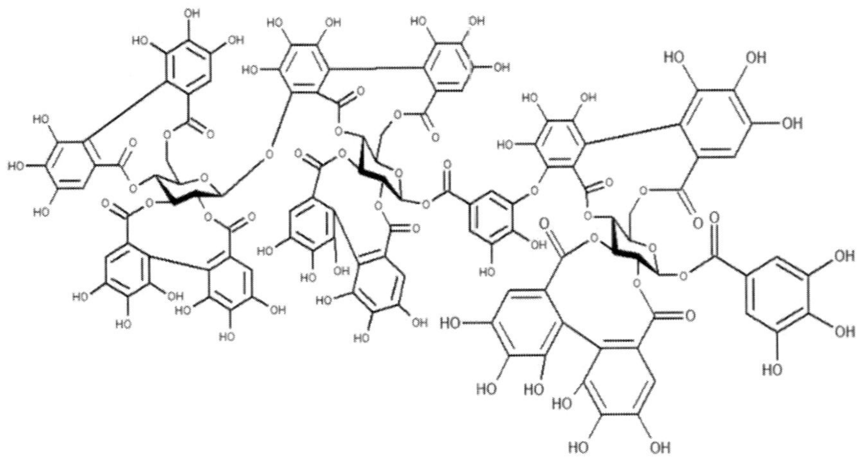

Figure 8. Raspberry ellagitannin containing 6 ellagic acid-type components and two additional monomeric phenolics, for a total of 14 gallic acid units.

CONCLUSIONS

Although the available data are still limited, epidemiological studies show that eating habits can significantly influence the incidence of neurodegenerative diseases such as dementia (Including AD) and PD. For example, data on AS incidence obtained from the so-called “Agees Personnes Quid” (PAQUID) studies showed that people who drink three to four glasses of wine a day (or red fruit juice equivalent) for three years have a decrease by 80% of dementia and AD, if compared to those who drink less or do not drink at all.

A conspicuous part of the protective effects of the Mediterranean diet is due to the presence of antioxidants in foods and beverages obtained from red fruits. Many epidemiological studies have also found an inverse association between the intake of vitamin E from red fruits and the appearance of PD. The protective effect of dietary antioxidants against neurodegenerative and cardiovascular diseases and different types of cancer is supported by data obtained from animals. Furthermore, the supplementation of the diet with red fruits can have beneficial effects on the age-related decline of neuronal and cognitive functions. The ability of these compounds to counteract the damaging effects of ROS and the relevance of their antioxidant action is therefore based on sound science.

REFERENCES

[1] M. Valko, C.J. Rhodesb, J. Moncola, M. Izakovica and M. Mazura, *Chem-Biol Interact,* 160, 1 (2006).

[2] JM. McCord, *Am. J. Med.* 108, 652 (2000).

[3] R. Noor, S. Mittal, J. Iqbal, *Med. Sci. Monit.* 8, 210 (2002).

[4] B. Ames, *J. Biol. Chem.* 278, 4369 (2003).

[5] Squier T.C. and D.J. Bigelow, *Front. Biosci.* 5: D504 (2000).

[6] Grune T, R. Shringarpure, N. Sitte and K. Davies, *J. Gerontol. A Biol. Sci. Med. Sci.* 56: B459 (2001).
[7] P. Jansen-Dürr and H.D. Osiewacz, *EMBO Rep.* 3, 1127 (2002).
[8] L. Lyras, R.H. Perry, E.K. Perry, P.G. Ince, A. Jenner, P. Jenner and B. Halliwell, *J. Neurochem.* 71, 302 (1998).
[9] J.M. Thole, T.F. Kraft, L.A. Sueiro, Y.H. Kang, J.J. Gills, M. Cuendet, J.M. Pezzuto, D.S. Seigler and M.A. Lila, *J. Med. Food,* 9, 498 (2006).
[10] K. de la Torre-Carbot, J.L. Chávez-Servín, O. Jaúregui, A.I. Castellote, R.M. Lamuela-Raventós, T. Nurmi, H.E. Poulsen, A.V. Gaddi, J. Kaikkonen, H-F. Zunft, H. Kisewetter, M. Fitó, M-I. Covas, C. López-Sabater, *J. Nutr.* 140, 501 (2010).
[11] J-K. Moon, J-R. Kim, Y.-J. Ahn and T. Shibamoto, *J. Agric. Food Chem.* 58, 6672 (2010).
[12] Y. Allouche, F. Warleta, M. Campos, C. Sanchez-Quesada, M. Uceda, G. Beltarn and J.J. Gaforio, *J. Agric. Food Chem.* 59, 121 (2011).
[13] N. Bendinelli, G. Masala, C. Saieva, S. Salvini, C. Calonico, C. Sacerdote, C. Agnoli, S. Grioni, G. Frasca, A. Mattiello, P. Chiodini, R. Tumino, P. Vineis, D. Palli and S. Panico, *Am. J. Clin. Nutr.* 93, 275 (2011).
[14] C.E. Guerrero-Beltrána, M. Calderón-Olivera, J. Pedraza-Chaverria and Y.I. Chirino, *Exp. Toxicol. Pathol.* Dec 1 2010, Epub ahead of print, http://www.sciencedirect.com/science/article/pii/S0940299310001831 (2010).
[15] C.E. Guerrero-Beltrána, P. Mukhopadhyaya, B. Horvátha, M. Rajesha, E. Tapiad, I. García-Torrese, J. Pedraza-Chaverrib and P. Pachera, *J. Nutr. Biochem.* Jun 16 2011, Epub ahead of print, http://www.sciencedirect.com/science/article/pii/S0955286311000775 (2011).
[16] M. Ruiz-Canela and M.A.Martínez-González, *Maturitas,* 68, 245 (2011).
[17] R. Blomhoff, *Curr. Opin. Lipidol.* 16, 47 (2005).
[18] T. Finkel and N.J. Holbrook, *Nature,* 408, 239 (2000).
[19] G. Lighthall, B.P. Hamilton and M. Hamlyn, *J. Hypertens,* 22, 1487 (2004).
[20] C. Xu, S. Goodz and E.M. Sellers, Advanced Drug Delivery Review 54, 1245 (2002).
[21] P. Giem, W.L. Beeson, G.E. Fraser GE, *Neuroepidemiology,* 12, 28 (1993).
[22] S. Nechuta, W. Lu, Z. Chen, Y. Zheng, K. Gu, H. Cai, W. Zheng and X.O. Shu, *Cancer Epidemiol. Biom. Prev.* 20, 262 (2011).
[23] E.Y. Dimova, A. Samoylenko and T. Kietzmann, *Antioxid Redox Sign* 4, 777 (2004).
[24] J.L. Rosner and G. Storz, *Curr. Top Cell. Regul.* 35, 163 (1997).
[25] M.B. Toledano and W.J. Leonard, *Proc. Natl. Acad. Sci. USA,* 88, 4328 (1991).
[26] K. Shirasu, R.A. Dixon and C. Lamb, *Curr. Opin. Immunol.* 8, 3 (1996).
[27] H. Kamata and H. Hirata, *Cell Signal,* 11, 1 (1999).
[28] T. Finkel, *Curr. Opin. Cell. Biol.* 10, 248 (1998).
[29] D. Hecht and Y. Zick, *Biochem. Biophys. Res. Commun.* 188, 773 (1992).
[30] G.L. Schieven, R.S. Mittler, S.G. Nadler, J.M. Kirihara, J.B. Bolen, S.B. Kanner and J.A. Ledbetter JA, *J. Biol.Chem.* 269, 20718 (1994).
[31] K.S. Yao, M. Clayton, P.J. O'Dwyer, *J. Natl. Cancer Inst.* 87, 117 (1995).
[32] H. Kamata, C. Tanaka, H. Yagisawa, S. Matsuda, Y. Gotoh, E. Nishida and H. Hirata, *J. Biol. Chem.* 271, 33018 (1996).
[33] S.G. Rhee, S.W Kang, L.E. Netto, M.S. Seo and E.R. Stadtman, *Biofactors,* 10, 2 (1999).

[34] M. Saitoh, H. Nishitoh, M. Fujii, K. Takeda, K. Tobiume, Y. Sawada, M. Kawabata, K. Miyazono, H. Ichijo, *EMBO J.* 17, 2596 (1998).
[35] W.C. Barrett, J.P DeGnore, Y.F. Keng, Z.Y Zhang, M.B. Yim and P.B. Chock, *J. Biol. Chem.* 274, 34543 (1999).
[36] P. Chiarugi, T. Fiaschi, M.L. Taddei, D. Talini, E. Giannoni, G. Raugei and G. Ramponi, *J. Biol. Chem.* 276, 33478 (2001).
[37] T.J. Preston, W.J. Muller and G. Singh, *J. Biol. Chem.* 276, 9558 (2001).
[38] E.R. Stadtman and C.N. Oliver, *J. Biol. Chem.* 266, 2005 (1991).
[39] D.H. Flint, J.F. Tuminello and M.H. Emptage *J. Biol. Chem.* 268, 22369 (1993).
[40] W.F. Beyer Jr, I. Fridovich. *Anal. Biochem.* 161: 559 (1987).
[41] K. Asada and M. Takahashi, *Photoinhibition*, Elsevier, Amsterdam (1997).
[42] K. Abe, L.H. Pan, M. Watanabe, H. Konno, T. Kato and Y. Hoyama, *Neurosci. Lett.* 19, 124 (1997).
[43] M.F. Beal, R.J. Ferrante, S.E. Browne, R.T Matthews, N.W. Kowall and R.H. Brown Jr, *Ann. Neurol.* 42, 644 (1997).
[44] M.A. Smith, P.L. Richey Harris, L.M. Sayre, J.S. Beckman and G. Perry, *J. Neurosci.* 17, 2653 (1997).
[45] K. Hensley, M.L. Maidt, Z. Yu, H. Sang, W.R. Markesbery and R.A. Floyd, *J. Neurosci.* 18, 8126 (1998).
[46] P.F. Good, P. Werner, A. Hsu, C.W. Olanow and D.P. Perl, *Am. J. Pathol.* 149, 21 (1996).
[47] P.F. Good, A. Hsu, P. Werner, D.P. Perl and C.W. Olanow, *J. Neuropathol. Exp. Neurol.* 57, 338 (1998).
[48] Yoritaka, N. Hattori, K. Uchida, M. Tanaka, E.R. Stadtman and Y. Mizuno, *Proc. Natl. Acad. Sci. USA,* 93, 2696 (1996).
[49] M.L. Selley, *Free Radic. Biol. Med.* 25, 169 (1998).
[50] B.P. Yu, *Physiol. Rev.* 74, 13 (1994).
[51] M.F. Beal, *Neuroscientist,* 3, 21 (1997).
[52] Y. Ishikawa, T. Satoh, Y. Enokido, C. Nishio, T. Ikeuchi and H. Hatanaka, *Brain Res.* 824, 71 (1999).
[53] R.F. Castilho, M.W. Ward and D.G. Nicholls, *J. Neurochem.* 72, 1349 (1999).
[54] D.G. Graham, *Molec. Pharmacol.* 14, 633 (1978).
[55] J.N. Keller, Z. Pang, J.W. Geddes, J.G. Begley, A. Germeyer, G. Waeg and M.P. Mattson, *J. Neurochem.* 69, 273 (1997).
[56] Q. Guo, L.S. Sebastian, B.L. Sopher, M.W. Miller, C.B. Ware, G.M. Martin, M.P. Mattson, *J. Neurochem.* 72, 1019 (1999).
[57] W.R. Markesbery and J.M. Carney, *Brain Pathol.* 9, 133 (1999).
[58] W.R. Markesbery, *Free Radic. Biol. Med.* 23, 134 (1997).
[59] S. Przedborski and V. Jackson-Lewis, *Marcel Dekker Inc., New York*, 273 (2000).
[60] B.I. Giasson, R. Jakes, M. Goedert, J.E. Duda, S. Leight, J.Q. Trojanowski and V.M-Y. Lee, *J. Neurosci. Res.* 59, 528 (2000).
[61] A.B. Manning-Bog, A.L. McCormack, J. Li, V.N. Uversky, A.L. Fink and D.A. Di Monte, *J. Biol. Chem.* 277, 1641 (2002).
[62] M.K. Lee, W. Stirling, Y. Xu, X. Xu, D. Qui, A.S. Mandir, T.M. Dawson, N.G. Copeland, N.A. Jenkins and D.L. Price, *Proc. Natl. Acad. Sci. USA*, 99, 8968 (2002).

[63] J.T. Greenamyre, T.B. Sherer, R. Betarbet and A.V. Panov, *IUBMB Life* 52, 135 (2001).

[64] K.K. Griendling and D.G. Harrison, *Circ. Res.* 85, 562 (1999).

[65] K.K. Griendling , D. Sorescu, B. Lassegue and M. Ushio-Fukai, *Arterioscler. Thromb. Vasc. Biol.* 20, 2175 (2000).

[66] J. Chandra, A. Samali and S. Orrenius, *Free Rad. Biol. Med.* 29, 323 (2000).

[67] T. Finkel and N.J. Holbrook, *Nature*, 408, 239 (2000).

[68] M. Eyries, T. Collins and L. M. Khachigian, *Endothelium,* 11, 133 (2004).

[69] Reitman, I. Friedrich, A. Ben-Amotz and Y. Levy, *Isr. Med. Assoc. J.* 4, 590 (2002).

[70] S. Furukawa, T. Fujita, M. Shimabukuro, M. Iwaki, Y. Yamada, Y. Nakajima, O. Nakayama, M. Makishima, M. Matsuda and I. Shimomura, *J. Clin. Invest.* 114, 1752 (2004).

[71] S.O. Olusi, *Int. J. Obes.* 26, 1159 (2002).

[72] H. Xu, G.T. Barnes, Q. Yang, G. Tan, D. Yang, C.J. Chou, J. Sole, A. Nichols, J.S. Ross, L.A. Tartaglia and H. Chen, *J. Clin. Invest.* 112, 1821 (2003).

[73] H. Tilg and G.S. Hotamisligil, *Gastroenterology,* 131, 934 (2006).

[74] M. Cave, I. Deaciuc, C. Mendez, Z. Song , S. Joshi-Barve and C. McLain. *J. Nutr. Biochem.* 18, 184 (2007).

[75] M. Wortham, L. He, M. Gyamfi, B.L. Copple and J.Y. Wan, *Dig Dis Sci.* 53, 2761 (2008).

[76] B.H. Halliwell and J.M.C. Gutteridge, *Free Rad. Biol. Med., 2nd edn*, Clarendon Press, Oxford, UK.

[77] E. Niki , N. Noguchi, H. Tsuchihashi and H. Gotoh, *Am. J. Clin. Nutr.* 62, 1322S (1995).

[78] Mortensen, L.H. Skibsted and T.G. Truscott, *Arch. Biochem. Biophys.* 385, 13 (2001).

[79] G. Block, C. Jensen, M. Dietrich, E.P. Norkus, M. Hudes, L. Packer, *J. Am. Coll. Nutr. 23,* 141 (2004).

[80] M.W. Davey, M. van Montagu, D. Inze, M. Sanmartin, A. Kanellis, N. Smirnoff, I.J.J Benzie, J.J. Strain, D. Favell and J. Fletcher, *J. Sci. Food Agr.* 80, 825 (2000).

[81] A-M. Lampi, A. Kamal-Eldin and V. Piironen, *Functional Food - Biochemical and Processing Aspects*, CRC Press LLC, Boca Raton, FL, USA, 1 (2002).

[82] M.J. Stampfer and E.B. Rimm. *Am. J. Cli. Nutr.* 62, 1365 (1995).

[83] W. Stahl and H. Sies, *Mol. Aspects Med.* 24, 345 (2003).

[84] E. Giovannucci, A. Ascherio, E.B. Rimm, M.J. Stampfer, G.A. Colditz abnd W.C. Willett, *J. Natl. Cancer Inst.* 87, 1767 (1995).

[85] L. Kohlmeier, J.D. Kark, E. Gomez-Gracia, B.C. Martin, S.E. Steck, A.F.M. Kardinaal, J. Ringstad, M. Thamm, V. Masaev, R. Rieserma, J.M.Martin-Moreno, J.K. Huttunen, F.J. Kok, *Am. J. Epidemiol.* 146, 618 (1997).

[86] S. Goulinet and M.J. Chapman, *Arterioscler. Thromb. Vasc. Biol.* 17, 786 (1997).

[87] Y. Arai, S. Watanabe, M. Kimira, K. Shimoi, R. Mochizuki and N. Kinae, *J. Nutr.* 130, 2243 (2000).

[88] J.B. Harborne, *Encyclopedia of Plant Physiology, Volume 8, Secondary Plant Products*, Springer-Verlag, Berlin, 329 (1980).

[89] Z. Liu , W. Li, J. Sun, C. Liu, Q. Zeng, J. Huang, B. Yu and J. Huo, *Asia Pac. J. Clin. Nutr.* 13, 204 (2004).

[90] L.H.Yao, Y.M. Jiang, J. Shi, F.A. Tomás-Barberán, N. Datta, R. Singanusong and S.S. Chen, *Plant Food Hum. Nutr.* (Formerly Qualitas Plantarum) 59, 113 (2012).
[91] N. Skottova, R. Vecera, K. Urbanek, P. Vana, D. Walterova and L. Cvak, *Pharmacol. Res.* 47, 17 (2003).
[92] G. Mojzisovà and M. Kuchta, *Physiol. Res.* 50, 529 (2001).
[93] P.M. Kris-Etherton, K.D. Hecker and A. Bonanome, S.M. Coval, A.E. Binkoski, K.F. Hilpert, A.E. Griel and T.D. Etherton, *Am. J. Med.* 113, 71 (2002).
[94] S. Parthasarathy, E. Wieland and D.A. Steinberg, *Proc. Natl. Acad. Sci. USA,* 86, 1046 (1989).
[95] E.N. Frankel, J. Kanner, J.B. German, E. Park, J.E. Kinsella, *Lancet,* 43, 454 (1993).
[96] Y.J. Surh, Nat Rev Cancer 3, 768 (2003).
[97] T. Richard, A.D. Pawlus, M-L. Iglésias, E. Pedrot, P. Waffo-Teguo, J-M. Mérillon, J-P. Monti, *Annals of the New York Academy of Sciences* 1215, 103 (2011).
[98] M. Minakawa, A. Kawano, Y. Miura and K. Yagasaki, *J. Clin. Biochem. Nutr.* 48, 237 (2011).
[99] M. Jang, L. Cai, G.O. Udeani, K.V. Slowing, C.F. Thomas, C.W. Beecher, H.H. Fong, N.R. Farnsworth, A.D. Kinghorn, R.G. Mehta, R.C. Moon and J.M. Pezzuto, *Science,* 275, 218 (1997).
[100] L. Casas, C. Mantell, M. Rodríguez, A. Torres, F.A. Macías and E.J. Martínez de la Ossa. *J. Supercrit. Fluid,* 45, 37 (2008).
[101] M. Manfra, M. De Nisco, A. Bolognese, V. Nuzzo, A. Sofo, A. Scopa, L. Santi, G.C. Tenore and E. Novellino E, J Sci Food Agr 91, 2749 (2011).
[102] R.S. Jackson, *Wine Science: Principles, Practice, Perception. Second Edition.* Academic Press, San Diego, CA, USA (2000).
[103] P.Y. Ribéreau-Gayon, Y. Glories, A. Maujean and D. Dubourdieu, *Handbook of Enology, Volume 2: The Chemistry of Wine and Stabilization and Treatments*, John Wiley and Sons Ltd., Chichester, UK (2006).
[104] N. Terrier, C. Poncet-Legrand and V. Cheynier. *Wine Chemistry and Biochemistry.* Springer, New York, USA, 463 (2009).
[105] D.M. Goldberg and G.J. Soleas, *Wine. A scientific exploration,* Taylor and Francis, New York, USA 160 (2003).
[106] K.T. Howitz, K.J. Bitterman, H.Y. Cohen, D.W. Lamming, S. Lavu, J.G. Wood, R.E. Zipkin, P. Chung, A. Kisielewski, L.L. Zhang, B. Scherer and D.A. Sinclair, *Nature,* 425, 191 (2003).
[107] J.G. Wood, B. Rogina, S. Lavu, K. Howitz, S.L. Helfand, M. Tatar and D. Sinclair, *Nature,* 430, 686 (2004).
[108] J.M. Kong, L.S. Chia, N-K. Goh, T-F. Chia and R. Brouillard, *Phytochemistry,* 64, 923 (2003).
[109] Sofo, V. Nuzzo, G. Tataranni, M. Manfra, M. De Nisco and A. Scopa, *J. Plant Physiol.* Accepted for publication. In press.
[110] H. Wang, G. Cao, and R.L. Prior, *J. Agr. Food Chem.* 5, 304 (1997).
[111] S. de Pascual-Teresa, D.A. Moreno and C. García-Viguera, *Int. J. Mol. Sci.* 11, 1679 (2010).
[112] G. Mazza, *Acta Hortic.* 754, 577 (2007).
[113] E.I. Meja-Meza, J.A. Yanez, C.M. Remsberg, J.K. Takemoto, N.M. Davies, B. Rasco, C. Clary, *J. Food Sci.* 75, H5 (2010).

[114] M. Pinent, M.C. Bladé, M.J. Salvadó, M.J., Arola and L. Ardévol, *J. Agric. Food Chem.* 53, 5932 (2005).
[115] B. Jayaprakasam, S.K. Vareed, L.K. Olson and N.G. Nair, *J. Agr. Food Chem.* 53, 28 (2005).
[116] M.D. Rivero-Pérez, P. Muñiz and M.L. González-Sanjosé, M.L., *Food Chem. Toxicol.* 46, 2815 (2008).
[117] B. Radovanović and A. Radovanović, *Molecules,* 15, 4213 (2010).
[118] A.B. Bautista-Ortína, J.I. Fernández-Fernández, J.M. López-Roca, E. Gómez-Plaza, *J. Food Compos. Anal.* 20, 546 (2007).
[119] C. Santos-Buelga and V. de Freitas, *Wine Chemistry and Biochemistry. Springer, New York,* 529 (2009).
[120] J. Carreño, L. Almela, A. Martínez and A. Fernández-López, *Ebensmittel-Wissenschaft und-Technologie,* 30, 259 (1997).
[121] M.O. Downey, N.K Dokoozlian, M.P. Krstic, *Am. J. Enol. Viticult.* 57, 257 (2006).
[122] Hall, D.W. Lamb, B.P. Holzapfel and J.P. Louis, *Prec. Agric.* 12, 103 (2010).
[123] P.K. Boss and C. Davies. *Grapevine Molecular Physiology and Biotechnology*, Springer, Netherlands, 263 (2009).
[124] R. Boulton, *Am. J. Enol. Viticult.* 52, 67 (2001).
[125] Zimman and A.L. Waterhouse, *Am. J. Enol. Viticult.* 55, 139 (2005).
[126] Y. Zhang, A.Y. Chen, M. Li, C. Chen and Q. Yao, *J. Surg. Res.* 148, 17 (2008).
[127] V. Sharma, C. Joseph, S. Ghosh, A. Agarwal, M. Kumar Mishra and E. Sen, *Mol. Cancer Ther.* 6, 2544 (2007).
[128] C. López-Sánchez, F.J. Martín-Romero, F. Sun, L. Luis, A.K. Samhan-Arias, V. García-Martínez and C. Gutiérrez-Merino, *Brain Res.* 1182, 123 (2007).
[129] J. Lee, E. Jung and J. Lee, *Obesity,* 17, 226 (2009).
[130] G. Roby, J.F. Harbertson, D.A. Adams, M.A. Matthews, *Austr. J. Grape. Wine Res.* 10, 100 (2004).
[131] F. Roby and M.A. Matthews, *Austr. J. Grape. Wine Res.* 10, 74 (2004).
[132] E.P. Lansky and R.A. Newman, *J. Ethnopharmacol.* 109, 177 (2007).
[133] J.D. Su, T. Osawa, S. Kawakishi and M. Namili, *Phytochemistry,* 27, 1315 (1988).
[134] Y. Kashiwada, G.I. Nonaka, I. Nishioka, J.J. Chang and K.H. Lee, *J. Nat. Prod.* 55, 1033 (1992).
[135] S.Y. Schubert, E.P. Lansky and I. Neeman, *J. Ethnopharmacol.* 66, 11 (1999).
[136] Faria and C. Calhau, *Crit. Rev. Food Sci. Nutr.* 51, 626 (2011).
[137] Malik, F. Afaq, S. Sarfaraz, V.M. Adhami, D.N. Syed, and H. Mukhtar, *Proc. Natl. Acad. Sci. USA,* 102, 14813 (2005).
[138] D.N. Syed, A. Malik, N. Hadi, S. Sarfaraz, F. Afaq and H. Mukhtar H, *Photochem. Photobiol.* 82, 398 (2006).
[139] V.M. Adhami, N. Khan and H. Mukhtar, *Nutr. Cancer,* 61, 811 (2009).
[140] S.A. Khan, *Pak. J. Pharm. Sci.* 22, 346 (2009).
[141] A.J. Pantuck, J.T. Leppert, N. Zomorodian, W. Aronson, J. Hong, R.J. Barnard, N. Seeram, H. Liker, H. Wang, R. Elashoff, D. Heber, M. Aviram, L. Ignarro and A. Belldegrun, *Clin. Cancer Res.* 12, 4018 (2006).
[142] D.J. Loren, N.P. Seeram, R.N. Schulman and D.M. Holtzman, *Pediatric Res.* 57, 858 (2005).
[143] M. Hadipour-Jahromy and R. Mozaffari-Kermani , *Phytotherapy Res.* 24, 182 (2010).

[144] F. de Nigris, S. Williams-Ignarro, L.O. Lerman, E. Crimi, C. Botti, G. Mansueto, F.P. D'Armiento, G. De Rosa, V. Sica, L.J. Ignarro and C. Napoli, *Proc. Natl. Acad. Sci. USA,* 102, 4896 (2005).

[145] C.B. Stowe, *Compl. Ther. Clin. Pract.* 17, 113 (2011).

[146] M. Aviram and L. Dornfeld, *Atherosclerosis,* 158, 195 (2001).

[147] R.E. Hartman, A. Shah, A.M. Fagan, K.E. Schwetye, M. Parsadanian, R.N. Schulman, M.B. Finn and D.M. Holtzman, *Neurobiol. Dis.* 24, 506 (2006).

[148] K. Azadzoi, R. Schulman, M. Aviram, M. Siroky, *J. Urol.* 174, 386 (2005).

[149] A.J. Edwards, B.T. Vinyard, E.R. Wiley, E.D. Brown, J.K. Collins, P. Perkins-Veazie, R.A. Baker, B.A. Clevidence, *J. Nutr.* 133, 1043 (2003).

[150] P. Lucotti, E. Setola, L.D. Monti, E. Galluccio, S. Costa, E.P. Sandoli, I. Fermo, G. Rabaiotti, R. Gatti and P. Piatti, *Am. J. Physiol. Endocrinol. Metab.* 291, E906 (2006).

[151] J.K. Collins, G. Wu, P. Perkins-Veazie, K. Spears, P.L. Claypool, R.A Baker and B.A. Clevidence, *Nutrition,* 23, 261 (2007).

[152] University of Kentucky, Watermelon reduces atherosclerosis, animal study finds. *ScienceDaily,* http://www.sciencedaily.com/releases/ 2011/10/111027125153.htm (2011).

[153] Texas AandM University, Watermelon May Have Viagra-effect. *ScienceDaily,* http://www.sciencedaily.com/releases/2008/06/ 080630165707.htm (2008).

[154] S. Agrawal, R. Misra and A. Aggarwal, *Clin. Rheumatol.* 26, 201 (2007).

[155] G.W. Comstock , A.J. lberg, H.Y. Huang, K. Wu, A.E. Burke, S.C. Hoffman, E.P. Norkus, M. Gross, R.G. Cutler, J.S. Morris, V.L. Spate and K.J. Helzlsouer , *Cancer Epidem. Biomar. Prev.* 6, 907 (1997).

[156] B.B. Freeman and K. Reimers, *Am. J. Lifestyle Med.* http://ajl.sagepub.com/content/ early/2010/11/10/1559827610387488.abstract (2010).

[157] T.H. Rissanen, S. Voutilainen, K. Nyyssönen, R. Salonen, G.A. Kaplan, J.T. Salonen, *Am. J. Clin. Nutr.* 77, 133 (2003).

[158] S.M. Henning, N.P. Seeram, Y. Zhang, L. Li, K. Gao, R-P. Lee, D.C. Wang, A. Zerlin, H. Karp, G. Thames, J. Kotlerman, Z. Li and D. Heber, *J. Med. Food,* 13, 116 (2010).

[159] F. Giampieri, S. Tulipani, J.M. Alvarez-Suarez, J.L. Quiles, B. Mezzetti and M. Battino, *Nutrition,* 28, 9 (2012).

[160] J. Salas-Salvado, M. Bullo, A. Pérez-Heras and E. Ros, *Brit. J. Nutr.* 96, S45 (2006).

[161] Burton-Freeman, A. Linares, D. Hyson, T. Kappagoda, *J. Am. Coll. Nutr.* 29, 46 (2010).

[162] S.J. Zunino, M.A. Parelman, T.L. Freytag, C.B. Stephensen, D.S. Kelley, B.E. Mackey, L.R. Woodhouse and E.L. Bonnel, *Brit. J. Nutr.* 9, 1 (2011).

[163] J. Montonen, P. Knekt, T. Härkänen, R. Järvinen, M. Heliövaara, A. Aromaa and A. Reunanen, *Am. J. Epidemiol.* 161, 3 (2005).

[164] Y. Gilgun-Sherki, E. Melamed and D. Offen, *J. Neurol.* 251, 261 (2004).

[165] N.P. Seeram, L.S. Adams, Y. Zhang, R. Lee, D. Sand, H.S. Scheuller and D. Heber, *J. Agric. Food Chem.* 54, 9329 (2006).

[166] Y. Zhang, N.P. Seeram, R. Lee, L. Feng and D., *J. Agric. Food Chem.* 56, 670 (2008).

[167] D.E. Wedge, K.M. Meepagala, J.B. Magee, S. Hope Smith, G. Huang and L.L. Larcom, *J. Med. Food,* 4, 49 (2001).

[168] S. Fortalezas, L. Tavares, R. Pimpão, M. Tyagi, V. Pontes, P.M.Alves, G. McDougall, D. Stewart, R.B. Ferreira and C.N. Santos, *Nutrients,* 2, 214 (2010).

[169] S.H. Häkkinen, S.O. Kärenlampi, H.M. Mykkänen, I.M. Heinonen and A.R. Törrönen, *Eur. Food Res. Technol.* 212, 75 (2000).
[170] M. Kosar, E. Kafkas, S. Paydas and K. Hüsnü, *J. Agric. Food Chem.* 52, 1586 (2004).
[171] M. da Silva Pinto, F.M. Lajolo, M.I. Genovese, *Food Chem.* 107, 1629 (2008).
[172] R. Bobinaité, P. Viškelis, P. Rimantas Venskutonis, *Food Chem.* 132, 1495 (2012).
[173] L.F. D'Antuono, M.L. Maltoni, S. Foschi, S. Elementi, E. Borgini, G. Baruzzi, W. Faedi, *Riv. Fruttic. Ortofloric.* 66, 64 (2004).
[174] H-G. Yan, W-H. Zhang, J-H. Chen and Z-E. Ding, *Afri. J. Biotechnol.* 11, 3623 (2012).
[175] MP. Khknen and M. Heinonen , *J. Agric. Food Chem.* 51, 628 (2003).
[176] A.M. Amorini, G. Fazzina, G. Lazzarino, B. Tavazzi, D. Di Pierro, R. Santucci, F. Sinibaldi, F. Galvano and G. Galvano, *Free Radic. Res.* 35, 953 (2001).
[177] A.M. Amorini, G. Lazzarino, F. Galvano, G. Fazzina, B. Tavazzi and G. Galvano , *Free Radic. Res.* 37, 453 (2003).
[178] L. Ziberna, F. Tramera, S. Mozec, U. Vrhovsekd, F. Mattivid, S. Passamontia, *Free Radic. Res. Med.* 52, 1750 (2012).
[179] J. Azevedo, I. Fernandes, A. Faria, J. Oliveira, A. Fernandes, V. de Freitas, N. Mateus, *Food. Chem.* 119, 518 (2010).
[180] X. Luo, S. Fang, Y. Xiao, F. Song, T. Zou, M. Wang, M. Xia and W. Ling, *Mol. Cell. Biochem.* 362, 211 (2012).
[181] K.J. Joshipura, F.B. Hu, J.E. Manson, .J. Stampfer, E.B. Rimm, F.E. Speizer, G. Colditz, A. Ascherio, B. Rosner, D. Spoegelman, W.C. Willett, Ann Intern Med 134, 1106 (2001).
[182] P.M. Kris-Etherton, K.D. Hecker, A. Bonanome, S.M. Coval, A.E. Binkoski, K.F. Hilpert, A.E. Griel and T.D. Etherton, *Am. J. Med.* 113, 71S (2002).
[183] N.R. Song, H. Yang, J. Park, J.Y. Kwon, N.J. Kang, Y.S. Heo, K.W. Lee and H.J. Lee, *Food Chem.* 133, 658 (2012).
[184] E. Karimi, E. Oskoueian, R. Hendra, A. Oskoueian and H.Z.E. Jaafar, *Molecules,* 17, 1203 (2012).
[185] H. Li, Z. Deng, H. Zhu, C. Hu, R. Liu , J.C. Young, R. Tsao, *Food Res. Int.* 46, 250 (2012).
[186] N. Nalini, S. Aranganathan and J. Kabalimurthy, *Toxicol. Mech. Methods*, Epub ahead of print, http://informahealthcare.com/doi/abs/ 10.3109/15376516.2012.673092 (2012).
[187] S-L. Hwang, P-H. Shih and G-C. Yen, *J. Agric. Food Chem.* 60, 877 (2012).
[188] A.A. Alshatwi, G. Shafi, T.N. Hasan, A.A. Al-Hazzani, M.A. Alsaif, M.A. Alfawaz, KY Lei and A. Munshi, *Asian Pac. J. Cancer. Prev.* 12, 1555 (2011).
[189] C.A. Camargo, M.C.C. Gomes-Marconed, N.C. Wutzki and H. Aoyama, *Anticancer Res.* 32, 129 (2012).
[190] Ghorbani, M. Nazari, M. Jeddi-Tehrani and H. Zand, *European J. Nutr.* 51, 39 (2012).
[191] J. Reyes-Carmona, G.G. Yousef, R.A. Martinez-Peniche and M.A. Lila, *J. Food Sci.* 70, S497 (2005).
[192] Elisia, C. Hu, D.G. Popovich and D.D. Kitts, *Food Chem.* 101, 1052 (2007).
[193] W-Y. Huang, H-C. Zhang, W-X. Liu and C-Y. Li, *Journal of Zhejiang University - Science B* 112, 94 (2012)
[194] S.Y. Wang and H. Lin, *J. Agric. Food Chem.* 48, 140 (2000).
[195] F. Aqil, A. Gupt, R. Munagala, J. Jeyabalan, H. Kausar, R.J. Sharma, I.P. Singh and R.C. Gupta, *Nutr. Cancer,* 64, 428 (2012).

[196] M. Kähkönen, P. Kylli, V. Ollilainen, J-P. Salminen and M. Heinonen, *J. Agric. Food Chem* 60, 1167 (2012).
[197] L. Kaume, W.C. Gilbert, C. Brownmiller, L.R. Howard and L. Devareddy, *J. Funct. Foods*, Available online 12 March 2012, http://www.sciencedirect.com/science/article/pii/S175646461200031X (2012).

In: Cultivars
Editor: Katya Carbone

ISBN: 978-1-62417-101-7

Chapter 4

PHYTOCHEMICAL PROFILES AND ANTIRADICAL CAPACITY OF GRAPE SEED EXTRACTS FROM DIFFERENT ITALIAN CULTIVARS: REUSING OF WINERY BY-PRODUCTS

Francesca Cecchini, Simonetta Moretti, Barbara Giannini and Katya Carbone*
Agricultural Research Council – Research Unit for Enology in Central Italy (CRA-ENC)
Rome, Italy

ABSTRACT

Over the last decade, much research has been focused on the potential health benefits of antioxidants and many synthetic and natural compounds have been evaluated for their antioxidant profile. Secondary compounds of higher plants have been demonstrated in vitro experiments to protect against oxidative damage by inhibiting or quenching free radicals and reactive oxygen species (ROS). The term "free radicals" refers to a wide family of highly reactive chemical species such as ROS, reactive nitrogen species (RNS) and reactive chlorine species (RClS). Animal studies have shown that dietary phytochemical antioxidants are capable of removing these harmful molecules. In this context, the valorization of by-products of plant origin, as a source of functional compounds for use in the food industry or in other areas, such as the cosmetic or the pharmaceutical ones, is a promising sector, especially for the wine industry, where the winery by-products still contain large quantities of polyphenolic compounds with high antioxidant power. Grape seeds belong to this kind of waste. They contain, as secondary metabolites, between 5 to 8% polyphenols depending on the variety. These compounds are mainly flavonoids, including simple molecules as gallic acid as well as more complex ones as procyanidin dimers, trimers and highly polymerized ones.

The antioxidant power of procyanidin molecules has been demonstrated to be 50 times greater than vitamin E. In this chapter, we will review the nutraceutical properties of seed extracts from different grape cultivars (cvs), analyzed over the last few years in

* Corresponding author. Tel. + 39 06 9630222, Email address: katya.carbone@entecra.it

our laboratories with the aim of highlighting the influence of cultivars on the healthy properties of these waste products. The results point out statistically significant differences ($p \leq 0.05$) in the content of total polyphenols, flavans and antiradical activity between the seed extracts of the different cvs under examination and also differences regarding the polyphenol profiles. Reported data underline that the wine industry by-products, especially in the case of white grape varieties that do not undergo maceration during the winemaking process, can be made into some really useful health supplements and the genotype key role to improve their healthy power.

1. Introduction

Vitis vinifera, L. (common grape vine) is a species of *Vitis*, whereas the term subsp. *vinifera* is restricted to cultivated forms (table 1).

Table 1. *Vitis vinifera* botanical classification

SCIENTIFIC CLASSIFICATION	
Kingdom:	Plantae
Division:	Magnoliophyta
Class:	Magnoliopsida
Order:	Vitales
Family:	Vitaceae
Genus:	*Vitis*
Species:	*V. vinifera*

(http://en.wikipedia.org/wiki/Vitis_vinifera).

It is one of the earliest domesticated fruit crops and it has been widely cultivated and appreciated for its fruit and wine. In fact, cultivars of *Vitis vinifera* form the basis of the majority of wines produced around the world [1].

The fruit is a berry, known as a grape. It is a non-climateric fruit growing on the perennial woody vines of the genus *Vitis*. Grapes are natives of Western Asia and Central Europe and are botanically known as *Vitis vinifera*. This species has more than 10,000 varieties, which have been generated from its wild progenitor, *Vitis vinifera* ssp. *sylvestris*, by vegetative propagation and by cross breeding. Grapes are the most important raw material for making wine. Approximately 70% of world grape production is used for this purpose, less than 30% as fresh fruit and about 1 – 2% as dried fruit. Italian wine production in 2011 amounted to 42.2 million hectoliters of wine, which was second only to that of France, while it was the leading country in exports [2].

Grapes have been also shown to be a rich source of high-added-value bioactive compounds, named phytochemicals [3].

These compounds possess strong antioxidant properties that enable them to scavenge free radicals, donate hydrogen, chelate metals, break radical chain reactions, and quench singlet oxygen in vitro and in vivo [4]. All these properties enable them to act in the prevention of oxidative-stress-related diseases (i.e. cancer) [5]. There is growing interest in using naturally occurring compounds as potential cancer chemo preventative agents in human populations. In this regard, a significant correlation between dietary intake and many types of cancer has been shown in epidemiological data generated throughout the world, while in animal experiments, many dietary substances have been documented to have anti-cancer properties.

Besides this fact, in recent years, researchers are also making great efforts to investigate the antioxidant capacities of grape wastes as grape pomace, seeds, stems and grape canes [6], which are good candidates for further development such as nutraceutical supplements or antioxidant remedies. In this chapter will review the nutraceutical properties of seed extracts from different grape cultivars, analyzed in the last few years in our laboratories in order to highlight the influence of cultivars and environmental parameters on the healthy properties of these waste products.

2. Phytochemical Composition of Grape Berry

From a structural point of view, the berry is composed of three different types of tissue (figure 1): pulp, skin, and seeds, with the sheer bulk of wine being derived from the pulp.

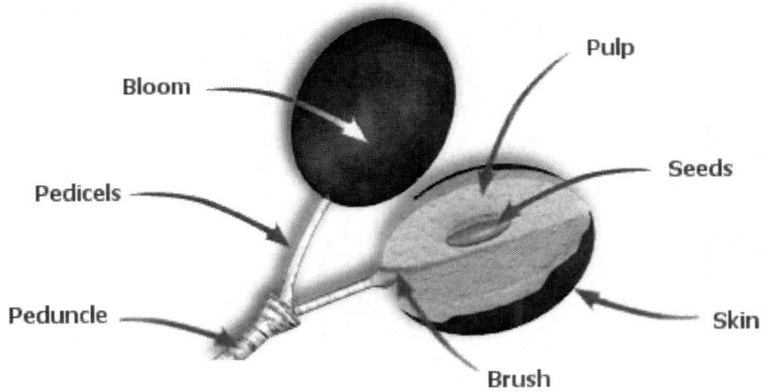

Illustration by http://www.youcellar.com/en/bunch-grape-vine.php

Figure 1. Berry's structure.

These tissues vary considerably in composition, and therefore by extension, they contribute differently to the overall wine composition.

Furthermore, the distribution of these tissues inside the berry is considered to be closely related to the genotype. For example, phenotypic analysis of clusters from Cabernet Sauvignon (CBS) and Malvasia del Lazio (MVL) showed significant differences not only between berry weights, but also between those of seeds [7].

In this work, the average mass of the berries of CBS was significantly lower (about 44%) compared to that of MLV, despite the two varieties had the same number of seeds *per* berry (table 2).

Table 2. Distribution of tissues in the analyzed grape berries (mean ± SE)

Tissue	CBS	MLV
berry[1]	1.88 ± 0.04[a]	3.38 ± 0.25[b]
seeds[2]	2.14 ± 0.04	2.13 ± 0.19
seeds[3]	5.57 ± 0.51[b]	2.35 ± 0.10[a]
pulp[3]	58.66 ± 10.75	73.82 ± 4.69
skins[3]	19.31 ± 4.27	14.50 ± 0.98

[1]mean fresh weight of a berry. [2]number of seeds per berry. [3]mean tissue fresh weight per 100 g of grape. Values with different letters in single rows are significantly different ($p < 0.05$).

It is well known that the berry size reached at maturity stage depends, other things being equal (i.e. climate, wine-making, etc.), upon the number of seeds present in it and that this is a genetic trait of the plant. Barbagallo et al. [8] reported, for grapes belonging to the same cv, an increase in the number and weight of seeds with the increasing weight of berries. On the contrary in this case, comparing different genotypes, Carbone et al. [7] observed a reverse trend: the white grapes, of larger sizes, showed the same number of seeds of the red grape genotype, but significantly lighter.

In this regard it is worth remembering that during the vinification of white grape cvs, skins and seeds represent the winery by-products. To our opinion these results seem to be very interesting as these grape tissues contain many health-promoting components such as the so-called “phytochemicals”, a term that refers to every naturally occurring chemical substance in plants with biological activity [3, 9]. Grape phytochemicals (i.e. phenylpropanoids, isoprenoids, alkaloids) arise from different metabolic pathways. The main compounds arise from acetyl-CoA and shikimic acids, via the intermediates mevalonate and aromatic amino acids, respectively. Among them, phenylpropanoid compounds are the most popular ones. Phenolics are ubiquitous secondary metabolites in plants. They include a large group of biologically active components, from simple phenol molecules to polymeric structures with molecular mass above 30kDa [10].

The content of phenolic compounds in plants varies with the species, the variety, the considered organ, the physiological stage and the pedoclimatic conditions [7, 11, 12]. It is also known the variation of the content of phenolic compounds in plants in response to fungal attacks [13].

They are present both in insoluble form mainly in cell walls and in soluble one within the plant cell vacuoles. Polyphenols play a vital role in determining the wine's color and flavor. They are involved in browning reactions in grapes and wine and also play a key role in the aging and maturation of wines, but may also act as antioxidants, with mechanisms involving both free-radical scavenging, metal chelation, singlet oxygen quenching and their effects on cell signalling pathway and on gene expression. Phenols in foods have gained much attention thanks to these properties and their possible beneficial implications for human health, a consequence of their proven biological activity in the prevention of degenerative diseases (i.e. cardiovascular disease and cancer) [14]. Phenols represent the third most abundant constituent in grapes and wines after carbohydrates and organic acids. They are broadly distributed inside grapes and are mainly located in the seeds and the skins. On the contrary, the juice contains a very small amount (3 to 5% of total phenols). The total extractable phenolics in grape are present at only 10% or less in pulp, 60-70% (5 to 8% by weight) in the seeds and 28-35% in the skins. In grape, phenylpropanoid biosynthetic route leads to simple phenols or phenolic acids, flavonoids, stilbenes and proanthocyanidins (figure 2) [14].

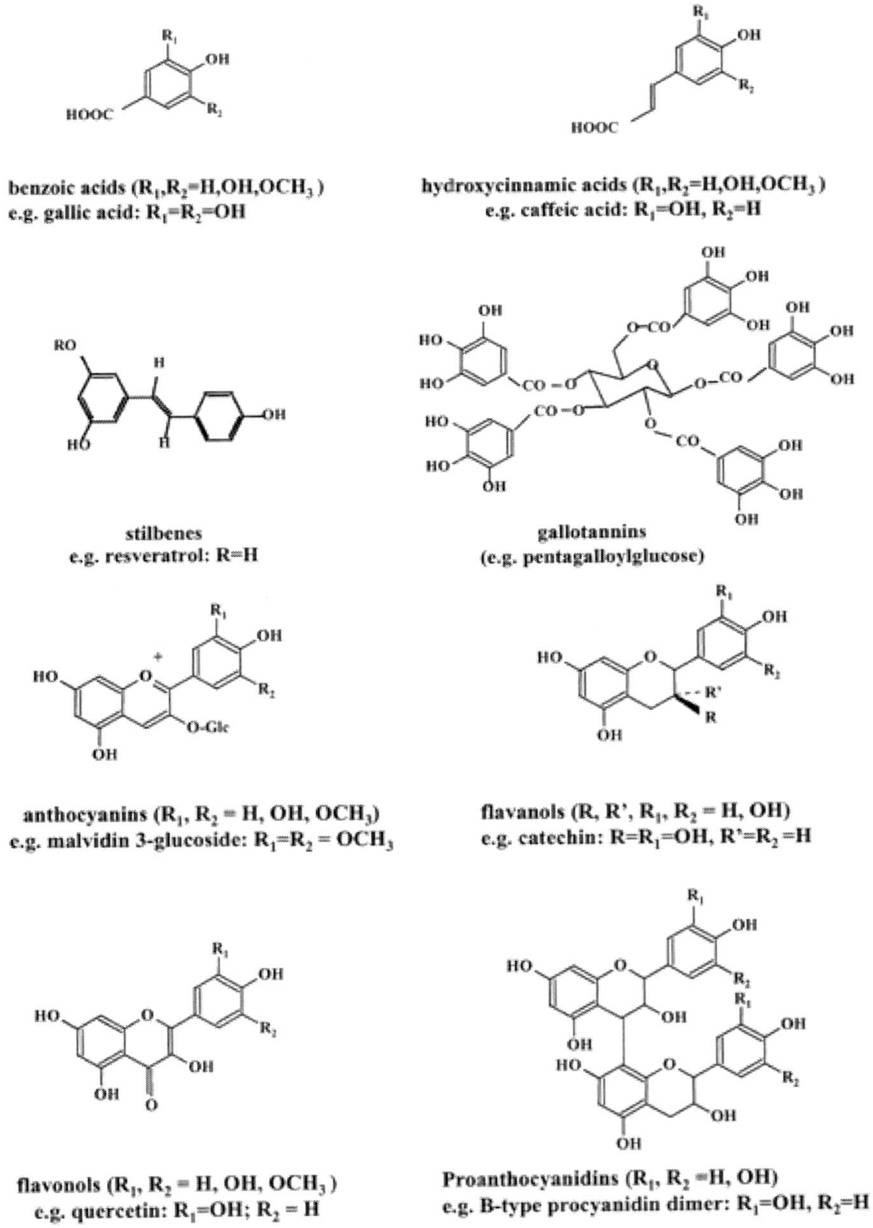

Figure 2. Main classes of phenolic compounds in grapes.

Simple phenols include hydroxybenzoic and hydroxycinnamic acids, while the term flavonoid refers to a broad class of compounds structurally different (i.e. flavonols, flavanols, anthocyanins). Among them, flavanols (also named flavans) provide catechin and epicatechin, the monomeric units of proanthocyanidins (also named condensed tannins), which are recognizing as being responsible for the majority of the potential health benefits attributed to grapes. Proanthocyanidins are characterized by a polymerization degree (PD) ranging mainly between 3 and 11, up to 17 and more. Grape skins are rich source of anthocyanins (red grapes), hydroxcinnamic acids, flavanols and flavanol glycosides, whereas flavanols are mainly present in seeds. Moreover, seeds contain lipids, proteins, carbohydrates and 5-8% polyphenols depending on the variety. Phenolic compounds in grape seeds are mainly flavonoids, including gallic acid, the monomeric flavan-3-ols: (+)-catechin, (-)-epicatechin, gallocatechin, and epicathechin 3-O gallate, procyanidin dimers, trimers, and more highly polymerized proanthocyanidins and they are known for their high antiradical power.

The type and content of grape polyphenols depend mainly on the cultivar and tissue considered [7] and therefore by extension, they contribute differently to the overall wine phenolic composition. White wines are usually produced from juice with little skin and seed contact. Their phenolic content is on average low, in the range of 100 to 500 mg/L gallic acid equivalent (GAE). Red wines, on the other hand, are customarily produced with skin and seed contact. Depending on the length of contact time, the phenolic content of a red wine generally varies between 800 to 3500 mg/L GAE.

3. Phytochemical Composition of Grape Seeds

As introduced in the former section, the phenolic compounds in grapes and grape products can be divided into two groups: (i) phenolic acids and related compounds and (ii) flavonoids. The phenolic compounds in grape seeds are mainly flavonoids (i.e. flavan-3-ols monomers, dimmers and trimers). Among them, grape seeds are a rich source of monomeric phenolic compounds, such as (+) catechin, (-) epicatechin and (-) epicatechin-3-O-gallate.

These flavans easily condensed into oligomeric procyanidins and polymeric compounds (condensed tannins). Procyanidins are often divided into two series: (i) B-series for dimeric procyanidins and (ii) C-series for the trimeric ones (figure 3). Five different dimers (procyanidin B1, B2, B3, B4 and B5) and two trimers (C1 and C2) were identified from grape skin and seeds [15]. These compounds are composed of cathechins and epicatechins (figure 3). All of the acylated procyanidins found in grape seeds are esters of gallic acid. The type and the concentration of these compounds depend on many factors, such as the variety of grapevine, environmental and viticultural factors. In this regard, Montalegre et al. [16] analyzed the phenolic composition of ten grape *Vitis vinifera* varieties grown in a warm climate.

In this study, polyphenolic compounds in the seeds of white grape cvs ranged from 730 to 1390 ppm (on fresh weight) while that of the red genotypes ranged from 330 to 870 ppm (on fresh weight).

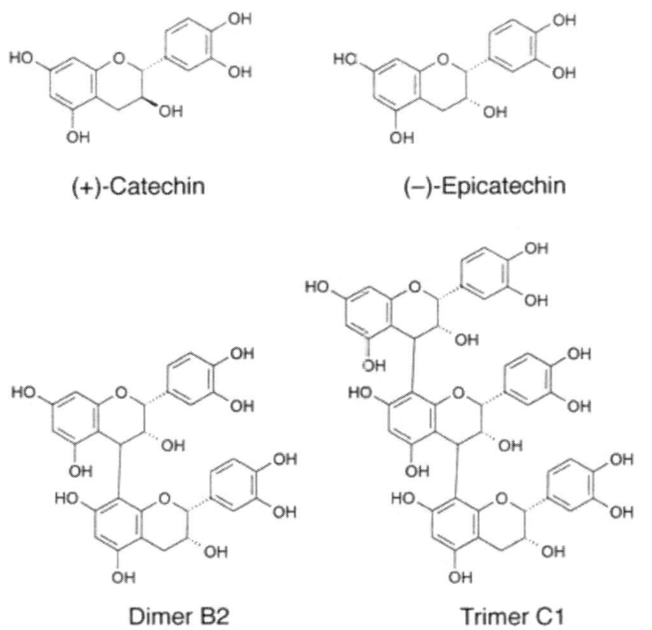

Figure 3. The structures of procyanidin monomer and oligomers.

These findings support the highest flavanol concentration of white grape seeds compared to that of red ones, already highlighted in other studies. Teresa et al. [17] reported the chemical composition of *V. vinifera* grape seeds, identifying 17 chemical compounds of which 11% represented by (+) catechin, 6% by dimer B2, 10% by (-) epicatechin, 7% by B1-3-O-gallate and 9% by (-) epicatechin-3-O-gallate. Highly polymerized procyanidins are also present in grape seeds, such as tetrameric, trimeric and dimeric ones [18].

4. BIOLOGICAL SIGNIFICANCE OF GRAPE SEED PHENOLICS FOR HUMAN HEALTH

From ancient to modern times, herbs and other plants have been used as medicinal agents, especially in China, India and Egypt. At present, a large number of plants used in the traditional medicine has become a part of our health care system. Natural products offer large structural diversity and modern separation techniques, combined with structure elucidation and combinatorial synthesis, have led to the revitalization of plant products as sources of new drugs. A total of 91 plant-derived compounds in clinical trials as of September 2007 are described in the review by Saklani and Kutty [19]. Besides, Mishra and Tiwari [20] reported a total of 19 natural product (NP) based drugs, which were approved for marketing worldwide in between the year 2005 to April 2010, among which 7 being classified as NPs, 10 semi-synthetic NPs and 2 NP-derived drugs. Among these natural substances employed for the prevention and treatment of multiple disease conditions, grape seed extracts are promising in delaying development of several cancers and neurodegenerative diseases. Grape seeds are by-products of grapes formed during the industrial production of grape juice and wines. They represent an example of discarded products with high beneficial effects on human health due to their free radical scavenging capability. Free radicals have been implicated in over a hundred disease conditions in humans, including arthritis, hemorrhagic shock, atherosclerosis, advancing age, ischemia and reperfusion injury of many organs, Alzheimer and Parkinson's disease, gastrointestinal dysfunctions, tumor promotion and carcinogenesis, and AIDS. Antioxidants found in grape seeds are potent scavengers of free radicals and serve as inhibitors of neoplastic processes. Grape seed scavenging power is superior to other well-known antioxidants, such as vitamin C (20 times more), E (50 times more) and beta-carotene. Particularly, grape seed proanthocyanidin extract (GSPE) possesses a broad spectrum of therapeutic properties. In GSPE, the main polyphenolic monomers and oligomeric proanthocyanidins are catechin, epicatechin, epicatechin gallate and procyanidin B2. Among these, procyanidin B2 has been shown to be the most effective compound in trapping oxygen free radicals [21]. The ability of the GSPE to act as an antioxidant depends primarily on the chemical structure of its constituents. The position of hydroxyl groups and other features are important for their free radical scavenging activities. Besides, GSPE is also known to be a pro-oxidant at concentrations that are significantly higher than the commonly used dose [22]. The ROS-scavenging by flavonoids may affect ROS-dependent signaling and the biological effects will represent the sum total of these different properties. During the last decade an increasing number of studies have been conducted on the anticarcinogenic effects of grape seed extracts (GSEs). Roy et al. [23] indicated that the induction of apoptosis maybe one of the mechanisms for the anticancerogenic effect of grape seed proanthocyanidins. Recently, Wang et al. [24] reported that polyphenols extracted from grape seeds are able to inhibit amyloid-beta (Abeta) aggregation, reduce Abeta production and protect against Abeta neurotoxicity *in vitro*, significantly attenuating the development of Alzheimer's disease (AD)-type brain pathology. This new finding, corroborated by other studies [25], holds a significant promise as a preventive method or treatment in Alzheimer's disease patients. The healthy effects of grape seed proanthocyanidins are mainly attributed to their antioxidant properties, although they also can modulate gene expression at the transcriptional level. Pajuelo et al. [26] pointed out that GSPE affects mainly the skeletal muscle and brown adipose tissue (BAT) mitochondria, increasing their oxidative capacity rapidly after acute supplementation. These results indicate that chronic consumption of a moderate dose of GSPE can correct an energy imbalance in a situation of diet-induced obesity, thereby improving the mitochondrial function and thermogenic capacity of the BAT [27]. Moreover, it has been demonstrated that the intake of GSE may be a feasible therapeutic strategy for prevention of a high-fructose diet-induced insulin resistance and oxidative stress. This finding points to an extremely simple and low-cost potential approach for averting both Type 2 diabetes and obesity [28]. Among the other diseases, ischemic heart one represents one of the leading causes of death in industrialized countries. It has been demonstrated that a significant reduction in coronary artery disease incidence has been linked to the consumption of flavonoids in the plants, suggesting that they may enhance tolerance to cardiac ischemia-reperfusion injury. In this context, grape seed proanthocyanidins can potentially offer cardiac protection against ischemic heart disease [29]. All these experimental evidences highlight the great nutraceutical potential of grape seed extracts, focusing the attention of the scientists first and of the industries later on these important by-products of winery.

5. FACTORS THAT INFLUENCE THE NUTRACEUTICAL CONTENT OF GRAPE SEEDS

Many papers point out that there is a significant correlation between the antioxidant properties and the phytochemical profile of grape seeds [30 - 33]. However, these properties not only depend on the total polyphenol content, but also on the type and concentration of individual compounds. In fact, the content of polyphenols, the composition of phenolic complex and antioxidant activity or radical scavenging capacity of grape and wine are clearly affected by many extrinsic and intrinsic factors, such as cultivar, climatic conditions, the effect of geographic origin of the grapes, soil, chemistry, fertilization, and the degree of fruit ripeness. In particular, these studies show that the observed difference in phenolic content and radical scavenging capacity or antioxidant activity is clearly influenced by the genetic characteristic of the variety, as well as by biotic and abiotic stress response of the plant. Differences were also found between clones of the same cultivar [30, 31].

In these studies, all analyzed varieties/clones cultivated under the same conditions of microclimate, soil and agronomic practices showed significant differences in antioxidant/antiradical activity. These outcomes confirm the genotype effect on the synthesis of active biomolecules with antioxidant activity.

In the next paragraphs, we will discuss about the influence of different parameters (i.e. genotype, climate, terroir, ripeness) on the nutraceutical power of grape seeds. Reported outcomes are derived from studies performed in our laboratories.

5.1. Materials and Methods

Herein, we report the methods we used in our experimental trials on the grape seeds from different cultivars.

Unless otherwise specified, the grape seeds were obtained from freshly harvested berries and then manually separated from the pulp. Hence, they were extracted at 30°C, for 6 days. In order to simulate the maceration phase during the alcoholic fermentation, the extraction was performed in a model wine (125 mL) containing: 5.0 g of tartaric acid, 22 mL of NaOH 1N, 2.0 g of $Na_2SO_2O_5$, 120 mL of 96% ethanol and distilled water up to 1000 mL. Obtained grape seed extract (GSE) was then analyzed as follow.

5.2. Total Polyphenols

The amount of total polyphenols (TP) in the GSE was determined according to the Folin-Ciocalteau method [34]. Catechin was employed as a calibration standard and results were expressed as catechin equivalent (CE) (mg CE/g of fresh seeds). The absorbance was measured using a spectrophotometer at the wavelength of 750 nm. All the measurements were made in triplicate.

5.3. Total Flavan 3-ols

The total flavan 3-ols (FL) content in GSE was determined by the vanillin assay [35], using catechin as the standard and results were expressed as mg CE/g of fresh seeds. The absorbance was measured using a spectrophotometer at the wavelength of 500 nm. All the measurements were made in triplicate.

5.4. Antiradical Activity

The antiradical activity (AA) was measured by the free stable radical 2,2-diphenyl-1-picryl-hydrazyl (DPPH•) (Sigma-Aldrich, Fine Chemicals, Missouri USA) spectrophotometric assay [36], slightly modified. Briefly: 0.1 mL of sample, at various concentrations, was added to 3.9 mL of DPPH• solution (4.73×10^{-2} g/L) and to 3.9 mL of methanol (spectrophotometric blank).

The kinetics of absorbance of DPPH• for different sample concentrations were monitored at constant temperature (23.0 ± 0.5°C, wavelength 515 nm), every 10 minutes until the plateau was reached (steady state). The initial concentration of DPPH• was controlled by using a calibration curve made by measuring the absorbance at 515 nm of standard samples of DPPH•at different concentrations. For each GSE concentration tested, the concentration of remaining DPPH• at the steady state was determined, and values transferred in a graph showing the concentration of residual DPPH•

AA was defined as mg of fresh (or dry) tissue necessary to reduce the initial DPPH• concentration by 50%, which is a value generally defined as "Efficient Concentration" (EC_{50}): the higher the antiradical activity, the lower this value is. AA was also expressed as ARP ($1/EC_{50}$). All the measurements were made in triplicate.

5.5. HPLC Analysis

Monomeric catechins and procianydins B1, B2, B3, B4 were analysed by HPLC analysis [31]. The identification of (+)-catechin, (-)-catechin, procyanidins B1 and B2 was made by comparison of retention times with those of pure standards, as well as by UV-visible spectra. In the absence of pure standards, the identification of procyanidins B3 and B4 was made by comparing the retention times and UV-visible spectra with those reported in the literature.

6. Genotype Effect

Table 3 summarizes the results of seed extracts from Nero d'Avola (a red grape cultivar of Southern Italy), Verdicchio and Malvasia (white grape cultivars of Central Italy) [7, 37].

Table 3. Phytochemical content and antiradical activity of Nero d'Avola, Verdicchio and Malvasia seeds

Cultivars	TP[1]	FL[1]	EC_{50}[2]
Nero d'Avola	31,79 ± 1.50[b]	8.67 ± 0.09[c]	1.67 ± 0.02[c]
Verdicchio	32.13 ± 1.89[b]	10.59 ± 0.10[b]	1.90 ± 0.01[d]
Malvasia	63.93 ± 2.73[a]	22.08 ± 1.54[a]	0.40 ± 0.02[b]

[1]TP and FL content expressed as mg of catechin equivalent (CE)/g of fresh tissue. [2]AA expressed as mg of fresh tissue that reduce of 50% the DPPH• (EC_{50}). The data are reported as means of three replications ± standard deviation. Values in column followed by different letters differ significantly ($p \leq 0.05$).

These cultivars were grown at the experimental vineyard of the Research Unit for Enology of Central Italy (CRA-ENC), located in Velletri (Rome) in 2009 year. It is well known that grape skin phenolic content is strictly related to the skin's color, being higher in red grape (mainly due to the anthocyanins), while there is no a direct relationship between seed phenolic content and grape color. In our study, the analysis of variance showed significant differences ($p \leq 0.05$) between TP, FL and AA in the seeds of different cultivars (table 3). TP and FL content were higher in white grape seeds, even if we found no significant differences between TP of Verdicchio and Nero d'Avola. Nevetherless, GSE antiradical activity of Nero d'Avola was significantly higher than that of Verdicchio and significantly lower than that of Malvasia.

This fact is probably due to the different degree of polymerization of flavanols of red grape as dimeric, trimeric, oligomeric and polymeric procyanidins take account of most of the superior antiradical capacity of these grape seeds.

We also carried out another study on the GSE of 15 native white (n = 10) and red (n = 5) genotypes grown in Central Italy (data not shown), which recently raised interest for their known oenological and organoleptic properties.

Results ponted out significant differences between TP (values ranged from 16 to 60 mg CE/g of seeds), FL (values ranged from 13.84 to 60.00 mg CE/g of seeds) and AA (expressed as EC_{50}, ranged from 0.22 to 1.38 mg of seeds). The study was conducted in two different years (2006 and 2008) and although we obtained different values in part also due to different climatic conditions, the results clearly highlighted that the GSEs from white grapes have TP and FL content and AA higher than those of red grapes [38].

Moreover, during these years, our studies have focused not only on defining differences in phytochemical profiles between different grape genotypes, but also on the differences between clones of the same cultivar. A clone is a genetically identical group of individuals derived originally from a single individual by asexual propagation (i.e. plant cuttings, or grafting) [39]. Some of the viticultural characteristics that allow to distinguish the clones of one another include berry and cluster size, morphology, yield, fruit chemistry, colour, phenolics, flavour, the time of budbreak and the ripening period, vine vigour, and others. All these parametres help to determine the antiradical power of the clone. The following is a summary of data from our studies carried out on grapes coming from different clones of the same cultivar, which were conducted during the period 2002 to 2006 [30, 31, 40]. In these studies we have obtained interesting results concerning the characterization and comparison of seven clones (named A5, A8, A9, A10, A19, A20, A21) of the Cesanese d'Affile cv. (*Vitis vinifera* L.), a native red grape variety of the Central Italy, appreciated for its high oenological quality and for its varietal specificity. These studies, which let us to obtain the homologation of the clones, allowed us to point out some relationships between the different morphological features, component structures and antiradical activity of grapes and wines.

Clones were grown in the experimental vineyards located in Lazio region (Italy) under the same pedoclimatic conditions and with the same training system (cordon spur).

The grape of seven clones harvested at different times and at the same ripening (between 21 and 22 g/L of reducing sugars) showed differences in morphological and chemical characteristics and antiradical activity. Table 4 lists the morphological characteristics of different clones of Cesanese d'Affile cv (year: 2006).

Table 4. Morphological characteristics of seven clones of Cesanese d'Affile[1]

Clones	Berry (g)[2]	Skin/berry (g)[3]	Seeds (g)[4]	Seeds *per* berry[5]
A5	1.25 ± 0.017^{a}	0.59 ± 0.001^{d}	0.043 ± 0.001^{c}	1.19 ± 0.014^{a}
A8	1.62 ± 0.071^{cd}	0.38 ± 0.017^{c}	0.041 ± 0.001^{b}	1.59 ± 0.071^{c}
A9	1.67 ± 0.084^{c}	0.39 ± 0.020^{c}	0.044 ± 0.001^{c}	1.52 ± 0.028^{ab}
A10	1.50 ± 0.024^{bc}	0.33 ± 0.001^{b}	0.040 ± 0.002^{a}	1.85 ± 0.071^{d}
A19	1.58 ± 0.004^{cd}	0.38 ± 0.001^{c}	0.051 ± 0.001^{e}	1.80 ± 0.001^{d}
A20	1.48 ± 0.075^{b}	0.39 ± 0.020^{c}	0.051 ± 0.001^{e}	1.42 ± 0.190^{ab}
A21	1.21 ± 0.063^{a}	0.27 ± 0.014^{a}	0.046 ± 0.002^{d}	1.33 ± 0.071^{a}

[1]Data are reported as means of three replications ± standard deviation. [2]Berry weight. [3]g of skin *per* berry. [4]Seed weight. [5]Number of seeds *per* berry. Values in column followed by different letters differ significantly ($p \leq 0.05$).

In 2006, A5 and A21 clones showed the lowest mean weight of berry and the lowest number of seeds *per* berry. In previous years, though with different values, A5 clone confirmed the same trend, while clone A21, in 2002, showed a greater berry weight and a greater number of seeds *per* berry, when compared to all others (data not shown). In particular, clone A21 showed a greater sensitivity to climatic variations. Significant differences were also found in anthocyanin and hydroxycinnamic acid content of the skins between the analyzed clones. Figure 4 shows average values of TP and AA, expressed as EC_{50} (mg seeds/mL of GSE), for the year 2006.

For A5 clone, polyphenols, as well as procyanidins B1, B2, B3 and B4 and AA were always significantly lower than those of other clones in all the years tested. Moreover, A5 and A21 clones, having the smallest berries, were found more efficient in terms of AA of the skins (data not shown) and less efficient in terms of AA of the seeds. In A5 clone, TP content and AA of the seeds were always significantly lower than all other clones in all the years tested. Obtained results confirm the biodiversity of clones belonging to the same cultivar. It follows that the contribution of the morphological, chemical and phytochemical characteristics of the clone affects the antiradical power of seeds.

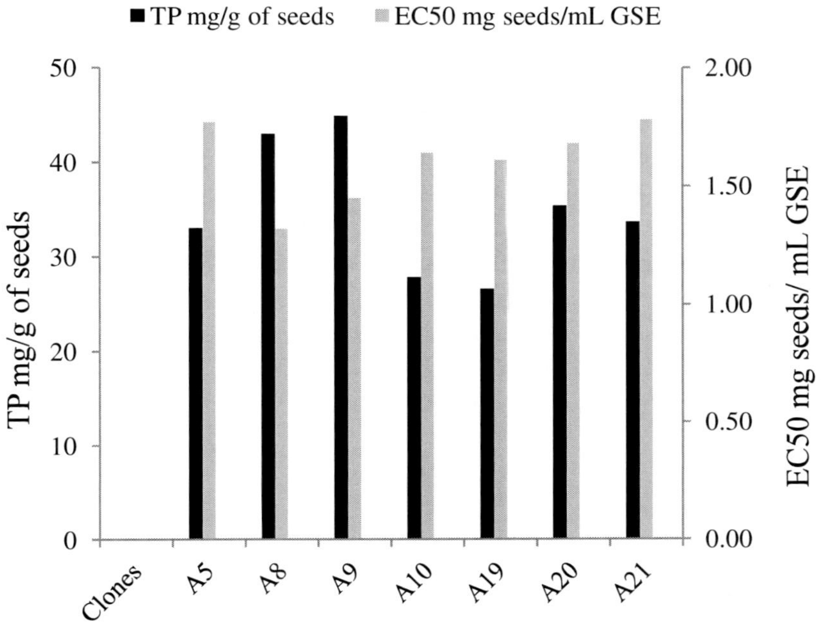

Figure 4. Cesanese d'Affile clone mean values refer to year 2006 of TP and AA.

7. Environment Effect

It is well known that gene expression is influenced by the surrounding environment [41 - 44] and the degree of grape ripening [48]. Consequently, although the genetic traits of the cultivar determine its phytochemical profile and related antiradical power, the environment plays a key role in modulating this biosynthetic activity. Herein, with the term environment we mean all of the following factors: climate, ripening, and terroir. In the next subheadings, we will discuss about the influence of these parameters on the nutraceutical power of grape seeds. Reported outcomes are derived from studies performed in our laboratories.

7.1. Climatic Conditions

Temperature, humidity and light are important physical parameters for grapes. Hot and dry climate is ideal, while areas with high rainfall are not suitable. The climate has an effect on the vine phenology mainly given by the sum of temperatures, and on the vine water status by the vineyard water balance. During berry ripening, temperature is a major factor that determines grape quality for winemaking, which necessitates optimal sugar, organic acid and color content. In added, many studies show how the climatic conditions influence both the synthesis of biomolecules in different grape tissues [46, 47], and their antiradical activity [47]. In fact, same cultivars grown in different pedoclimatic zones show a different behavior in polyphenol synthesis and as a result a different antiradical activity. Revilla et al. [48] found significant differences between catechin, epicatechin and procyanidin (B1, B2, B3, B4 and C1) content in seeds of Muscat of Hambourg cv sampled in El Encin (Spain) during the period 1992 to 1993, despite their similar degree of maturity. Because the quality of the harvest varies from year to year, vintage has been considered as an important factor for the metabolic composition of grape berries, influencing sugar, acidity, nitrogen and phenolic compound balance.

In this context, we studied the climatic effect on the antiradical activity of the GSE of 15 native cultivars in Central Italy, evaluating only the effect of climatic variation from year to year between 2006 and 2008 (vintage effect), being equal the other experimental conditions (data not shown). Seed extracts showed significant differences in TP, FL and AA in the two years. In 2006, the climatic conditions were more favorable and TP, FL and AA of GSE were significant higher than in 2008, for all the cultivars tested [38]. Moreover, data showed that GSE from white cultivars exhibited the highest AA than that of red ones.

7.2. Ripening Effect

Berry development is characterized by two successive sigmoid curves, separated by a lag phase corresponding to the third stage of development (figure 5) [49]. Stage I is characterized by the first period of berry growth, from bloom to approximately 60 days afterward. During this period, the pericarp and seed cell numbers increase and approach to the full size. Also during the first growth period, the berry expands in volume as solutes accumulate.

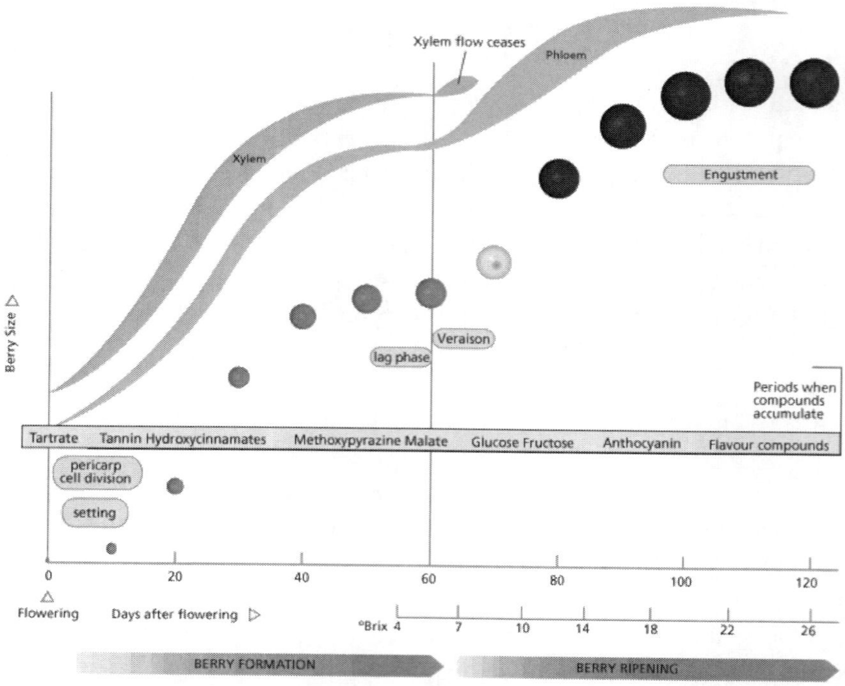

Illustration by J. Koutroumanidis; http://www.practicalwinery.com/images/JulyAugust02/julaug02p15.gif

Figure 5. Ripening curve of the berry.

There are several solutes that accumulate in the berry during this period, all of which reach an apparent maximum around véraison. Stage II is characterized by change in berry size; seed embryos develop with a concomitant hardening of their coat [50]. Stage III, termed the vérasion, is characterized by berry softening and antocyanin accumulation for red variety, while seeds are fully developed.

During grape ripening significant changes in polyphenol amount, composition and molecular weight occur. Kennedy et al. [51] found, over course of ripening and at different level of vine water status, that flavan-3ol monomers and procyanidins of Cabernet Sauvignon GSE declined during fruit ripening. In our laboratories, to assess the changes in antiradical activity during fruit ripening, two Italian cultivars (Nero d'Avola and Verdicchio) were sampled from vèraison to surmaturation (table 5). In both varieties from the third sampling (late July), berry weight grows proportionally to the accumulation of sugars and decreases proportionally in post-maturation (late October).

Table 5. Change of reducing sugars in Nero d'Avola and Verdicchio cvs during ripening

		Reducing sugars g/L[1]	
Sampling	*Date*	*Nero d'Avola*	*Verdicchio*
1	17-July	15 ± 0.08	5 ± 0.07
2	24- July	37 ± 0.12	19 ± 0.09
3	31-July	68 ± 0.07	23 ± 0.12
4	07- Aug.	96 ± 0.15	44 ± 0.15
5	12- Aug.	116 ± 1.03	67 ± 1.03
6	19- Aug.	160 ± 1.40	106 ± 0.99
7	28- Aug.	167 ± 0.98	131 ± 1.13
8	02- Sep.	180 ± 1.37	185 ± 1.20
9	09- Sep.	178 ± 1.40	186 ± 1.25
10	23- Sep.	200 ± 1.03	241 ± 1.86
11	07- Oct.	200 ± 1.15	254 ± 1.90
12	26- Oct.	201 ± 1.20	275 ± 2.10

[1]Reducing sugars were assayed by OIV method.

The incidence % of the skins on berry weight changed significantly between the two varieties from july-october.

AA of GSE of Nero d'Avola, expressed as ARP ($1/EC_{50}$), increased linearly from pre-vèrason to ripining (figure 6). AA reached the maximum value at the sixth sampling (ARP = 0.41; 160g/L reducing sugars) and remained constant until the tenth sampling (200 g/L of reducing sugars). TP content increased up to 5th sampling, with a maximum value of 40.69 mg /g of seeds, corresponding to 116.5 g/L of reducing sugars. AA in GSE of Nero d'Avola was more correlated with sugar concentration ($r = 0.76$, $p \leq 0.01$) and less correlated with TP ($r = 0.44$, $p \leq 0.05$), suggesting a different trend for AA and TP during ripening.

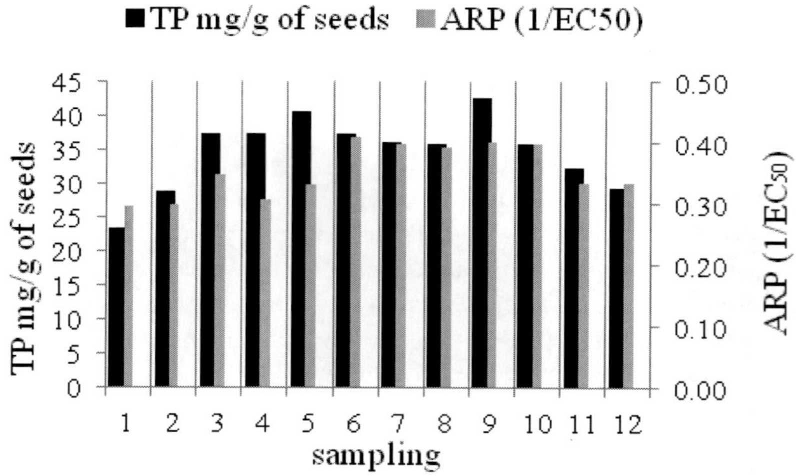

Figure 6. Changes in TP and AA during ripening of Nero d'Avola cultivar.

For Verdicchio instead, antiradical activity of GSE increased from pre-vèrasion up to 8th sampling (ARP = 0.48; 185 g/L reducing sugars; figure 7).

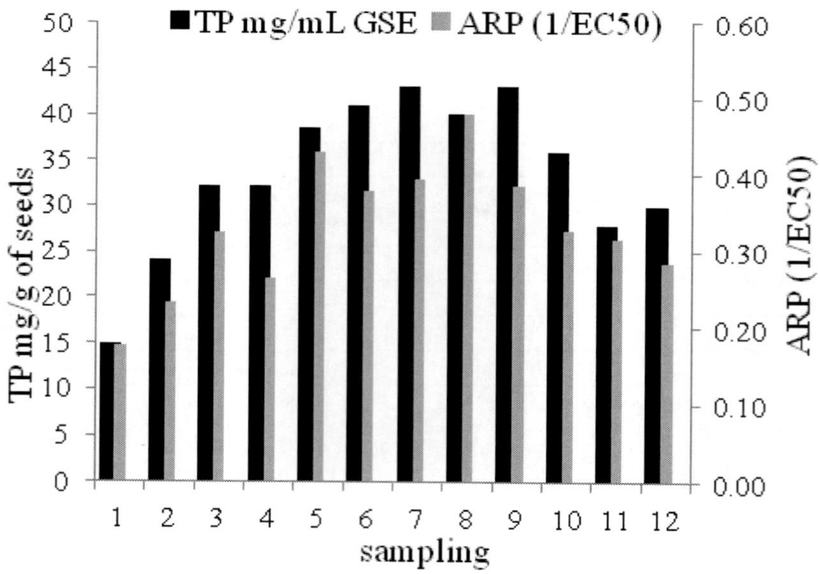

Figure 7. Changes in TP and antioxidant activity during ripening of Verdicchio cultivar.

TP content showed a similar trend during ripening, so in this case, there was a high significant correlation (r = 0.80, p≤ 0.01) between total polyphenols and AA. In both cultivars, polyphenols decreased at the end-stage of maturation (September -October). Our results are in agreement with those of Revilla et al. [55], who showed that grape cultivars (Tempranillo, Garnacha, and Cabernet Sauvignon) grown in Central Spain had the highest concentration of catechins and procyanidins during the early stages of berry development.

In our case, the total content of catechins and procyanidins in seeds decreased dramatically during September. The amount of these compounds in seeds was nearly fivefold more at the end of August compared with the levels at the beginning of October. Also in post-maturation due to the seed lignification, both TP, and the radical scavenging activity decreased.

7.3. Terroir Effect

Terroir can be defined as an interactive ecosystem, in a defined zone. It takes in account the climate, soil and vine (rootstock and cultivar) [52]. Minerals and water supply to the vine, and microclimate influence the development of vine and grape maturation.

Besides, climate influences vine phenology, vine development, grape ripening and disease pressure mainly through temperatures, rainfall, potential evapotranspiration and solar radiation.

Solar radiation influences the photosynthesis and interferes with grape phenolics. Climate and soil interact for the vine phenology, vine water and mineral status. [45].

In our preliminary study, in order to evaluate the effect of terroir on the nutraceutical properties of grapes, we studied the antiradical activity of Dolcetto red grape grown in two different Italian regions, Piemonte and Lazio, located in Northern and Central Italy, respectively [53].

Climate, soil composition (clay-gravel, loam and sandy) and altitude are very different between these regions, accounting for the environmental impact on the grape quality. Grapes were harvested at the same degree of maturation (table 6).

Table 6. Oenological parameters of Dolcetto cv

Region	SS[1] (g/L)	pH	TA[2] (g/L)
Piemonte	197 ± 0.8	3.28 ± 0.02	7.10 ± 0.52
Lazio	190 ± 1.2	3.32 ± 0.01	6.80 ± 0.41

[1]SS: soluble solids; [2]TA: titratable acidity as tartaric acid.

In this study, before polyphenol extraction, skins, pulp and seeds were manually separated, and dried for 24h in an oven at 105°C until they reached a constant weight, and phytochemical and AA results were expressed on dry basis (DM). Obtained data highlighted significant differences in grape berry morphology between the grapes from the two regions, with the exception of the number of seeds *per* berry (table 7).

On average, seeds accounted for 3.15% and 3.87%, skins for 21.80% and 22.58% and pulp for 38.34% and 42.47% of the total weight of fresh berries from Piemonte and Lazio respectively. Piemonte grape TP content and total antocyanins (only for skin; data not shown) were significantly higher ($p \leq 0.05$) than that from Lazio, in all analyzed tissues (table 8).

Data did not showed significant differences ($p \leq 0.05$) in FL content between grapes in both skins and seeds. AA (expressed as EC50 mg/DM) was significantly higher in skins and significantly lower in the pulp from Lazio region, when compared with skins and pulp of from Piemonte one. Besides, AA was not significantly different between seeds from both regions.

Table 7. Grape berry morphology in Dolcetto cv from Piemonte and Lazio regions [a]

T[1]	Berry (g)[2]	Skin/berry (g)[3]	Pulp/berry (g)[4]	Seeds/berry (g)[5]	Seeds (g)[6]	Seeds[7]
P	1.33 ± 0.04^a	0.29 ± 0.02^a	0.51 ± 0.02^a	0.042 ± 0.01^a	0.024 ± 0.01^a	1.75 ± 0.03^a
L	1.86 ± 0.05^b	0.42 ± 0.03^b	0.79 ± 0.04^b	0.072 ± 0.02^b	0.040 ± 0.02^b	1.79 ± 0.02^a

Data are reported as means of three replications ± standard deviation and referred as g of fresh weight. [1]T: terroir. [2]Berry weight. [3]g of skins *per* berry. [4]g of pulp *per* berry. [5]g of seeds *per* berry. [6]Seed weight. [7]Number of seeds *per* berry. Values in column followed by different letters differ significantly ($p \leq 0.05$).

Table 8. Terroir effect on phytochemical content and antiradical activity of seeds

Terroir	TP[1]	FL[1]	EC_{50}[2]
Piemonte	53.50 ± 3.20^b	8.52 ± 0.51^a	0.35 ± 0.01^a
Lazio	27.15 ± 2.12^a	9.25 ± 0.58^a	0.38 ± 0.01^a

[1]Data are reported as means ± standard deviation and referred as mg CE/g of DM. [2]EC_{50} is expressed as mg of dry seeds that reduce of the 50% the DPPH• concentration. Different letters in column indicate significant differences ($p \leq 0.05$).

Obtained results suggest that the different geographical origin of grape belonging to the same genotype strongly influences the morphology of the berry, its phenolic composition and AA. This in turn could have some effect on the antiradical capacity of this cultivar during winemaking process. Besides this, from a nutraceutical point of view, seeds of this cultivar do not show appreciable differences due to their geographical origin.

CONCLUSIONS

In this chapter, we have reported a summary of the results on the nutraceutical power of grape seeds obtained in our laboratories in the last few years. In defining the nutraceutical potential of these by-products, we have taken into account both their phytochemical profile, expressed in terms of both their content of polyphenols and flavanols, and their antiradical power. Reported data underline that the wine industry by-products, especially in white grape varieties that do not undergo maceration during the winemaking process, can be made into some undeniably useful health supplements. The genotype plays a key role in improving their healthy power as well as the environmental conditions (i.e. climate, terroir, ripening, etc.). Seeds from white grape cultivars may offer unique nutraceutical values for their peculiar contents in bioactive phytochemicals, suggesting their uses into a wide range of traditional and specialty products, naturally enriched with health-promoting compounds. Overall, the results presented in this chapter might be of importance in the identification of potential raw material for innovative healthy food production as well as for drugs or cosmetic products.

REFERENCES

[1] S. Myles, A. R. Boyko, C. L. Owens, P. J. Brown, F. Grassi, M. K. Aradhya, B. Prins, A. Reynolds, J. M. Chia, D. Ware, C. D. Bustamante and E. S. Buckler, Preprint, 2010 at http://www. Pnas.org/cgi/doi/10.1073/pnas.1009363108.
[2] Organisation Internationale de la Vigne et du Vin (O.I.V.). (2012). Paris, France.
[3] M. Iriti and F. Faoro, *Medical Hypotheses,* 67, 833 (2006).
[4] J. Dai and R. J. Mumper, *Molecules,* 15, 7313 (2010).
[5] C. A. Rice-Evans, N. J. Miller and G. Paganga, *Trends Plant Sci.,* 2, 152 (1997).
[6] Zhang, Y. Fang, H. Wang, H. Li and Z. Zhang, *Molecules,* 16, 2846 (2011).
[7] K. Carbone, B. Giannini, and F. Cecchini, *Riv. Vitic. Enol.*, 4, 47 (2011).
[8] M. G. Barbagallo, S. Guidoni and J. J. Hunter, *Italus Hortus.*, 17, 512 (2010).
[9] H. Zielinski, H. Kozlowska, *J. Agric. Food Chem.*, 48, 2008 (2000).
[10] E. Dreosti, *Nutrition*, 16, 692 (2000).
[11] Scalbert and G. Williamson, *J. Nutr.*, 130, 2073S (2000).
[12] L. Yu, S. Haley, J. Perret, M. Harris, *Food Chem.*, 86, 11 (2004).
[13] Carbone, B. Giannini, M.C. Serra and F. Cecchini. "Influenza dell'*Uncinula necator* sulle caratteristiche fenotipiche, sul profilo polifenolico e sul potenziale antiradicalico di uve Dolcetto", X congresso italiano di Scienza e Tecnologia degli Alimenti, May 2011 (Milan, Italy) pp12.
[14] Pearson, C. H. Tan, J. B. German, P. A. Davies and M. A. Gershwin, *Life Sci.,* 66, 1913 (1999).
[15] Shi, J. Yu, J. E. Pohorly, Y. Kakuda, *J. Med. Food*, 6, 291 (2003).
[16] R. R. Montealegre, R. R. Peces, J. L. Chacon Vozmediano, J. Martinez, J. Guascuena, E. G. Romero, *J. Food Compos. Anal.,* 19, 687 (2006).
[17] E. B. Teresa, G. F. Yolanda, C. R. Julian, S. B. Celestino, *J. Agric. Food Chem.*, 40, 1794 (1992).
[18] T. Fuleki and J. M. Ricardo da Silva, *J. Agric. Food Chem.,* 45, 1156 (1997).
[19] Saklani and S. K. Kutty, *Drug Discovery Today*, 13, 161 (2008)
[20] B. Mishra and V. K. Tiwari, Opportunity, Challenge and Scope of Natural Products in Medicinal Chemistry, Research signpost, Kerala, India (2011).
[21] J. M. R. Da Silva, N. Darmon, Y. Fernandez and S. Mitjavila, *J. Agric. Food Chem.,* 39, 1549 (1991).
[22] Z. H. Shao, T. L. Vanden Hoek, J. Xie, K. Wojcik, K. C. Chan, C. Q. Li, K. Hamann, Y. Qin, P. T. Schumacker, L. B. Becker and C. S. Yuan, *Cardiovasc. Toxicol.,* 3, 331 (2003).
[23] M. Roy, M. S. Manjeshwar, C. A. Elmets and S. K. Katiyar, *Neoplasia*, 7, 24 (2005).
[24] Y. J. Wang, P. Thomas, J. H. Zhong, F. F. Bi, S. Kosaraju, A. Pollard, M. Fenech and X. F. Zhou, *Neurotox Res.,* 15, 3 (2009).
[25] P. Liu, L. J. Kemper, J. Wang, K. R. Zahs, K. H. Ashe and G. M. Pasinetti, *J. Alzheimer's Dis.*, 26, 657 (2011).
[26] D. Pajuelo, S. Díaz, H. Quesada, A. Fernández-Iglesias, M. Mulero, A. Arola-Arnal, M. J. Salvadó, C. Bladé and L. Arola, *J. Agric. Food Chem.,* 59, 4279 (2011).
[27] D. Pajuelo, S. Díaz, H. Quesada, A. Fernández-Iglesias, M. Mulero, A. Arola-Arnal, M. J. Salvadó, C. Bladé and L. Arola, *Br. J. Nutr.,* 107, 170 (2012).

[28] W. Suwannaphet, A. Meeprom, S. Yibchok-Anun and S. Adisakwattana, *Food Chem. Toxicol.,* 48, 1853 (2010).
[29] C. Z. Wang, S. Mehendale, T. Calway and C. S. Yuan, *Am. J. Chin. Med.,* 39, 661 (2011).
[30] A.Garofolo, B. Giannini, S. Favale and M. Savino, *Riv. Vitic. Enol.,* 4, 35 (2006).
[31] A.Garofolo, B. Giannini and S. Favale, *Riv. Vitic. Enol.,* 2-3, 11 (2009).
[32] C. Xu, Y. Zhang, L. Cao and J. Lu, *Food Chem.,* 4, 1557 (2010).
[33] D. De Beer, E. Joubert, W.C.A. Gelderblom and M. Manley, *J. Agric. Food Chem.,* 51, 901 (2003).
[34] V. L. Singleton and J. A. J. Rossi, *Am. J. Enol. Vitic.,* 16, 144 (1965).
[35] R. Di Stefano and M. C. Cravero, *Riv. Vitic. Enol.,* 2, 37 (1991).
[36] W. Brand-Williams, M. E. Cuvellerand and C. Berset, *Food Sci. Tech.,* 28, 25 (1995).
[37] B. Giannini, F. Cecchini and S. Moretti (2011). "Nutraceutic property of Italian red and white grape cultivar and wine", The 2nd International Conference Series on Wine Active Compounds, pp32-322 .
[38] Moretti, M. C. Comandini, R. Valori and F. Cecchini (2011). The 17th International Symposium of GiESCO, Aug. 2011 (Asti-Alba, Italy) pp115-118.
[39] H. T. Hartmann, D. E. Kester, and F. T. Davies, Plant Propagation Principles and Practices. Englewood Cliffs, New Jersey (1990).
[40] F. Cecchini, B. Giannini, S. Moretti and M. Garofolo, *J. Biotech.,* 150, 307 (2010).
[41] F. Broussaud, V. Cheynier, C. Asselin and M. Moutounet, *Am. J. Enol. Vitic.,* 50, 277 (1999).
[42] V. Cheynier, M. Moutounet and P. Sarni-Manchado, Oenologie, fondementes scientifiques et tecnologiques. Lavoisier TecandDoc, Paris (1998).
[43] H. Ojeda, C. Andary, E. Creaba, A. Carbonneau and A. Deloire, *Am. J. Enol. Vitic.,* 53, 261 (2002).
[44] O. Downey, N. K. Dokoozlian and M. P. Krstic, *Am. J. Enol. Vitic.,* 57, 257 (2006).
[45] C. Van Leeuwen, P. Friant, X. C. O. Tregoat, S. Koundouras and D. Dubourdieu, *Am. J. Enol. Vitic.,* 55, 207 (2004).
[46] R. Rodrìguez Montealegre, R. Romero Paces, J. L. Chacòn Vozmediano, J. Martìnez Gauscuena and E. Garcìa Romero, *J. Food Comp. Anal.,* 19, 687-693 (2006).
[47] J. Lachman, M. Šulk, K. Faitovà and V. Pivec, *Int. J. Wine Res.,* 1, 101 (2009).
[48] E. Revilla, J. M. Escalona, E. Alonso and V. Kovac, Food Flavors: Generation, Analysis and Process Influence, Elsevier Science Publishers, Amsterdam (1995).
[49] B. G. Coombe, *Acta Horticulturae.,* 34, 261 (1973).
[50] B. G. Coombe, *Plant Physiol.,* 35, 241 (1960).
[51] J. A. Kennedy, M. A. Matthews and A. L. Waterhouse, *Phytochemistry.* 55, 77 (2000).
[52] G. Seguin. Symposium on Marittime Climate Winegrowing (1998). Department of Horticultural Sciences, Cornell University, Genova, NY pp36-53.
[53] F. Cecchini, B. Giannini M. Morassut and M. Moretti (2011). "Effect of the terroir on the nutraceutical proprieties in Vitis vinifera L. cv dolcetto", The 17th International Symposium of GiESCO, Aug. 2011 (Asti-Alba, Italy) pp123-126.

In: Cultivars
Editor: Katya Carbone

ISBN: 978-1-62417-101-7

Chapter 5

COMMON BEANS: ANTIOXIDANT CAPACITY AND HEALTH CONNECTION

R. Campos-Vega[1,2]*, B. Dave Oomah [3] and G. Loarca-Piña[1]

[1]Programa en Alimentos del Centro de la República (PROPAC), Research and Graduate Studies in Food Science, School of Chemistry, Querétaro State University, México

[2]Kellogg de México, Carretera al Campo Militar, San Antonio de la Punta, Querétaro, México.

[3]National Bioproducts and Bioprocesses Program, Pacific Agri-Food Research Centre, Agriculture and Agri-Food Canada, Summerland, British Columbia, Canada

ABSTRACT

Common beans are major sources of bioactive compounds in addition to providing essential nutrients. Recent research has highlighted the importance of bioactive compounds in human health and nutrition. Some of these beneficial physiological effects can be due to the antioxidant activities of bean polyphenols. Polyphenols are the most abundant antioxidants in our diet. They exhibit a wide range of pharmacological and medicinal properties, including anti-inflammatory, anticarcinogenic, vasodilator actions mostly attributed to their free radical scavenging, metal chelating, and antioxidant activities. In addition, indigestible common bean fraction can be an important source of bioactive compounds like condensed tannins known to exhibit significant antioxidant capacity. The transportation of dietary antioxidants through the gastrointestinal tract may be an essential function of indigestible fraction, an effect that has received little attention so far. An evaluation of the available literature reveals that the level of bioactive compounds in common beans is influenced by some factors like post-harvest storage, processing, cooking, fermentation and extrusion among others. This chapter reviews the effects of bioactive compounds such as phenolic compounds, anthocyanins, isoflavones and their antioxidant activity on health benefits attributed to common beans. Factors influencing their antioxidant capacity like processing and metabolism are also discussed.

* Corresponding author: telephone (442) 211 1400 Ext 1720; Email address: chio_cve@yahoo.com.mx

INTRODUCTION

Common beans (*Phaseolus vulgaris* L.) are well known as a good source of nutrients and bioactive compounds that contribute to health benefits and have therapeutic properties. [1, 2, 3, 4]. The potential health benefits of beans have also been attributed to the presence of secondary metabolites such as phenolic compounds that possess antioxidant properties [5]. Phenolics are known as natural antioxidants and due to qualitative diversity they are often divided into major groups: phenolic acids, flavonoids, isoflavones, tannins, and others [6]. Nowadays, polyphenolics are being studied for their potential role in the prevention and treatment of several chronic diseases including certain forms of cancer, osteoporosis, and heart disease, also for their ability to relieve menopausal symptoms [7].

Cardiovascular disease (CVD) is caused by disorders of the heart and blood vessels, and includes coronary heart disease (heart attacks), cerebrovascular disease (stroke), elevated blood pressure (hypertension), peripheral artery disease, rheumatic heart disease, congenital heart disease and heart failure. The major causes of cardiovascular disease are consumption of tobacco, physical inactivity, and an unhealthy diet [8]. By 2030, almost 23.6 million people will die from CVDs, mainly from heart disease and stroke [8, 9]. Diets rich in fruits and vegetables like beans can have beneficial health effects due to flavonoid contribution. Flavonoids are potentially beneficial in a wide range of diseases, from cardiovascular disease to cancer and neurodegenerative conditions. Many studies have shown the antioxidant power of particular flavonoids and flavonoid-rich extracts. However, nowadays, it is widely accepted that if flavonoids have any preventive or curative activity through their ingestion, this effect must involve, not only their antioxidant potential, but also the modulation of multiple cellular pathways that are crucial in the pathogenesis of those diseases [10]. The Zutphen study [11] of elderly men in the Netherlands found that the flavonoid intake was inversely associated with heart disease mortality and incidence of heart attack over a five-year period. Those who had the highest consumption of flavonoids had 60% less mortality from heart disease than the low flavonoid consumers.

ANTIOXIDANT CAPACITY

Antioxidants refer to compounds possessing free radical scavenging, transition metal-chelating activities, and/or singlet oxygen-quenching capacity [12, 13]. For example, phenolic phytochemicals inhibit the formation of free radicals, thereby minimizing the deterioration of significant biomolecules in the human body [14, 15]. Many studies have shown that the cell wall contains a high concentration of phenolics. Aparicio-Fernandez et al., [16] reported that phenolic contents in Black Jamapa bean were higher in the seed coat than in the cotyledon. Similar findings were reported for black, red and white beans [17, 18, 19, 20]. Furthermore, pure flavonoid compounds present in common bean seed coats have significant antioxidant activity (AA) [21] and seed coat color may be related to phenolic concentration [22]. Beans with dark colored seed coats, such as black bean, contained a higher concentration of phenolics than those with light-colored seed coats, such as white bean [23, 24]. Generally, hydrophobic substances contribute more to the seed coat phenolics content than hydrophilic substances [25]. Also, total phenolic content (TPC), alone or in combination with other phenolic constituents, has been proposed as a potential marker for selecting/discriminating AA in beans [26].

Genotype influence on Antioxidant Capacity

The anthocyanin content, TPC and AA of 29 common beans from different origins and seed coat color, were recently determined by Akond et al. [27]. The seeds included genotypes with distinct shape, color and seed weight (white, cream, purple, red and black, with striped, rhomboid spotted and circular mottled variations). Anthocyanin content varied significantly among genotypes and market classes, ranging from 0.05 to 0.47 mg g^{-1}. The bean genotypes with a TPC ranging from 5.87 to 14.14 mg gallic acid equiv/g also exhibited significant variation (17 to 37%) in AA against 2,2-diphenyl-1-picrylhydrazyl radical (DPPH$^{\cdot}$).

A study investigated the phenolic content and antiradical activities of the beans and hull fractions of white kidney, red pinto, Swedish brown and black kidney beans [19]. Total phenolic content of bean hulls and whole seed extracts ranged from 6.7 to 270 and 4.9 to 93.6 mg catechin equivalents / g extract, respectively. Trolox equivalent antioxidant capacity (TEAC) assay, based on scavenging of 2, 2, 9-azinobis-(3-ethylbenzothiazoline-6-sulfonate) radical anion, revealed that the antioxidant capacity of red, brown, and black whole seed extracts was in the same order of magnitude with minimal variation. TEAC values of red and brown whole seed extracts were superior to that of black whole seed extract. On the basis of the total phenolic content and TEAC values, it can be deduced that colored beans possess superior antioxidative activity compared with white beans. The superoxide and hydroxyl radical scavenging activities of the bean extracts were concentration dependent, whereas the concentration effects were limited for the hydrogen peroxide scavenging capacity. All extracts prevented human low-density lipoprotein (LDL) cholesterol oxidation by 61.4% to 99.9% at 2 to 50 ppm level as catechin equivalents. The study postulates that the antiradicals found in these bean types may be effective in reducing the occurrence of cancers [19].

Bean hull or seed coat colors are effective markers of antioxidant capacity. This was demonstrated in a study evaluating the antioxidant potential of aqueous acetone (80%) extracts of four bean varities with different hull colors (white, brown, red, and black) [25]. These bean extracts effectively retained supercoiled DNA against 2, 2′-azobis (2-methylpropionamidine) dihydrochloride (AAPH)-induced radical damage at 5, 10, 50, and 100 ppm. Delphinidin, cyanidin, and procyanidins B2, C1, C2, and X were identified in red, brown, and black bean hull extracts, whereas white bean hull extract contained only a small amount of flavonoids. Vanillic, caffeic, *p*-coumaric, ferulic, and sinapic acids were the main phenolic acids identified in bean hull extracts.

Process Influence on Antioxidant Capacity

Analysis of a raw bean is important for the nutritional value of the food, but common beans are consumed cooked which modifies their characteristics as well as the bioavailability of the nutrients. Three bean genotypes were recently evaluated [28] for AA, total phenolics, tannin and phytate contents in: raw, cooked with and without soaking water. The broths, grains and soaking water were analyzed. Cooking without soaking preserved most of the nutrients in the grains. There were losses of total phenolics, tannin and phytate in all samples of the cooked beans. Flor de Mayo M38, Pinto Villa and Bayo Victoria, three common bean cultivars developed in Mexico, were tested for phenol content and free radical scavenging activity before and after autoclaving [29]. Long cooking times enhanced diffusion of phenols from seed coats to cooking water and then to cotyledons as evidenced by the remarkable activity of the cooking water similar to crude seed coat extracts despite their lower (80%) phenolic content. Soaking can significantly decrease (26-39 %) the bioavailability of total phenols (TPs) of dry common beans (CB) as observed in soaking dry pinto beans (PB) in cold water [30]. Total flavonoid content was well retained in PB cooked without soaking but was not detected in CB after *in vitro* digestion. CB soaked in hot water and cooked with the addition of $NaHCO_3$ showed the greatest inhibitory effect on DPPH radical ($p<0.05$). *In vitro* digestion increased the AA of both CB and PB. Industrial dehydration generally reduces bioactive compounds (phenolics) and AA (ORAC) [31].

The hard-to-cook (HTC) phenomenon in beans has a negative effect on the content of free and bound phenolic antioxidants and antioxidant capacity [32].Folin-Ciocalteu, Trolox equivalent antioxidant capacity, and HPLC methods were used to quantify the content of phenolic acids and antioxidant capacity. Results showed that the HTC phenomenon did not equally affect the content and antioxidant capacity of phenolic acids in different bean cultivars. Black beans were most affected, with reduction in the contents of free and acid hydrolyzable phenolic acids by 35 and 36%, respectively, and the AA by 18 and 25%, respectively. This study showed that the HTC phenomenon affected a potential nutritive characteristic of dry beans.

The effects of solid state fermentation on the physicochemical, nutritional and antioxidant properties of common bean flour have been studied [33]. Tempeh flour had more phenolics (2.2x) than the bean flour and exhibited antiradical activity (43%) and AA (38%) correlated with total phenolics content.

The addition of bean modified the starch digestibility and antioxidant characteristics of tortilla, obtaining a product with nutraceutical characteristics according to Grajales-García et al. [34]. Extractable polyphenols and proanthocyanidins were higher in the tortilla with bean than control tortilla. This pattern produced higher antioxidant capacity of tortilla with bean (17.6 μmol Trolox eq/g) than control tortilla (7.8 μmol Trolox eq/g).

Navy and red bean flours (BF) added to cornstarch at 30% were extruded to produce fortified puffed snacks [35]. At this level, crude protein increased 12-fold, while total phenols, DPPH• and ORAC *in vitro* AA also increased. Red bean fortification yielded extracts with higher levels of phenols and both DPPH• and ORAC AA compared to navy beans. In navy and red bean extracts, total phenols, DPPH•, and ORAC AA were reduced by 10, 17, and 10%, and by 70, 62, and 17% after extrusion, respectively.

Compound Influence on Antioxidant Capacity

Protein hydrolysis yields peptides whose bioactivity is most often associated with antioxidant activity. A study of protein hydrolysates (alcalase and pepsin–pancreatin hydrolysates) showed antioxidant capacity after 80 and 120 min of hydrolysis, respectively [36]. In this context, a protein isolate, the storage protein phaseolin and a lectin extract from

Phaseolus vulgaris L. var. Jamapa, were hydrolyzed with pepsin and pancreatin to investigate the possible release of peptides with antioxidant and metal chelating properties [37].

Antioxidant activity in Caco-2 human colon cancer cells exposed to a free radical generator, and iron and copper chelating activities were determined using colorimetric methods. The highest AA, 71% inhibition, was found in the hydrolyzed protein isolate. Copper and iron chelating activities were highest in the lectin and phaseolin hydrolysates, 53% and 81%, respectively. Thus, experimental data indicates that antioxidant peptides are abundant in pepsin–pancreatin hydrolysates, which may represent a valuable health-promoting property in common bean.

Tissue Influence on Antioxidant Capacity

Hulls obtained by mechanical abrasive dehulling from four bean cultivars were extracted with two solvents, aqueous acetone (70%) and water, and the extracts evaluated for AA in relation to their phenolic contents [38]. TPC and AA of bean hulls, measured using oxygen radical absorbance capacity (ORAC) values were 6–8-fold those of corresponding whole beans. Aqueous acetone (70%) extracted over twice the amount of total phenolics from hulls that exhibited significantly higher AA than water.

HEALTH CONNECTION

Cardiovascular Disease

Xu et al. [39] investigated the AA of the hydrophilic extracts from 9 selected legumes based on copper-induced human LDL oxidation model *in vitro*. The AA was assessed based on the formation of conjugated dienes (lag time of oxidation) and thiobarbituric acid reactive substances (TBARS) as the early and late stage markers of LDL oxidation. The results showed that the extracts of black beans (*Phaseolus vulgaris* L.), lentils (*Lens culinaris*), black soybeans (*Glycine max*), and red kidney beans (*Phaseolus vulgaris* L.) had significantly ($P < 0.05$) longer LDL oxidation lag times (129, 124, 108, and 111min, respectively) than the LDL control group (95min). Meanwhile, the AA of these legumes against LDL-lipid peroxidation in the above assays was found to correlate significantly ($P < 0.01$) with their phenolic substances. Therefore consuming legumes like black beans and red kidney beans may have potential in preventing the development of atherosclerosis from the perspective of inhibiting LDL oxidation.

The pigments isolated from bean seed coat exhibit strong antioxidant activity. These anthocyanin pigments play an important role in chemically protecting the bean from oxidative damage. When the beans are ingested the pigments may also provide protection to the consumer against oxidative damage of cell membrane lipids, and supplement the work of vitamins C and E [40].

Vascular inflammation is a common link between arthrosclerosis and hypertension [41]. Polyphenolics from beans can attenuate hypertension in rats, and reduce macrophage infiltration associated with inflammatory signaling [42]. Macrophage infiltration is an immunopathogenic mechanism involved in the pathogenesis of hypertension [43]. Recently, polyphenol-containing adzuki beans were shown to suppress heart rate elevation in spontaneously hypertensive rats, and also decreased macrophage number in the heart and kidney. The anti-oxidant and anti-inflammatory action of polyphenols can contribute to reductions in macrophage number by inhibition of NF-*k*B activation, thereby reducing the production of signaling chemokine [42].

Cancer and Mutagenesis

Bean could be an important source of polyphenolic compounds with potential biological use as antioxidant and anticancer agents. The antimutagenic activity of a phenolic rich extract from the seed coats of black Jamapa beans (*Phaseolus vulgaris*) was tested against aflatoxin B_1 (AFB_1) in the Kado microsuspension assay [44]. Newly harvested beans showed higher antimutagenic activity against AFB_1 mutagenicity than stored beans, suggesting that freshly harvested beans provide the maximum advantage of components with biological activity.This research group also reported a strong correlation between proanthocyanidin concentration in methanolic extract of Jamapa bean (BME) and antiradical capacity, suggesting that these compounds contribute significantly to antiradical activity [45]. The treatment of human cervix adenocarcinoma (HeLa) cells with 35 μg BME ml^{-1} for 24 h increased the expression of Bax and Caspase-3, pro-apoptotic proteins (6.13 and 1.2 times for Caspase-3 and Bax, respectively). Additionally, a correlation analysis showed that phenolic extract had high antimutagenic and AA and also inhibited cytochrome p450 enzymatic activity. The results suggest that the primary mechanism of action of phenolic compounds in beans against AFB_1 mutagenicity may be extra-cellular in the microsuspension assay [46].

Hulls from dry edible beans contain large amounts of phenolic compounds known to have high AA [46].

In *in vitro* and *in vivo* research trials, anthocyanins and other antioxidant components present in common beans have demonstrated marked ability to reduce cancer cell proliferation and inhibit tumor formation. The capacity of anthocyanin pigments to interfere with the process of carcinogenesis seems to be linked to multiple potential mechanisms of action, including inhibition of cyclooxygenase enzymes and potent antioxidant potential [47]. Hou et al. [48] revealed that anthocyanins inhibit tumorigenesis by blocking activation of a mitogen-activated protein kinase pathway. This report provided the first indication of a molecular basis for the anticarcinogenic properties of anthocyanins. In this context, Dong et al. [49] isolated and identified 24 compounds from black bean seed coat extracts, including 12 triterpenoids, 7 flavonoids, and 5 other phytochemicals. Among the isolated compounds, 9 compounds (1, 2, 6-8, 14, 15, 19, and 20) showed potent inhibitory activities against human liver cancer cells (HepG2) proliferation; and potent antiproliferative activities against Caco-2 cell growth; and 7 compounds (5, 7-9, 11, 19, and 20) had potent antiproliferative activities against MCF-7 cell growth in a dose-dependent manner. Six flavonoids (compounds 14-19) showed potent AA [49].

In animal models, whole bean consumption or bean hull extract (BHE) supplementation reduces the incidence of colon cancer induced by azoxymethane in rats [4] prevents DNA damage or liver injury in mice [2, 50]. Moreover, BHE supplementation may have beneficial effects on bone health in mice by decreasing bone resorption [51].

Previously, there were limited evidence to suggest an inverse relationship between bean consumption and colon cancer among the Mexican population. A rat feeding study [52] was undertaken to evaluate the effects of different common beans cultivars, at the average per capita consumption existing in Mexico, on colon cancer. Rats were fed a control diet or diets containing 2% colored beans ("Negro 8075", "Ayocote Morado", "Pinto Zapata", and "Ayocote Negro") and white-beige beans ("Flor de Mayo Anita", "Blanco Tlaxcala", and "Marcela") for 4 weeks (wks) before, during (8 wks) and 10 wks after treatment with the carcinogen dimethylhydrazine. The colons were removed to evaluate tumor volume, incidence and multiplicity. The incidence of adenocarcinomas in the colon was lower in rats fed the "Negro 8075" (72%) , "Ayocote Morado" (90%), and "Pinto Zapata" bean (67 %) diets than those fed the control diet or other beans (100%). Total tumor multiplicity was also significantly lower in rats fed the "Negro 8075" bean (1), "Pinto Zapata" bean (0.8) and "Ayocote Morado" bean (1.1) diets compared to the rats fed the control diet (3). Similar results were observed for the tumor volume. The reduction in colon carcinogenesis in rats fed the colored beans might be attributed to the content of phenolic compounds, among other substances.

López-Reyes et al., [53] assessed the anti-fibrotic effects of methanolic black bean extract antioxidants in a carbon tetrachloride (CCl_4) liver injury model in rats. Qualitative and quantitative histological analysis showed that administration of 70 mg/kg (body weight) of black bean extract reduced hepatic fibrosis index by 18% compared to positive controls ($P = 0.006$), resulting from a decrease in type I (- 44.3%, $P < 0.05$) and type IV (- 68.9%, $P < 0.05$) collagen gene expression compared to CCl_4-injured and quercetin treated rats. This study provides evidence that this methanol black bean extract ameliorates liver fibrosis and types I and IV collagen gene expression, in the animal model.

The potential health enhancing properties of condensed tannins from black beans as inhibitors of angiogenesis was discussed by Bawadi et al. [54]. After 24 h exposure to human epithelial colorectal adenocarcinoma cells (Caco-2), human breast adenocarcinoma cells (MCF-7 and Hs578T lines), and human prostate cancer cells (DU 145), they observed that a water-soluble extract (24 μM) of black bean condensed tannins suppressed fetal bovine serum stimulated cell migration, the secretion of matrix metalloproteinase-2 (MMP-2 or gelatinase A), matrix metalloproteinase-9 (MMP-9 or gelatinase B), and vascular endothelial growth factor ($VEGF_{165}$) receptor expression by the cancer cells in the conditioned media.

Diabetes

Aqueous extract of *Phaseolus vulgaris* pods possess antihyperglycemic activity [55]. This activity may be associated with the antioxidant property of *Phaseolus vulgaris* pods [56], demonstrated in rats with streptozotocin-induced diabetes. Oral administration of *Phaseolus vulgaris* pod extract (200 mg/kg body weight) for 45 days significantly reduced thiobarbituric acid reactive substances and hydroperoxides. The extract also causes significant increase in reduced glutathione, superoxide dismutase, catalase, glutathione peroxidase and glutathione-S-transferase in the liver and kidneys of rats with streptozotocin-induced diabetes. The extract also decreased plasma hydroperoxides, vitamin E, and ceruloplasmin considerably. The decreased serum levels of high-density lipoprotein cholesterol, antiatherogenic index (AAI), plasma insulin, vitamin C, and glutathione in the diabetic rats reverted to normal. The results showed the antioxidant and antihyperlipidemic properties of *Phaseolus vulgaris* pods in addition to its antidiabetic action [57].

Phenolic-Gut Microflora Metabolism

The gut microbiota exerts an enormous impact on the nutritional and health status of the host via modulation of the immune and metabolic functions. The microbiome provides additional enzymatic activities involved in the transformation of dietary compounds. Food bioactive compounds also exert significant effects on the intestinal environment, modulating the gut microbiota composition and probably its functional effects on mammalian tissues. This evidence is changing the way the biological roles of functional food components are being investigated since their metabolites and effects may depend on the gut microbiota and even change from one individual to another [58].

The colon is thus a rich source of potentially active phenolic acids that may impact gut health both locally and systemically. Both the small and large intestine (colon) contain absorption sites for phenolic acids, but low post-prandial concentrations in plasma indicate minimal absorption early in the gastrointestinal tract and/or rapid hepatic metabolism and excretion. Therefore, any bioactivity that contributes to gut health may predominantly occur in the colon.

Several phenolic acids affect the expression and activity of enzymes involved in the production of inflammatory mediators of pathways thought to be important in the development of gut disorders including colon cancer [59].

The bioavailability and effects of polyphenols greatly depend on their transformation by specific components of the gut microbiota via esterase, glucosidase, demethylation, dehydroxylation and decarboxylation activities [60]. Many dietary polyphenols are glycosides that are transformed into aglycones by commensal bacterial glycohydrolases, thereby modifying their bioavailability and affecting positively or negatively their activities and/or functional effects on mammalian tissues [61, 62]. The microbiota metabolites of polyphenols are better absorbed in the intestine, and their entero-hepatic circulation ensures that the residence time in plasma for the metabolites, is extended compared to that of their parent compounds, and finally are excreted via urine.

The gut microbiota is essential for the production of active isoflavone metabolites with estrogen-like activity; additionally, the metabolites produced exhibit different anti-inflammatory properties [63]. Similarly, the flavonoid quercetin generated by gut microbial enzymes exerts a higher effect in the down-regulation of the inflammatory responses than the glycosylated form present in vegetables (quercetin or 3-rhamnosylquercetin) [64].

Different studies have been carried out to understand gut microbiota transformations of particular polyphenol types and identify the responsible microorganisms. Bacteria in the gut can perform reactions that transform more complex plant phenolics such as anthocyanins, procyanidins, flavanones, flavonols, tannins and isoflavones into simple phenolic metabolites [60]. Phenolic acids (benzoic, phenylacetic and phenylpropionic acids) are the most abundant phenolic structures found in fecal water [65]. *Clostridium* and *Eubacterium* genera, which are phylogenetically associated, are common elements involved in the metabolism of many phenolics [66].

Table 1.

Related compound/extract/factor	Connection	Reference
Phenolics		
Phenol content	Free radical scavenging activity before and after autoclaving	[31]
Phenolics	Inhibit the formation of free radicals	[14]
Total phenolics, tannin and phytate	Antioxidant activity	[30]
Phenolic rich extract	Antimutagenic activity	[45]
Phenolic extract	Antimutagenic and antioxidant activity	[47]
Anthocyanin	Antioxidant activity	[27]
Phenolics	Gut metabolism	[63]
Protein hydrolysates (alcalase hydrolysates and pepsin–pancreatin)	Antioxidant capacity	[38]
Flavonoids and tannins		
Flavonoid	Antioxidant activity	[21]
Condensed tannins	Angiogenesis inhibition	[56]
Anthocyanins		
Anthocyanins and their breakdown metabolites	Atherosclerosis	[65]
Anthocyanins	Tumorigenesis inhibition	[49]
Proanthocyanidin polyphenol extract	Anti-oxidant and anti-inflammatory action	[44]
Extracts	Antioxidant activity, antihyperlipidemic and antidiabetic action	[58]
Bean hulls	Protect DNA damage, antioxidant potential	[25]
Phaseolus vulgaris pods		
Bean hulls extracts	Antioxidant activities in relation to their phenolic contents	[40]
Hydrophilic extracts	Cooper-induced human LDL oxidation; radical scavenging activity	[41]
Phenolics		
Methanolic extract	Antiradical capacity and pro-apoptotic activity	[46]
Methanolic black bean extract	Liver fibrotic effects	[55]
Hydrolysates		
Protein hydrolysates (pepsin–pancreatin hydrolysates)	Antioxidant and metal chelating properties	[39]
Process		
Soaking dry common beans and dry pinto beans	Antioxidant activity	[33]
Industrial dehydration	Antioxidant activity	[33]
Hard-to-cook phenomenon	Antioxidant capacity	[34]
Solid state fermentation	Antiradical activity and antioxidant activity	[35]
Tortilla with bean	Antioxidant capacity	[36]
Extrusion cooking, total phenols	Radical and oxygen radical absorbance capacity in *vitro* antioxidant activities	[37]

Gut microflora metabolize anthocyanins to phenolic acids and aldehydes. These metabolites may explain the relationship between anthocyanin consumption and reduced incidence of colon cancer according to a recent report [67]. All six major metabolites (3-*O*-methylgallic, gallic, syringic, protocatechuic, and vanillic acids, and 2,4,6-trihydroxybenzaldehyde), along with a Cabernet Sauvignon anthocyanin extract, were incubated with Caco-2 cells at 0−1000 μM concentrations over 72 h to determine cell proliferating effects and for 24 h to assess cytotoxicity effects and at 140 μM for 24 h to measure apoptotic induction. Gallic and 3-*O*-methylgallic acids inhibited cell proliferation and lacked cytotoxicity at low concentrations. The aldehyde metabolite and anthocyanin extract also inhibited cell proliferation at low concentrations and had low cytotoxicity at various concentrations.

The aldehyde was the best inducer of apoptosis among the four substances that effectively reduced cell proliferation. In addition, these same four treatments were degraded quickly in growth media (syringic, protocatechuic, and vanillic acids, and 2,4,6-trihydroxybenzaldehyde), suggesting the involvement of subsequent oxidation products in the reduction of cell viability. These results indicated that the anthocyanin microfloral metabolites gallic acid, 3-*O*-methylgallic acid, and 2,4,6-trihydroxybenzaldehyde reduce cell proliferation in Caco-2 cells more effectively than anthocyanins and may offer protection against colon cancer after their formation in the gut. Recently, [68] anthocyanins and their breakdown metabolites have been suggested to partly provide a protective effect against atherosclerosis that is multi-causal and involves different biochemical pathways. The effects of anthocyanins and their breakdown metabolites, protocatechuic, syringic, gallic, and vanillic acids were evaluated on different parameters involved in atherosclerosis, including inflammation, cell adhesion, chemotaxis, endothelial function, estrogenic/anti-estrogenic activity, and angiotensin-converting enzyme (ACE) inhibitory activity.

From the tested metabolites, only protocatechuic acid exhibited a slight inhibitory effect on nitric oxide (NO) production and TNF-α (tumor necrosis factor-alpha) secretion in LPS-INF-γ-induced (lipopolysaccharide- interferon-gamma) macrophages. Gallic acid decreased secretion of MCP-1 (monocyte chemotactic protein-1), ICAM-1 (Intercellular Adhesion Molecule 1), and VCAM-1 (vascular cell adhesion molecule 1) in endothelial cells.

All anthocyanins exhibited ACE-inhibitory activity. Delphinidin-3-glucoside, pelargonidin-3-glucoside, and gallic acid showed affinity for ERβ (estrogen receptor beta) and pelargonidin and peonidin-3-glucosides for Erα (estrogen receptor alpha).

Table 1 summarizes selected evidence on common beans antioxidant capacity and their health connection.

Conclusions

Although, there are numerous reports on the fate of the major antioxidants in common beans, there are still numerous key issues that need to be resolved and this will require further research. Namely, how exactly are these compounds absorbed and metabolized in humans? It has been postulated for decades that the consumption of common beans has beneficial health effects. However, only recently the biological properties and health effects of its constituents have been investigated. Recent studies have confirmed these assumptions. Common beans and their major phenolic components are strong antioxidants and good radical scavengers. The extracts of black beans and red kidney beans may have potential in preventing the development of atherosclerosis from the perspective of inhibiting LDL oxidation. It has also been shown that when beans are ingested the pigments may also provide protection to the consumer against oxidative damage of cell membrane lipids, and supplement the work of vitamins C and E. Recent studies have shown that bacteria in the gut can perform reactions that transform more complex plant phenolics such as anthocyanins, procyanidins, flavanols, flavonols, tannins and isoflavones into simple phenolic metabolites. It is unclear at this time if the beneficial properties of common beans are directly from its constituents or their metabolites. Further study in this area is required to determine the putatively beneficial properties of their metabolites.

REFERENCES

[1] P. B. Geil and J. W. Anderson, *J. Am. Coll. Nutr.*, 13, 549 (1994).
[2] L. Azevedo, J. C. Gomes, P. C. Stringheta, A. M. M. C. Gontijo, C. R. Padovani, L. R. Ribeiro and D. M. F. Salvadori, *Food Chem. Toxicol.*, 41, 1671 (2003).
[3] L. A. Bazzano, J. He, L. G. Ogden, C. Loria, S. Vupputuri, L. Myers and P. K. Whelton, *Arch. Intern. Med.*, 61, 2573 (2001).
[4] L. Hangen and M. R. Bennink, *Nutr. Cancer*, 44, 60 (2002).
[5] D. L. Luthria and M. A. Pastor-Corrales, *J. Food Compost Anal.*, 19, 205 (2006).
[6] R. Randhir and K. Shetty, *Innovat. Food Sci. Emerg. Tech.*, 8, 197 (2007).
[7] V. Habauzit and M. Christine, *Ther. Adv. Chronic Dis.* 3, 87 (2012).
[8] R. J. Williams, J. P. Spencer and C. Rice-Evans, *Free Radical Bio. Med.*, 36, 838 (2004).
[9] WHO. World Health Organization.. Cardiovascular diseases (CVDs). Fact sheet N° 317. September 2009. Available at: http://www.who.int/mediacentre/factsheets/fs317/en/index.html (Accessed on 12 April 2010).
[10] S. Pascual-Teresa, D. A. Moreno and C. García-Viguera, *Int. J Mol. Sci.*, 11, 1679 (2010).
[11] M. G. L. Hertog, E. J. M. Feskens, P. C. H. Hollman, M. B. Katan and D. Kromhout, *Lancet*, 342, 1007 (2003).
[12] C. A. Rice-Evans, N. J. Miller, P. G. Bolwell, P. M. Bramley and J. B. Pridham, *Free Radical Res.*, 22, 375 (1995).
[13] M. Luiz, A. Biasutti and N. A. Garcia, *Redox Rep.*, 7, 23 (2002).
[14] D. Bagchi, M. Bagchi, S. J. Stohs, D. Das, K., S. D. Ray, C. A Kuszynski, S. J. Shantaram, H. G. Pruess, *Toxicol*, 148, 187–197 (2000).
[15] C. A. Rice-Evans, N. J. Miller, G. Paganga, *Free Radical Bio. Med.*, 20, 933–956 (1996).
[16] X. Aparicio-Fernandez, G. G. Yousef, G. Loarca-Piña, E. De Mejia and M. A. Lila, *J. Agr. Food Chem.*, 53, 4615 (2005).
[17] N. M. A. Hassimoto, M. I. Genovese and F. M. Lajolo, *J. Agr. Food Chem.*, 53, 2928 (2009).
[18] T. Madhujith, R. Amarowicz and F. Shahidi, *J. Am Oil Chem Soc.* 81, 691 (2004).
[19] T. Madhujith and F. Shahidi, *J. Food Sci.*, 70, 85 (2005).
[20] L. G. Ranilla, M. I. Genovese and F. M. Lajolo, *J. Agr. Food Chem.*, 55, 90 (2007).

[21] C. W. Beninger and G. Hosfield, *J. Agr. Food Chem.*, 51, 7879 (2003).
[22] R. E. Simard and Z. Barampama, *Food Chem.*, 47, 159 (1993).
[23] A. Anton, K. A. Ross, T. Beta, R. G. Fulcher and S. D. Arntfield, *LWT-Food Sci. Tech.*, 41, 771 (2008).
[24] B. D. Oomah, C. Blanchard and P. Balasubramanian, *J. Agr. Food Chem.*, 56, 11312 (2008).
[25] T. Madhujith, M. Naczk and F. Shahidi, *J. Food Lipids*, 11, 220 (2004).
[26] B. D. Oomah, A. Cardador-Martínez and G. Loarca-Piña, *J. Sci. Food Agr.*, 85, 935 (2005).
[27] S. M. G. M. Akond, L. Khandaker, J. Berthold, L. Gates, K. Peters, H. Delong and H. Khwaja, *Am. J. Food Tech.*, 6, 385 (2011).
[28] S. T. Valdés, C. M. Medeiros Coelho, D. J. Michelluti and V. L. Cardoso Garcia Tramonte, *LWT. Food Sci. Technol.*, 44, 2104 (2011).
[29] N. E. Rocha-Guzmán, R. F. González-Laredo, F. J. Ibarra-Pérez, C. A. Nava-Berúmen and J. A. Gallegos-Infante, *Food Chem.*, 100, 31 (2007).
[30] H. G. Akillioglu and S. Karakaya, *Food Sci. Biotech.*, 19, 633 (2010).
[31] Y. Aguilera, Y. Estrella, V. Benitez, R. M. Esteban and M. A. Martín-Cabrejas, *Food Res. Int.*, 44, 774 (2011).
[32] C. M. Machado, M. G. Ferruzzi and S. S. Nielsen, *J. Agr. Food Chem.*, 56, 3102 (2008).
[33] M. Reyes-Bastidas, E. Z. Reyes-Fernández, J. López-Cervantes, J. Milán-Carrillo, G. F. Loarca-Piña and C. Reyes-Moreno, *Food Sci. Technol. Int.*, 16, 427 (2010).
[34] E. M. Grajales-García, P. Osorio-Díaz, I. Goñi, D. Hervert-Hernández, S. H. Guzmán-Maldonado and L. A. Bello-Pérez, *Int. J. Mol. Sci.*, 13, 286 (2012).
[35] A. Anton, R. G. Fulcher and S. D. Arntfield, *Food Chem.*, 113, 989 (2009).
[36] M E. Oseguera-Toledo, E. Gonzalez de Mejia, V. P. Dia and S. L. Amaya-Llano, *Food Chem.*, 127, 1175 (2011).
[37] J. Carrasco-Castilla, A. J. Hernández-Álvarez, C. Jiménez-Martínez, C. Jacinto-Hernández, M. Alaiz, J. Girón-Calle, J. Vioque and G. Dávila-Ortiz, *Food Chem.*, 131, 1157 (2012).
[38] B. D. Oomah, A. Corbé and P. Balasubramanian, *J. Agr. Food Chem.*, *58*, 8225 (2010).
[39] B. J. Xu, S. H. Yuan and S. K. C. Chang, *J. Food Sci.*, 72, 522 (2007).
[40] T. Tsuda, T. Osawa, K. Ohshima and S. Kawakishi, *J. Agr. Food Chem.*, 42, 248 (1994).
[41] J. Li and J. Chen, *Med. Hypothesis*, 64, 925 (2005).
[42] S. Sato, Y. Mukai, J. Yamate, J. Kato, M. Kurasaki, A. Hatai and M. Sagai, *Clin. Exp. Pharmacol., P.* 35, 43 (2008).
[43] D. J. Dzielak, *Hypertension*, 19, 36 (1992).
[44] X. Aparicio-Fernández, L. Manzo-Bonilla and G. F. Loarca-Piña, *J. Food Sci.*, 70, S73 (2005b).
[45] X. Aparicio-Fernández, R. Reynoso-Camacho, E. Castaño-Tostado, T. García-Gasca, E. González de Mejía, S. H. Guzmán-Maldonado, G. Elizondo, G. G. Yousef, M. A. Lila and G. Loarca-Pina, *Plant Foods Hum. Nutr.*, 63, 35 (2008).
[46] Cardador-Martínez, A. Albores, M. Bah, V. Calderón-Salinas, E. Castaño-Tostado, R. Guevara-González, A. Shimada-Miyasaka and G. Loarca-Piña, *Plant Foods Hum. Nutr.*, 61, 161 (2006).

[47] M. A. Lila, *J. Biomed. Biotech.,* 5, 306 (2004).
[48] D. X. Hou, K. Kai, J. J. Li, N. Terahara, M. Wakamatsu, M. Fujii, M. R. Young and N. Colburn, *Carcinogenesis*, 25, 29 (2004).
[49] M. Dong, X. He and R. H. Liu, *J. Agr. Food Chem.,* 55, 6044 (2007).
[50] K. H. Han, M. Fukushima, K. Ohba, K. Shimada, M. Sekikawa, H. Chiji, C. H. Lee and M. Nakano, *J.Nutr. Sci. Vitaminol., (Tokyo)*, 50, 380 (2004).
[51] J. Cao, B. R. Gregoire, X. Sheng and J. P. Liuzzi, *Food Res. Int.,* 43, 560 (2010).
[52] M. L. González-Dávalos, A. Palomino-Salinas, C. Ríos-Ugalde, M. Ramos-Gómez, F. G. Loarca-Piña, R. Reynoso-Camacho, E. González-Jasso, H. Guzmán-Maldonado and I. Torres, IFT Annual Metting. Session 89G, Nutraceutical and Functional Foods: Geleral III. (2005).
[53] G. López-Reyes, N. Arroyo-Curras, B. G. Cano, V. J. Lara-Díaz, G. E. Guajardo-Salinas, J. F. Islas, V. Morales-Oyarvide, L. A. Morales-Garza, F. J. Galvez-Gastelum, G. Grijalva and J. E. Moreno-Cuevas, *Ann. Hepatol.*, 7, 130 (2008).
[54] H. A. Bawadi, R. R. Bansode, A. II Trappey, R. E. Truax and J. N. Losso, *Cancer Lett.*, 218, 153 (2005).
[55] R. Roman-Ramos, J. L. Flores Sanoz and F. J. Alarcon Aquilar, Antihyperglycemic effect of some edible plants. *J. Ethnopharmacol.*, 48, 25 (1995).
[56] S. Venkateswaran and L. Pari, *Asia Pac. J. Clin. Nutr.,* 11, 206 (2002).
[57] S. Venkateswaran, L. Pari and G. Saravanan, *J Medic. Food.,* 5, 97 (2002).
[58] J. M. Laparra and Y. Sanz. *Pharmacol. Res.,* 61, 219 (2010).
[59] W. Russell and G. Duthie, *P. Nutr. Soc.,* 70, 389 (2011).
[60] M. Aura, *Phytochem. Rev.*, 7, 407 (2008).
[61] J. M. Gee and I. T. Johnson, *Curr. Med. Chem.*, 8, 1245 (2001).
[62] Scalbert, C. Morand, C. Manach and C. Remesy, *Biomed Pharmacother.,* 56, 276 (2002).
[63] J. S. Park, M. S. Woo, D. H. Kim, J. W. Hyun, W. K. Kim, J. C. Lee and H. S. Kim, *J. Pharmacol. Exp. Ther.,* 320, 1237 (2007).
[64] M. Comalada, D. Camuesco, S. Sierra, I. Ballester, J. Xaus, J. Gálvez and A. Zarzuelo, *Eur. J. Immunol.,* 35, 584 (2005).
[65] C. Cueva, M. V. Moreno-Arribas, P. J. Martín-Álvarez, G. Bills, M. F. Vicente, A. Basilio, C. López-Rivas, T. Requena, J. M. Rodríguez and B. Bartolomé, *Res. Microbiol.,* 161, 372 (2010).
[66] M. V. Selma, J. C. Espín and F. A. Tomás-Barberán, *J. Agr. Food Chem.,* 57, 6485 (2009).
[67] S. C. Forester and A. L. Waterhouse, *J. Agr. Food Chem.,* 58, 5320 (2010).
[68] M. Hidalgo, S. Martin-Santamaría, I. Recio, C. Sanchez-Moreno, B. Pascual-Teresa, G. Rimbach and S. Pascual-Teresa, *Genes Nutr.,* DOI: 10.1007/s12263-011-0263-5. (2012).

In: Cultivars
Editor: Katya Carbone
ISBN: 978-1-62417-101-7

Chapter 6

CHEMICAL PROPERTIES AND HEALTH BENEFITS OF ACTIVE CONSTITUENTS IN CULTIVAR SEED OILS

A. O. Cherif*
Laboratoire de Biochimie, des Lipides et des Protéines,
Département de Biologie, Faculté des Sciences de Tunis, Tunis, Tunisia

ABSTRACT

Phytochemicals in plant material have raised interest among scientist, producers and consumers for their roles in the maintenance of human health and in assessing protection against chronic degenerative disorders. Brazil nuts, peanut, olive oil, sesame oil and pumpkin oil etc. have been associated with many health benefits including mainly cholesterol-lowering effects, antioxidant activity and anti- proliferative effect. In this review bioactive constituents (phytosterols, flavonoid derivatives, tocopherols, squalene, stilbenes derivative (resveratrol) and bioactive fatty acids) and their associated health benefits will be discussed.

INTRODUCTION

Phytochemicals comprise a number of substances such as phenolics (flavonols, catechins), flavonoids, sterols (β-sitosterol, stigmasterol, campesterol, Δ5-avénasterol) and squalene. They are generally commercialized as nutraceuticals and as ingredients for functional foods. Phytochemicals are found in vegetables (e.g. tomatoes, olives, carrots) and fruits (e.g., lemon, berries, grapes), nuts (e.g., peanut, sesame, brazil nut), seeds (grape, pumpkin) and herbs (e.g., cocoa leaves, oregano) [1]. Many phytochemicals are potent effectors of biologic processes and have the capacity to influence disease risk via several complementary and overlapping mechanisms [2].

* Corresponding author: Laboratoire de Biochimie, des Lipides et des Protéines, Département de Biologie, Faculté des Sciences de Tunis, Tunis, 2092 El Manar. Tunisia.

In this review, five classes of phytochemicals compounds found in oil-seeds such as: bioactive fatty acids, phytosterols, squalene, flavonoids, stilbenes and their content in oil-seeds will be discussed. Focus will also be placed on the beneficial aspects of these nutraceuticals on human health.

Oilseeds

The composition of oilseeds has been studied extensively [3], but recently it has been thoroughly investigated with particular respect to the phytochemicals representing minor components (such as tocopherols, squalene, phytosterols and phenolic compounds). This interest is due to the activity of such compounds against cardiovascular diseases, lipid oxidation, protein cross-linking, DNA mutations and hemostasis function, which prevent the attack on biomolecules by free radicals. Most of these beneficial effects are due to: unsaturated fatty acids, squalene, flavonoids, stilbenes.

ACTIVE CONSTITUENTS

Bioactive Fatty Acids

Fatty acids are aliphatic monocarboxylic acids that act as the building blocks of lipids. They are either saturated, monounsaturated or polyunsaturated depending on the number of double bonds. Long-chain fatty acids (LCFA) are fatty acids with aliphatic tails of 16 or more carbons [4]. This group of fatty acids contains two or more double bonds. There are two principal families of PUFAs- the omega-3 and the omega-6 families [4].

Linoleic acid (LA) and α- linolenic acid (ALA) belong to the *n*-6 (omega-6) and *n*-3 (omega-3) series of polyunsaturated fatty acids (PUFA), respectively. They are defined as "essential" fatty acids since they are not synthetized in the human body and are mostly obtained from the diet. Food sources of ALA and LA are most vegetables oils, cereals and nuts [5]. From a biochemical point of view, both LA and ALA, have 18 carbon atoms in their acyl chain presenting two (LA) or three (ALA) C=C double bonds. The position of the first unsaturation counting from the methyl end of the fatty acid, the so-called omega-C, generated the name of the two different classes. The parent compound of the *n*-6 family, linoleic acid (LA) (C18:2, cis, cis Δ^9, Δ^{12}, *n*-6), is plentiful in nature, and is found in the seeds of most plants as well as cooking oils (with a high content) such as peanut oil, corn oil and sunflower oil. Alpha-linolenic acid (ALA) (C18:3, cis, cis, cis Δ^9, Δ^{12}, Δ^{15}, *n*-3), the parent compound to the omega-3 family, is far less common and is found primarily in soya bean, rapeseed and flaxseed oil. Both α- linolenic acid and linoleic acid can be elongated and desaturated to long-chain PUFAs: linoleic acid to arachidonic acid (AA), and α- linolenic acid to eicosapentaenoic acid (EPA) and docosahexaenoic acid (DHA) (Fig.1) [4].

An increased intake of ALA has been associated with a decrease in blood cholesterol levels, a reduction in the risk of heart attacks, and a growth reduction in breast, colon and prostate cancer [4]. LA has become increasingly popular in cosmetics because of its beneficial properties for the skin.

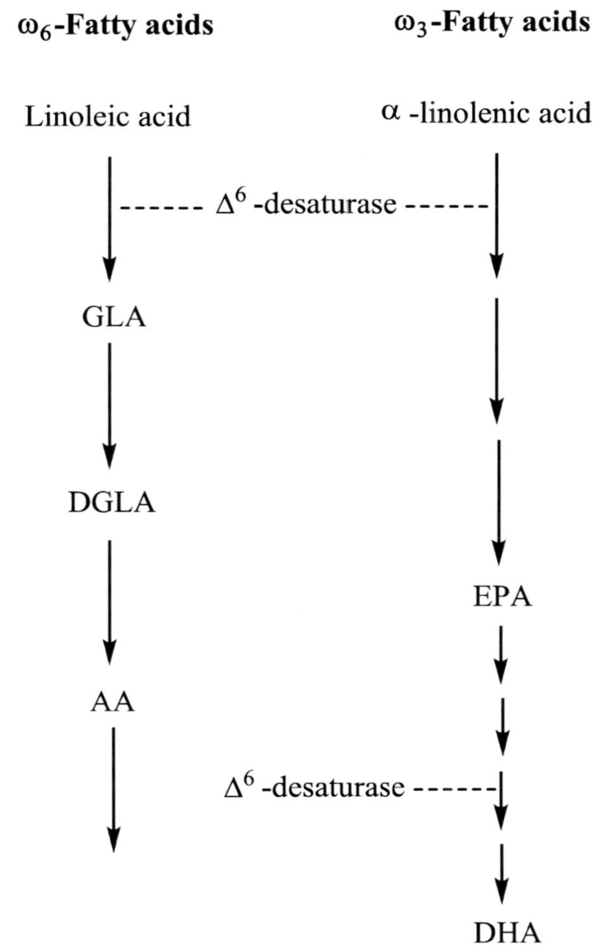

Figure 1. Linoleic acid and α-linolenic acid pathway.

γ- linolenic acid (GLA) and stearidonic acid (STA) have shown biologically beneficial effects. γ- linolenic acid as been hreported to feature anti-inflammatory effects as well as anti-cancer actions [4]. The long-chain n-3 PUFAs, such as EPA and DHA have the ability to prevent and treat hypertension, arthritis, inflammatory and autoimmune disorders as well as cancer [4]. In addition to all the beneficial properties described above, it has been reported that low levels of DHA result in low brain serotonin levels [4], that have been associated with Alzheimer's disease, and depression, among other diseases, and there is mounting evidence that DHA supplements may be effective in combating such diseases [4].

Arachidonic acid (AA) is also one of the essential fatty acids required by most mammals and is accumulated rapidly in the developing brain during the last trimester of gestation [4]. A deficit in AA results in reduced growth in infants [4]. AA acid supplements have also shown beneficial effects in human amnesic patients, improving the dysfunction due to organic brain damage or aging [4].

The conjugated linoleic acid (CLA) and pinolenic acid are also long-chain fatty acids featuring beneficial effects. Conjugated linoleic acid consists of a series of positional and geometric isomers of conjugated derivatives of linoleic acid. They are mainly *trans* fatty acids. Among the latter, only C18:2, *cis, trans* Δ^9, $\Delta^{11}$and C18:2, *trans, cis* Δ^{10}, Δ^{12} are considered biologically significant and active because of their potential biologically beneficial effects in attenuating many chronic diseases. Several CLAs are reported to be anti-carcinogenic, anti-atherogenic, anti-adipogenic, anti-diabetogenic and/ or anti-inflammatory [4]. On the other hand, studies on the pinolenic acid (C18:3, cis, cis, cis Δ^5 Δ^9, Δ^{12}) have shown beneficial functions in the human body: decreasing lipids, increasing HDL and prostacyclin, decreasing blood pressure, inhibiting aggregation of platelets as well as anti-inflammatory properties [4]. It has also stated that pinolenic acid may have LDL-lowering properties due to enhancing the hepatic LDL uptake [4].

PHYTOSTEROLS (DESM, DIMS, ET MMS)

Phytosterols (PS) make up the greatest proportion of the non-saponifiable components of vegetable oils [6]. They are 28or 29-carbon alcohols and resemble cholesterol in vertebrates in terms of stabilization of phospholipids bilayers in plant cell membranes. Their structure has a four-ring steroid nucleus, the 3β-hydroxyl group and often a 5, 6-double bond [7] (Fig. 2).

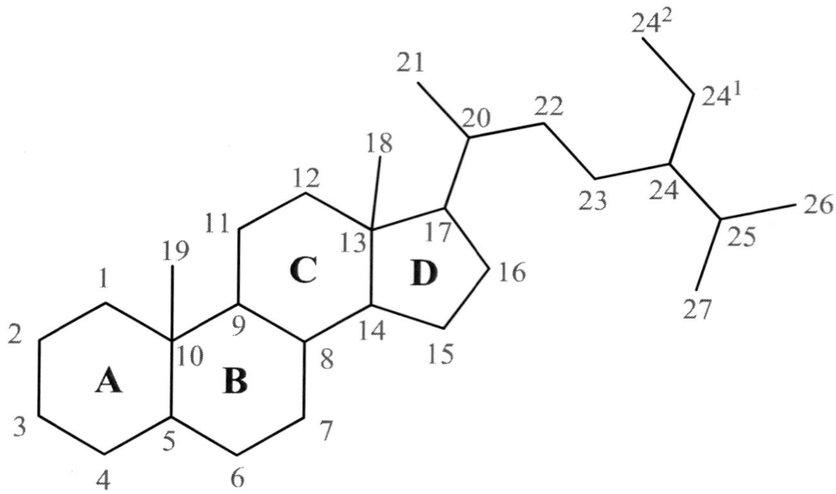

Figure 2. Steroid structure.

They have been categorized into three subclasses: on the basis of the number of methyl groups at the C_4 position: (4-DEMS) none (4-desmethylsterols or "sterols"), (4-MMS) one (4α-monomethyl-), and (4-DIMS) two (4,4'-dimethylsterols or triterpenes alcohols) [8-9].

The main compounds of the 4-desmethylsterol class are β-sitosterol , campesterol, stigmasterol and Δ^5- avenasterol (Fig.3) [10].

4,4'-dimethylsterols (triterpenes alcohols) and 4α-monomethylsterols are metabolic intermediates in the biosynthetic pathway leading to the end-product, 4-desmethylsterol, but they are usually present at lower levels than 4-desmethylsterols in most plant tissues (8-11). The major components of 4,4'-dimethylsterols are β-amyrin, α-amyrin, cycloartenol and 24-

methylencycloartanol (Fig. 3). In the case of 4α-monomethylsterols, mainly Obtusifoliol, gramisterol and citrostdienol are detailed (Fig. 3).

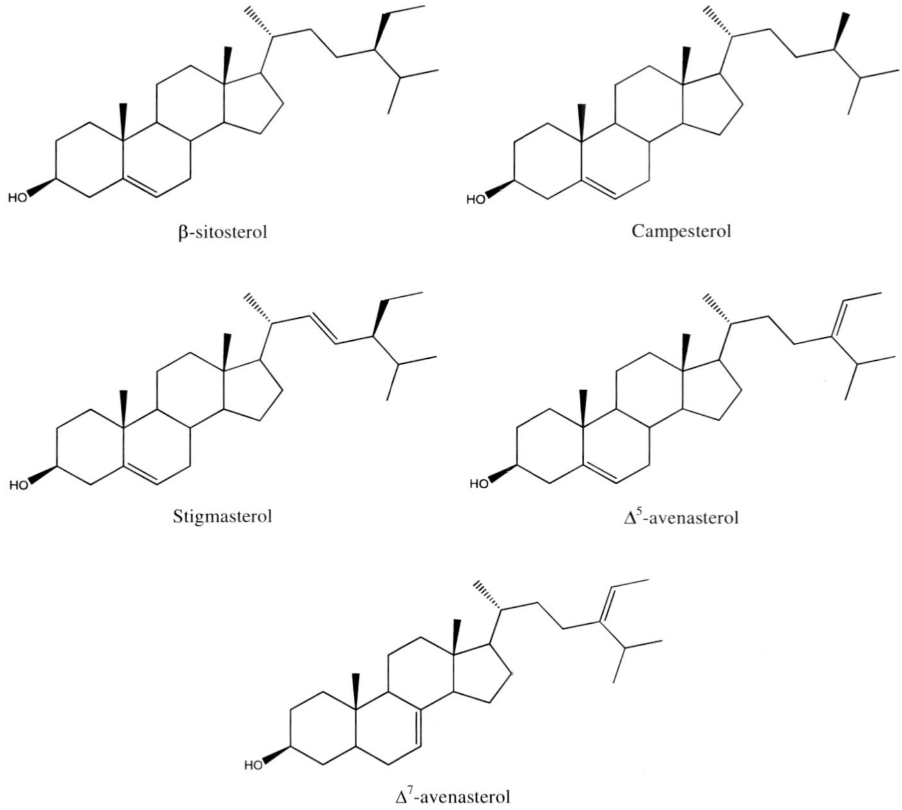

Figure 3. Detailed chemical structures of □-sitosterol, Campesterol, Stigmasterol, Δ5-avenasteroland and Δ7-avenasterol.

4-desmethylsterols make up the greatest proportion of non-saponifiable and they have important effects: they contribute to lowering serum cholesterol levels [12], and are also considered to have anti-inflammatory, anti-bacterial, anti-artherosclerotic, anti-oxidative, anti-ulcerative, and anti-tumoral properties in humans [8-13], as well as contributing to the oxidative and thermal stability and shelf-life of vegetables oils [14]. In addition, they are useful emulsifiers for cosmetic manufacturers and supply the majority of steroidal intermediates and precursors for the production of therapeutic steroids [15].

Triterpenes compounds are also important bioactive secondary metabolites, due to the wide range of their biological activities. They mainly display antimicrobial, cytotoxic, antitumoral, antiviral, anti-inflammatory- hepatoprotective, antifeedant and insecticidal activities [16].

SQUALENE

Squalene (2,6,10,15,19,23-hexamethyl-2,6,10,14,18,22-tetracosahexaene) is a highly unsaturated aliphatic hydrocarbon ($C_{30}H_{50}$) and the major constituent of hydrocarbons fraction (~80-90%) accruing at 800-12000 mg/kg. squalene is a key intermediate molecule between linear triterpenes and cyclic triterpenes (Fig.4). It contains six isoprene units and is the biochemical precursor of cholesterol and other steroids [17]. The synthesis of squalene begins with the conversion of acetyl coenzyme A to 3-hydroxyl-3-methylglutaryl coenzyme A (HMG CoA), followed by the reduction of HMG CoA to mevalonate and mediated by 3-hydroxy-3-methylglutaryl coenzyme A reductase (HMG CoA reductase). Mevalonate is then phosphorylated and eventually decarboxylated to form delta 3-isopentenyl diphosphate, which is the donor molecule for polyprenyl compounds. The successive addition of prenyl functional groups results in the formation of franesyl diphosphate. Two farnesyl diphosphate molecules undergo a reductive coupling to form squalene [17].

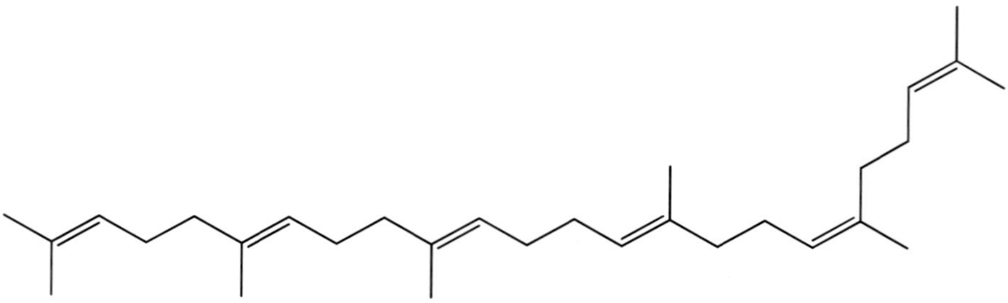

Figure 4. Squalene.

Squalene is widespread in nature, especially among olives, shark liver oil, wheat germ, and rice bran. Thus, in addition to being synthesized cells, it is consumed as an integral part of the human diet. In humans, squalene is synthesized in the liver and the skin, transported in the blood by very low density lipoproteins (VLDL) and low density lipoproteins (LDL), and secreted in large quantities by the sebaceous glands [18].

It has been extensively studied particularly in olive and pumpkin oils because of its important biological properties [8]. It is normally used in its natural or hydrogenated form (squalane) as a moisturizing or emollient agent in cosmetic preparations [19]. It is a metabolic precursor in cholesterol biosynthesis and is a potential oxidation inhibitor against PUFA (polyunsaturated fatty acids) contributing to the stability of various vegetable oils [20]. It is also regarded as one of the compounds responsible for the beneficial effects compared to the certain types of cancers [21] and is believed to be partially responsible for the low incidence of human cancer in the Mediterranean region where people's diets have a high olive content [17].

Squalene is frequently used in the preparation of stable emulsions for various applications, especially the delivery of vaccines, drugs, and other medicinal substances.

The amount of this hydrocarbon varies from 3.69 to 160.32 mg/100g of oil in Kenaf and Kalahari melon seed oils. The highest amount was found in pumpkin seed oil (590.69 mg/100g of oil) [18].

POLYPHENOLS

Polyphenols are the most studied compounds, they are further divided into flavonoids (including flavonols, flavones, catechins, flavanones, anthocyanidins, and isoflavones), phenolic acids, stilbenes, coumarins, and tannins [22]. They are also used as a source of potentially beneficial phytochemicals in the food and pharmaceutical industries [23].

Phenolic compounds are a heterogeneous class of secondary plant metabolites that can be divided into two main groups: flavonoid compounds, with a C6-C3-C6 structure (antocyanins, flavan-3-ols, flavonols, flavones, and flavanones) and nonflaonoid compounds, with C6-C1 and C6-C3 structures (hydroxybenzoic and hydroxycinnamic acids and stilbenes) [24].

Flavonoids and stilbenes in particular are involved in a defense mechanism against physical injuries and microbial contamination.

FLAVONOID DERIVATIVES

Flavonoids are the largest class of phenylpropanoids in plants. The basic flavonoid structure is two aromatic rings (one from phenylalanine and the other from three malonic acids) linked by three carbons C_6-C_3-C_6 carbon framework or more specifically a phenylbenzopyran function. Depending on the position of the linkage of the aromatic ring to the benzopyrano moiety, this group of natural products can be divided into three classes: flavonoids (2-phenylbenzopyrane derivatives, i.e. flavone), isoflavonoids (3-phenylbenzopyrane derivatives, i.e. isoflavone) and neoflavonoids (4-phenylbenzopyrane derivatives, i.e. neoflavone) (Fig. 5).

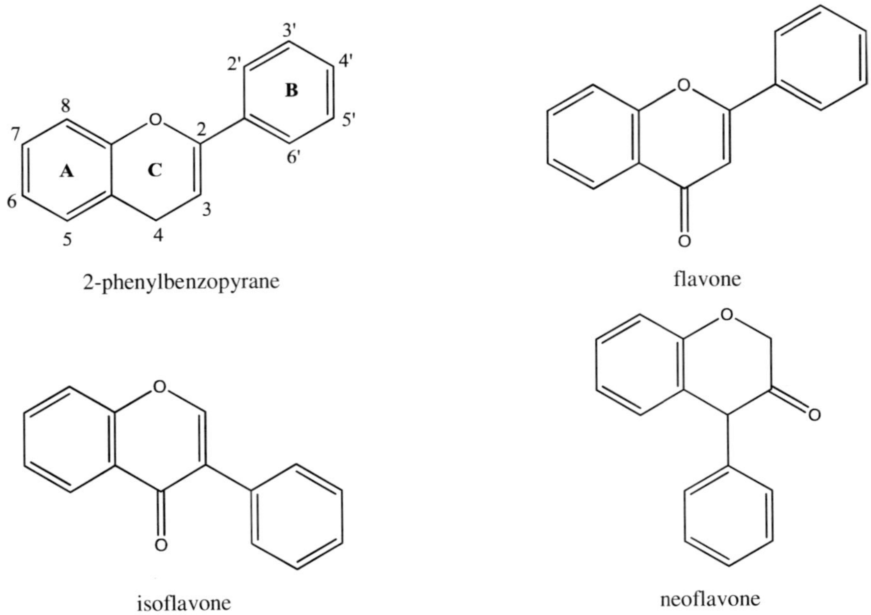

Figure 5. Main classes of flavonoids.

Flavonoids are primarily found in plants. Some flavonoids are pigments whereas others are not. There are about 6,000 known compounds in the flavonoid category. Certain flavonoids have been shown to reduce the risk of cancer and heart disease, while others lessen menopause symptoms or strengthen bones [25].

STILBENES DERIVATIVE (RESVERATROL)

Stilbenes are characterized by a 1,2-diphenylethylene backbone, usually derived from the basic unit *trans*-resveratrol (3, 5, 4'-trihydroxy-*trans*-stilbene) (Fig.6), although other structures are found in particular plant families. Ring A usually carries two hydroxyl groups in the *m*-position, whereas ring B is replaced by hydroxyl and methoxy groups in the *o*-, *m*- or *p*-position [26].

Stilbenes are synthesized by a wide range of plant species; however they are often in plants that are not routinely consumed for food or in the edible tissue. Peanuts (1.3-3.7 μg of resveratrol/g) and peanut butter are considered major dietary sources of stilbenes, along with grapes and their derivatives [26].

Stilbene synthesis has been associated with resistance to some common peanut diseases, in particular to fungal contamination. Also, the production of stilbenes is elicited by injuries, fungal contamination, insect damage, and other attacks [26]. However, stilbenes can be found in uninfected and uninjured plants, albeit in minor amounts [26].

Piceatannol and resveratrol, two of the phytoalexins produced by plants in response to fungal infection, mechanical damage, or ultra-violet irradiation [27], are very important compounds for health due to their antioxidative activities and other biological functions.

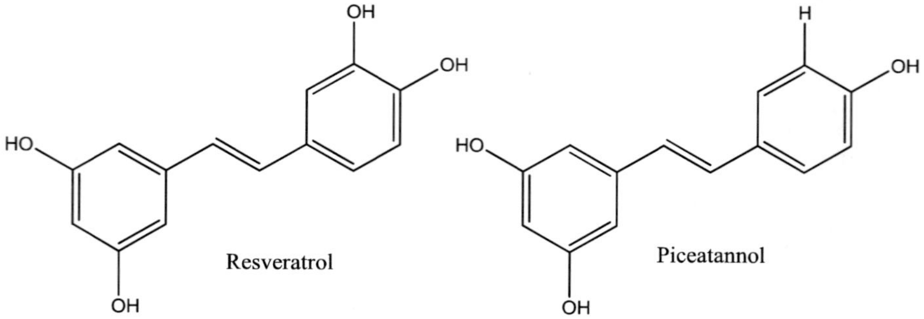

Figure 6. Structures of trans-resveratrol and trans-piceatannol.

Resveratrol (Res) in particular, is a naturally occurring phytoalexins and has recently attracted great interest, due to the "French paradox" , despite fat-rich diets, mortality from coronary heart disease is lower in France than in other countries due to the moderate consumption of red wine [28]. Epidemiological studies have suggested that Res is one of the active ingredients of red wine responsible for decreased coronary heart disease mortality [29]. Some studies have shown that Res modulates lipid metabolism, protects low-density lipoproteins against oxidative and free radical damage [28], and inhibits platelets activation and aggregation. It also has strong anticarcinogenesis effects and blocks the carcinogenesis stages of initiation, promotion, and progression [28].

Compared with Res, piceatannol has a similar chemical structure, but many potent biological activities. For instance, piceatannol, instead of Res, is a more efficient inducer of apoptosis [27] and called an anticancer compound. However, Res is a chemopreventive compound in cancer treatment.

Conclusion

Given the importance of dietary habit and food components to health, the provision of phytochemical information of a range of foods is vital to support the future work in assessing protection against chronic degenerative disorders. The "new" health benefits of these foods may prompt research into the assessment and determination of potential rich sources of phytochemical compounds in agricultural produce that could improve cultivar development, production practices, post-harvest storage and food processing. These food-based approaches would be essential for sustainable solutions to combat the alarming prevalence of chronic cancer, coronary heart diseases and diabetes.

References

[1] R.H. Liu. *J.Nutr.* 134, 3479S. (2004).
[2] (Lampe, 2007). W. J. Lampe and J. L. Chang. *Seminars in Cancer Biol.*. 17,347. (2007).
[3] Tuberoso, 2007 C. I.G. Tuberoso, A. Kowalczyk, E. Sarritzu and P. Cabras. *Food. Chem.* 103, 149. (2007).
[4] Rodriguez, 2010 A. Ruiz-Rodriguez, G. Reglero and E. Ibañez. *J. Pharm. Biomed. Ana.* 51, 305. (2010).
[5] G.L. *Russo. Biochemical pharmacology.* 77, 937. (2009).
[6] S. Azadmard-Damirchi, G.P. Savage and P.C. Dutta. *J. Am. Chem. Soc.* 82, 717. (2005).
[7] Lagarda M.J., Liatas G.C., Farré R. *J. Pharm. Biomed. Anal.* 41, 1486. (2006).
[8] R.A. Moreau, B.D. Whitaker and K.B. Hicks. *Prog. Lipid Res.* 41, 457. (2002).
[9] P. Benveniste. The Arabidopsis book. *http: // www.aspb.org/publications.* (2002).
[10] J.E. Pardo, M.A.Cuesta, A. Alvarruiz, J.D. Granell and M. Alvarez-Orti. *Food Chem.* 100, 977. (2007).
[11] C. Grunnwald. *Annu Rev Plant Physiol*, 26, 209. (1975).
[12] R. Engel and H. Schubert. *Innovative Food Sci. Emerg. Technol,* 6, 233. (2005).
[13] T.H.J. Beveridge, T.S.C. Li, and J.C.G Drover. *J. Agric. Food Chem.,* 50, 744. (2002).
[14] R. Przybylski and N.A.M. Eskin. *Inform.* 17, 187. (2006).
[15] A.O. Cherif (PhD Thesis). (2012)
[16] N. Alvarenga and A. Esteban. *Stud. Nat. Prod. Chem.* 30, 635. (2005).
[17] L.H. Reddy and P. Couvreur. *Advanced Drug Delivery Reviews.* 61, 1412. (2009).
[18] A.O. Cherif (PhD Thesis). (2012).
[19] A. Kaiya. *Yukagaku.* 39, 525. (1990).
[20] M. Malecka. *Nahrung.* 38, 135. (1994)

[21] T. J. Smith. 9, 1841. (2000)
[22] R.H. Liu. *J.Nutr*. 134, 3479S. (2004).
[23] Pradeep, 2011) S.R. Pradeep and M. Guha. *Food.Chem*. 126,1643. (1999).
[24] J. Lako, V.C. Trenerry, M. Wahlqvist, N.Wattanapenpaiboon, S. Sotheeswaran and R. Premier. *Food. Chem*. 101, 1727. (2007).
[25] J. H. Lee, I. Y. Baek, N. S. Kang, J. M. Ko, H. T. Kim et al. *Food Sci. Biotechnol.* 16, 29 (2007).
[26] R.M. Lopes, T. Agostini-Costa, M. A. Gimenes and D. Silveira. *J. Agric. Food. Chem.*59,4321. (2011).
[27] Lin, 2007 L.L.) Lin, C.Y. Lien, Y.C. Chen and K.L. Ku. *J. Chromatogr. B. Anal. Technol. Biomed. Life Sci.* 853, 175 . (2007).
[28] D.M. Golberg, S.E.Hahn and J.G. Parkes. *Clinica Chimica Acta*. 237, 155. (1995).
[29] Z. Lu, B. Cheng, Y. Hu, Y. Zhang and G. Zou. *Food Chem*. 17. (2009).

In: Cultivars
Editor: Katya Carbone

ISBN: 978-1-62417-101-7

Chapter 7

COMPARATIVE STUDY ON PHENOLIC PROFILE AND ANTIOXIDANT POTENTIAL OF SIX TUNISIAN AUTOCHTHONOUS OLIVE CULTIVARS

***Boutheina Gargouri*[1], *Ines Feki*[1], *Amir Ben Mansour*[2] *and Mohamed Bouaziz*[1,2]**

[1]Institut Supérieur de Biotechnologie de Sfax, Université de Sfax, Sfax, Tunisie

[2]Laboratoire d'Electrochimie et Environnement, Ecole National d'Ingénieur de Sfax, Université de Sfax, Sfax, Tunisie

ABSTRACT

Six olive cultivars, Chemlali, Chetoui, Zalmati, Chemchali, Dhokar and Gemri-Dhokar, growing in the south of Tunisia were investigated to identify their phenolic content, phenolic profile and evaluate antioxidant activity. Chemchali olives showed the highest mean content in total phenol (9.09 g kg^{-1}) followed by Chemlali cultivar (7.70 g kg^{-1}). In addition, the highest value of total flavonoids content was detected in Gemri-Dhokar cultivar extract (3.09 g kg^{-1}). The identification of phenolic compounds was based on separation by high performance liquid chromatography (HPLC) equipped with a diode array detector (DAD) followed by liquid chromatography-mass spectrometry analysis (LC-MS/MS). In all samples, hydroxytyrosol and oleuropein were the major compounds identified while rutin and luteolin 7-O-glucoside were the two main flavonoids. The antioxidant activity of olive extracts was evaluated by DPPH and ABTS assays. The highest antioxidant activity was detected in Chemchali and Chemlali cultivar extracts.

INTRODUCTION

In recent years, there has been growing interest in dietary consumption of virgin olive oil (VOO), attributable to its potential beneficial effects on human health. In part, these effects reflect the high level of monounsaturated fatty acids (MUFA) in VOO, which are more resistant to lipid peroxidation than polyunsaturated fatty acids (PUFA) [1]. Moreover, these effects appear to be related to the contribution of several minor constituents of VOO, including phenolic compounds as genuine antioxidants [2]. Phenolic compounds are minor components in olive oil; however, they are an important parameter for the evaluation of VOO quality as they contribute greatly to oil flavour and taste [3], as long as the oil is protected from autoxidation [4, 5].

VOO can be differentiated from all other vegetable oils due to its very particular phenolic composition [6]. Olive oil contains a large number of phenolic compounds, including derivatives of 2-phenylethanol (e.g. tyrosol and hydroxytyrosol), benzoic acid and cinnamic acid, flavonoids (e.g. luteolin and apigenin [7]), secoiridoids including aglycon derivatives of oleuropein, demethyloleuropein and ligstroside [8], and lignans [9]. The absolute concentration of phenols in olive oil is strongly affected by several factors, including the cultivar of the fruits used for its extraction, the degree of fruit ripeness, the climate in the area where olive trees are cultivated and the extraction procedures [10]. Several studies have shown that climatic factors such as temperature and precipitation have an effect on plant physiology, and consequently on the chemical characteristics of the oil extracted from the plant fruits [11, 12].

Tunisia is a very important country in the olive oil-producing world. It is the largest African exporter and the fourth largest exporter worldwide after Spain, Italy and Greece, with an average annual export of over 10,000 tonnes [13]. The olive tree (Olea europaea L.) is present in practically every region of the country, up to the border of the southern desert. Many olive varieties are present in Tunisia, but there are two that stand out. The Chemlali cultivar occupies more than two thirds of the total olive growing area and is cultivated in the center and south of the country. The Chétoui cultivar is the second main variety cultivated in Tunisia and is widespread in the north of the country in the plains as well in mountain regions, showing a high capacity for adaptation to various pedo-climatic conditions [14].
Beside these two principal varieties, which form the essential of the Tunisian olive grove, other secondary varieties like Chemcheli, Gerboui, Zalmati, and Oueslati are confined at more limited zones [15]. Moreover, several minor varieties are sustained in restricted areas in different parts of Tunisia such as Dhokar and Gemri-Dhokar, two rare olive trees growing in Tataouin zone (south of Tunisia).

The aim of this chapter is to identify and quantify phenolic compounds in six cultivars growing in Tunisia (Chemlali, Chetoui, Zalmati, Chemchali, Dhokar and Gemri-Dhokar) by using a reverse-phase HPLC coupled with UV-visible absorbance detection and mass spectrometry. Moreover, an evaluation of the antioxidant activity of olive fruit extract from six cultivars was carried out using the DPPH and ABTS assays.

Materials and Methods

Reagents

Pure HPLC solvents were used in all cases. 2, 2-Diphenyl-1-picrylhydrazyl (DPPH), 2, 6-di-tert-butyl-4-hydroxy-boxylic acid (BHT), 2, 4, 6-tri(2-pyridyl)-s-triazine (TPTZ), 2, 2’-

azinobis[3-ethylbenzothiazoline-6-sulfonate] (ABTS) and 6-hydroxy-2, 5, 7, 8-tetramethylchroman-2-carboxylic acid (Trolox), were purchased from Sigma. Folin-Ciocalteu reagent was obtained from Fluka. $FeCl_3.6H_2O$ and $CH_3COONax3H_2O$ were purchased from Merck. The double distilled water was used in the HPLC mobile phase.

Olive Samples

Sufficient amounts of olives were handpicked from all sides of different olive trees growing under rain-fed conditions in November 2008 and 2009. The samples analyzed and the mean average of the two years crops are shown in Table 1. The olive fruits (500g) were collected from olive tree plantations located in several localities in the north, south and centre of Tunisia. To obtain uniform amounts of fruits, collection was accomplished from different parts of each tree, so as to minimize the effect of watering and sun exposure. The samples were immersed in liquid nitrogen, and stored at −80 ◦C prior to analysis.

Maturity Index

The olive ripeness index was determined according to the method developed by the Agronomic Station of Jaen [16] based on the evaluation of the olive skin and pulp colors. Maturation index (MI) values range from 0 (100% intense green skin) to 7 (100% purple flesh and black skin).

Phenolic Compounds Extraction

The whole olive fruits were washed and crushed with a hammer crusher. The paste obtained (100 g) was extracted as described previously [17]. Briefly, a mixture of methanol and water (250 mL, 4:1 v/v) was added to the olive paste and the mixture was left to stand under agitation for 24 h. Subsequently, the solution was filtered using GF/F filter paper and washed with hexane (100 mL) to remove oil. The extract was concentrated by evaporation to dryness at 40 ◦C and the residue obtained was stored in glass vials at 0 ◦C in the dark until HPLC analysis.

Determination of Total Phenols

Total phenol content of different olive extracts was determined using the Folin–Ciocalteu technique [18]. Briefly, a 50 μL aliquot of the extracts was assayed with 250 μL Folin reagent and 500 μL sodium carbonate (20%, w/v). The mixture was vortexed and diluted with water to a final volume of 5 mL. After incubation for 30 min at room temperature, the absorbance was read at 765 nm. Total phenols were expressed as gallic acid equivalents (GAE), using a calibration curve of a freshly prepared gallic acid solution. For the gallic acid, the curve absorbance versus concentration is described by the equation $y = 0.0012x - 0.0345$ ($R^2 = 0.9997$).

Total Flavonoids

Total flavonoids were measured by a colorimetric assay developed by Zhishen et al. [19]. One milliliter aliquot of appropriately diluted sample or standard solutions of luteolin (20, 40, 60, 80 and 100 mg L^{-1}) was added to a 10 mL volumetric flask containing 4 mL double distillate H_2O. At zero time, 0.30 mL, 5% $NaNO_2$ was added to the flask. After 5 min, 0.30 mL 10% $AlCl_3$ was added. At 6 min, 2 mL (1 mol L^{-1}) NaOH were added to the mixture. Immediately, the reaction flask was diluted to volume with the addition of 2.40 mL of double distillate H_2O and thoroughly mixed. Absorbance of the mixture, pink in color, was determined at 510 nm versus prepared water blank. Total flavonoids of fruits were expressed on a fresh weight basis as mg 100 g^{-1} luteolin equivalents (LE). Samples were analyzed in triplicate.

Reverse Phase HPLC Condition

Before HPLC analysis, sample solution was prepared by dissolving 4 mg of extract in 1 mL of methanol-water (80:20 v/v). All samples were filtered through a 0.45 μm Minisart filter. HPLC analyses were performed with a Schimadzu apparatus equipped with a LC-10ATvp pump and a SPD-10Avp) UV-vis detector. The column was 4.6 x 250 mm (shim-pack, VP-ODS), and the temperature was maintained at 40°C. The flow rate was 0.5 mL/min. The mobile phase was 0.1 % phosphoric acid in water (A) and 70 % acetonitrile in water (B). The solvent gradient started with 20 % B and increased to 50 % in 30 min, followed by an isocratic plateau for 5 min. Finally, solvent B decreased to 20 % until the end of running time.

LC-MS Analysis

Mass spectrometry was used as the major tool for the assessment of the analysis of phenolic compounds. The LC–MS/MS experiments were carried out with an Agilent 1100 LC system consisting of degasser, binary pump, auto sampler, and column heater. The column outlet was coupled to an Agilent MSD Ion Trap XCT mass spectrometer equipped with an ESI ion source. Data acquisition and mass spectrometric evaluation was carried out on a personal computer with data analysis software (Chemstation). For the chromatographic separation a Zorbax 300Å Extend-C-18 Column (2.1 x 150 mm) was used. The column was held at 95% solvent A (0.1% formic acid in water) and 5% solvent B (0.1% formic acid in acetonitrile) for 1 min, followed by an 11 min step gradient from 5% B to 100% B, then it was kept for 4 min with 100% B, finally, the elution was achieved with a linear gradient from 100% B to 5% B in 2 min. The flow rate was 200 μL min^{-1} and the injection volume was 5 μL. The following parameters were employed throughout all MS experiments. For electrospray ionization with negative ion polarity the capillary voltage was set to 3.5 kV, the drying temperature to 350 °C, the nebulizer pressure to 40 psi, and the drying gas flow to 10 L min^{-1}. The maximum accumulation time was 50 ms, the scan speed was 26,000 m z^{-1} s^{-1} (ultra scan mode) and the fragmentation time was 30 ms.

Antioxidant Activity

DPPH Radical Scavenging Assay

DPPH (1, 1-diphenyl-2-picrylhydrazyl) radical-scavenging effect was evaluated according to the method employed by Bouaziz et al. [20]. Aliquots (50 μL) of various concentrations of the test compound in methanol were added to 2.5 mL of DPPH methanol solution (1.5×10^{-4} mol L^{-1}). After a 30 min incubation period at room temperature the absorbance was read against a blank at 517 nm. Scavenging of free radical DPPH in % (I %) was calculated using the following equation: $I\% = [(A_{blank} - A_{sample})/A_{blank}] \times 100$, where A blank is the absorbance of the control reaction (containing all reagents except the test compound), and A sample is the absorbance of the test compound. Test compound concentration providing 50% inhibition (IC_{50}, expressed in μgmL^{-1}) was calculated from the graph plotting inhibition percentage against extract concentration.

ABTS Radical Scavenging Assay

Trolox equivalent antioxidant capacity (TEAC) assay, which measures the reduction of the radical cation of ABTS by antioxidants, was performed as previously described by Laporta et al. [21]. Briefly, ABTS radical cation ($ABTS\bullet^{+}$) was produced by reacting an ABTS stock solution with 2.45 mmol L^{-1} potassium persulfate (final concentration) and allowing the mixture to stand in the dark at room temperature for 12 to 24 h before use. For the study of phenolic compounds the $ABTS\bullet^{+}$ solution was diluted with ethanol or water (lipophilic and hydrophilic assay, respectively) to an absorbance of 0.70 (± 0.02) at 734 nm. For the photometric assay 1 mL of the $ABTS\bullet^{+}$ solution and 100 μL antioxidant solution were mixed for 45 s and measured immediately after 5 min at 734 nm (absorbance did not change significantly up to 10 min). Compounds were assayed at five different concentrations determined within the linear range of the dose–response curve. A calibration curve was prepared with different concentrations of Trolox (0–20 $\mu molL^{-1}$). Results were expressed in $mmolL^{-1}$ of Trolox.

Results and Discussion

The mean weight and the fruit color (green, green with purple spots, purple, purple with black spots, and black) of the Chemlali, Chetoui, Zalmati, Chemchali, Dhokar and Gemri-Dhokar varieties, harvested at the beginning of November 2008 and 2009, are shown in Table 1. This ripening stage was characterized by a coloration of the fruit skin varying from green, purple and finally black. The Dhokar and Gemri-Dhokar fruits are black early compared to other varieties. This may be due to the desert climate where the olive tree is cultivated in shallow soils, at high temperature and low rainfall. The change in fruit color is related to the phenolic change and the increase in anthocyanin pigmentation with decrease of the chlorophyll concentrations [22, 23].

Table 1. Fruit Mass, Dominant Fruit Color and Maturation index in the six varieties

cultivar	fruit mass (g)	maturation index	fruit color
Chemlali	0,91 ± 0.10	3.80 ± 0.50	Purple–black
Chetoui	2,40 ± 0.85	3.00 ± 0.35	Green with red spots
Zalmati	1,20 ± 0.45	3.30 ± 0.60	Purple with black spots
Chemchali	2.40 ± 0.70	2.20 ± 0.43	Green with purple spots
Dhokar	1.56 ± 0.32	6.05 ± 0.73	Black
Gemri-Dhokar	1.01 ± 0.10	5.30 ± 1.00	Black

On the other hand, the six studied varieties showed different values of maturation index (MI). In fact, the Dhokar and Gemri-Dhokar cultivars showed a significantly higher MI value compared to the other varieties. The fruit of the latter cultivar are black in early date of the season; this is may be because the olive trees are cultivated in desert region known by high temperature and low rain. Indeed, these varieties are widespread in the south of Tunisia [24].

Phenolic and Flavonoid Contents in Olives Extracts

The average values of total polyphenol content in all cultivars are presented in Table 2. The total phenol content in the olive drupes from the six cultivars harvested at the beginning of November ranged from 1.45 to 9.09 g kg^{-1} (all data are reported on a fresh mass basis). Chemchali olives showed the highest mean content in total phenol (9.09 g kg^{-1}) followed by Chemlali cultivar with total phenols content of 7.70 g kg^{-1}. Lower mean values were found in Dhokar olives, they attained 1.45 g kg^{-1} of fresh olive fruit.

Table 2. Total phenolic content (g gallic acid equivalents/kg of fresh olive) and total Flavonoid content (g luteolin equivalents/kg of fresh olive) in the six varieties

Cultivar	Total phenols content (g kg^{-1} of fruit FW)	Total flavonoids content (g kg^{-1} of fruit FW)
Chemlali	7.70 ± 1.05	0.39 ± 0.05
Chetoui	6.05 ± 0.85	0.23 ± 0.08
Zalmati	5.45 ± 0.70	0.23 ± 0.07
Chemchali	9.09 ± 1.15	0.26 ± 0.05
Dhokar	1.45 ± 0.52	0.52 ± 0.08
Gemri-Dhokar	6.97 ± 0.67	2.09 ± 0.41

Table 3. Phenolic compounds detected in six Tunisian olive fruit cultivars with their HPLC retention times, UV max and mass spectral data

N°	Compound	Olive cultivar	RT (min)	λ max (nm)	[M-H]−	Main fragments by ESI-MS^2
1	Hydroxytyrosol	a,b,c,d,	1.9	220,227	153	123
2	Hydroxytyrosol glucoside	a, b, d, e, f	2.15	219, 273	315	153, 123
3	Unknown	a, d	2.2	228, 280	407	389, 375, 345, 357, 313
4	Oleoside	a, c, d, e, f	2.3	236	389	371,345, 209, 165
5	Nuzhenide	a, b, c, d, e, f	2.4	226, 277	685	523, 453, 421, 299
6	11-Methyloleoside	a, c	2.9	235	403	371, 342, 269
7	Oleuropein aglycone derivative	a, d, e, f	3.2	238, 284	377	197, 153
8	Rutin	a, b, c, d, e, f	4.1	218, 254, 352	609	463, 301, 271, 227
9	Luteolin7-O-glucoside	a, b, c	5.1	255, 267, 350	447	285, 327, 267
10	Luteolin7-O-rutinoside	c, d	5.2	255, 268, 351	593	447, 285
11	Luteolin4'-O-glucoside	a	5.3	247, 268, 337	447	285
12	Verbascoside	a, b, c, e, f	5.9	216, 291, 315	623	461, 315, 161
13	6'-β-Hexopyranosyl-oleoside	a, b, c, d	7.6	232, 284	551	533, 507, 389, 341, 281, 251, 221, 179
14	Ligstroside	b, d	7.9	217, 286, 315	523	453, 421, 361, 291
15	6'-β-Deoxyhexo-pyranosyl-oleoside	a, b, c, d	8.2	230, 280	535	517, 491, 390 345, 325, 307,
16	Oleuropein	a, b, c, d, e, f	8.3	228, 278	539	377, 307, 275, 345, 275, 153
17	Oleuroside	a	8.4	238, 277	539	403, 377, 307, 275, 345, 275,

Table 3. Continued

N°	Compound	Olive cultivar	RT (min)	λ max (nm)	[M-H]−	Main fragments by ESI-MS^2
18	Oleuropein aglycone	a, b, c, d, e, f	9.9	240, 284	377	307, 275, 345, 153
19	Oleuropein glucoside	a, b, c, d, e, f	10.8	238, 277	701	507, 539, 443, 377, 307, 275
20	Nzhenide 11-methyl oleoside	a, b, c, d, e, f	11.79	225, 278	1071	1117, 731, 685, 523, 453, 421
21	Unknown	a	13	218, 274	579	315, 297
22	Unknown	b, d	14.2	241	736	706, 650, 521

a: compound identified in Chemlali cultivar. **b**: compound identified in Chemchali cultivar. **c**: compound identified in Chetoui cultivar. **d**: compound identified in Zelmati cultivar. **e**: compound identified in Dhokar cultivar. **f**: compound identified in Gemri-Dhokar cultivar

Total flavonoids content was also measured in the extracts. The mean values of total flavonoids content in the six cultivars studied are presented in Table 2. It was expressed as luteolin equivalents, and ranged from 0.23 to 2.09 g kg^{-1} (all data are reported on a fresh mass basis) in the six studied cultivar. Gemri-Dhokar olives showed the significantly highest mean content in total flavonoids (2.09 g kg^{-1}). However, flavonoid concentrations were lower in Chemlali, Chetoui, Zalmati, Chemchali and Dhokar cultivars and ranged from 0.23 to 0.52 g kg^{-1}.

Samples were analysed by HPLC using mass spectrometry with electrospray as the ionization source (LC-MS/MS), in negative and positive polarity modes. However, negative ionization mode enabled higher signal to noise ratios than positive mode for all the detected compounds. The total ion current and UV chromatographic profiles obtained for the several olive fruit extracts analyzed were similar, although differences in concentrations were observed for the cultivars studied. The identification of phenolic compounds in Chemlali, Chetoui, Zalmati, Chemchali, Dhokar and Gemri-Dhokar olive fruit extracts was based on the search for pseudo molecular $[M–H]^-$ ions, using extracted ion mass chromatograms and comparing those with data in the literature [17, 25]. Data obtained from the LC–MS/MS analysis of olive extracts are summarized in Table 3.

Identification of Phenolic Compounds

For example, the ESI mass spectrum in negative mode of compound 5 (see Table 3) exhibited a base peak $[M-H]^-$ at m/z 685 and fragment ions at m/z 523, 453, 421 and 299 (Table 3). The MS^2 spectrum obtained by fragmentation of the ion m/z 685 presented the following m/z values: 299, 421, 453 and 523. The fragmentation of the Pseudo molecular ion $[M-H]^-$ of compound 5 at m/z 685 yielded a fragment at m/z 523 by neutral loss of 162 mass units. The ion m/z 299 was formed by the loss of the elenolic acid glucoside unit. The ions at m/z 421 and 453 are fragments of ion m/z 523. This suggested that 5 might be a nuzhenide. This compound was detected in the seeds and leaf of Olea europea [26, 27] and in other genera of oleaceae [28]. Recently, nuzhenide was identified in Portuguese olive fruit [29].
The ion m/z 377 was detected at two retention times (3.2 and 9.9 min). The two compounds eluted showing MS fragmentation patterns similar to those reported for oleuropein derivatives [30, 31]. In addition, The MS^2 spectrum of compound 18 obtained by fragmentation of the ion m/z 377 [M-H] presented a fragment at 307. This would be due to the loss of C_4H_6O, while the fragment at m/z 275 may have derived from rearranged fragments and the following m/z values: 345 and 153 as was previously shown [31]. However, compound 7 which eluted at 3.2 min exhibited the same mass spectra but in MS^2 spectra it showed other fragments (m/z 197 and 153) suggesting that it could be identified as oleuropein aglycone derivative in various tautomeric forms. The mass spectrum of compound 8 (retention time 4.1 min) exhibited a base peak at m/z 609 in negative ion mode and strong peaks at m/z 301. This observation can be a diagnosis of quercetin derivatives, and a fragment at m/z 463, due to the loss of 146 amu (atomic mass unit) corresponding to rhamnose. These results confirm the presence of rutin, having a molecular mass of 610 Da and already reported in other olive cultivars [28, 31]. The compound eluted at 7.6 min exhibited a mass spectrum with the main ion at m/z 551. This may correspond to the pseudo molecular ion $[M-H]^-$. The principal fragment originated from the loss of 44 amu, giving rise to the intense signal at m/z 507, and

the ionic species at m/z 389 representing the oleoside structure. Its formation was accomplished by the loss of a hexose moiety (162 amu) suggesting that the compound was a hexoside derivative of oleoside. Moreover, the presence of a fragment at m/z 341, which corresponds to a disaccharide, indicated that this hexose molecule should be linked to the sugar moiety of oleoside. This suggested that 13 might be an oleoside derivative and the obtained scheme of fragmentation in negative mode agreed with those described previously [25]. According to the above results, compound 13 was a 6'-β-hexopyranosyl-oleoside. This compound was identified in the four cultivars studied. The mass spectrum of compound 15 showed a high-intensity ion at m/z 535. The ESI-MS^2 spectrum of that ion (Table 3) showed some similarities to the spectral profile of compound 13. The main fragment was obtained by the loss of 44 amu (ion at 491). Furthermore, an ionic species corresponding to the oleoside ion (m/z 389 and m/z 325) were also detected. The loss of 146 amu and 162 amu from the pseudo molecular ion can be justified by the elimination of two sugars rhamnose and glucose, respectively. This compound has previously been detected in olive leaves and fruits by the use of ESI-MS and was identified as 6'-deoxyhexopyranosyloleoside [32]. The ion **20** detected in the mass spectrum in the negative mode of 11.79 min chromatographic peak at m/z 1117 and 1071 are shown in figure 1. The ion at m/z 1117, detected in full MS spectrum, is a formic acid adduct, [M-HCOO]-, of nuzhenide 11-methyl oleoside (molecular mass 1072). Nuzhenide 11-methyl oleoside has more 386 mass units than nuzhenide corresponding to an addition of one unit of 11-methyl oleoside to the latter molecule. The detection of this compound at two retention times could correspond to the presence of isomers. The MS2 spectrum obtained by fragmentation of the ion m/z 1071 yields the The ion at m/z 685 by the loss of 386 mass units corresponding to 11-methyl oleoside unit and the ion at m/z 523, by neutral loss of 162 mass of a glucose moiety units. The ions at m/z 421 and 453 are fragments of ion m/z 523 (Figure 1). According to Silva et al. [33] nuzhenide 11-methyl oleoside is located mainly in olive seed and is the main phenolic compound in this part of the fruit, which is also in accordance with other results obtained concerning the analysis of phenolic compounds in fruits and seeds of Italian olive tree cultivars by HPLC/DAD/MS using Electrospray (ESI) Mass Spectrometry (MS) and atmospheric pressure chemical ionisation (APCI) as the ionization sources [34]. The comparison of the ESI-MS data with data in the literature helped to identify other compounds (Table 3). Their fragmentation schemes were in agreement with those previously detected in Olea europaea fruits and leaves [17, 35, 36].

Oleuropein and Hydroxytyrosol Quantification

Phenolic compounds in olive fruits are important factors to evaluate the quality of virgin olive oil since they are partly responsible for its autoxidation stability. Phenolic compounds quantification was achieved using HPLC apparatus by the absorbance recorded in the chromatograms relative to external standards. Among the identified compounds, hydroxytyrosol and oleuropein were the major phenolic compounds. The analysis of Table 4 reveals considerable quantitative differences in oleuropein contents among the different cultivars. In all the analyzed cultivars, Chetoui and Chemlali cultivar showed highest amount of oleuropein reaching 5.86 and 4.02 g kg^{-1}, respectively (Table 4). While for Gemri-Dhokar and Dhokar cultivars, harvested on the same date, the oleuropein concentrations were low and attained 1.16 and 0.25 g kg^{-1} respectively.

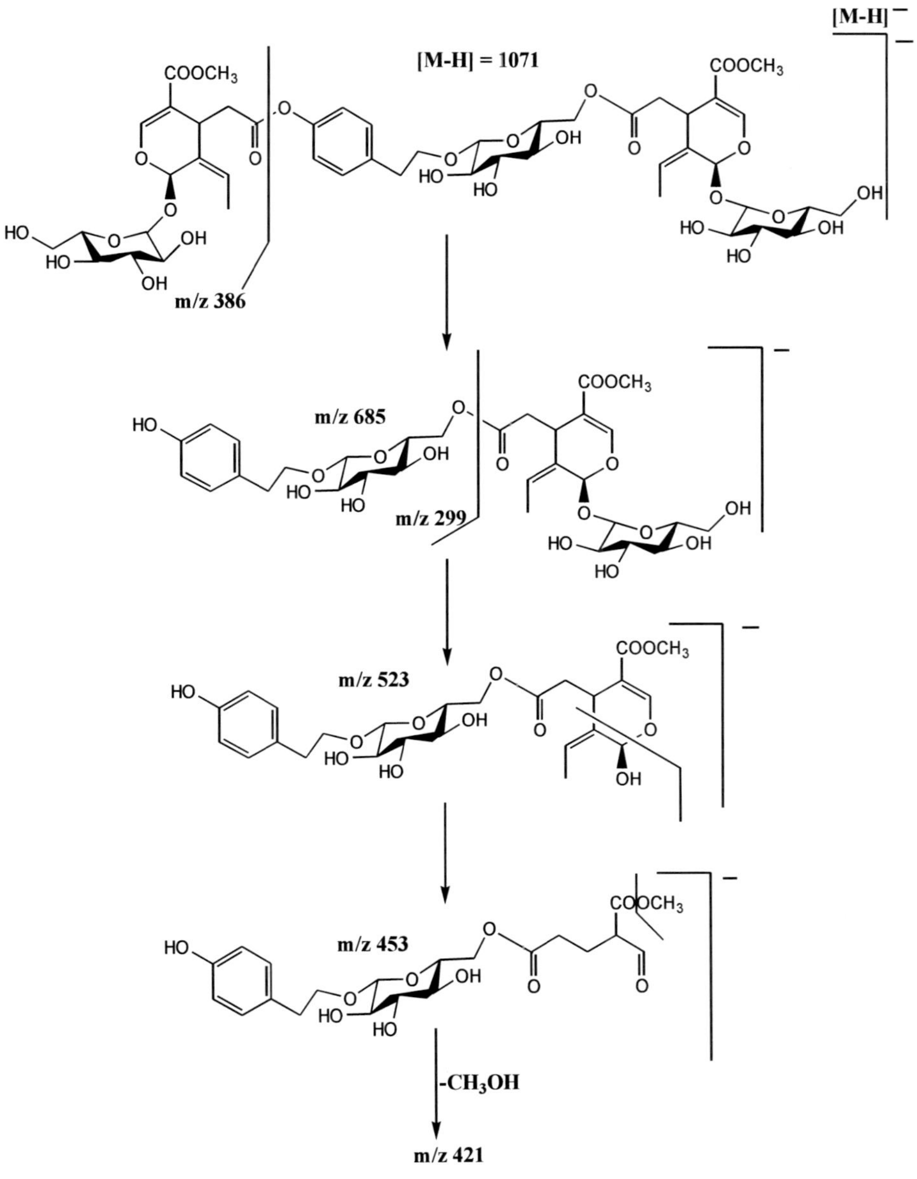

Figure 1. Fragmentation scheme for ions obtained by MS^2 of precursor ion m/z 1071.

On the other hand, the concentrations of hydroxytyrosol reached 0.62, 0.42, 0.41 and 0.25 g kg^{-1}, respectively in Zalmati, Chemlali, Chemchali and Chetoui cv. However, they were only 0. 12 and 0,08 g kg^{-1} of Gemri-Dhokar and Dhokar fruits, which were the lowest values.

Hydroxytyrosol is the principal degradation products of oleuropein, it was found in much smaller quantities relative to oleuropein and showed an inverse trend of evolution in all cultivars. This difference in hydroxytyrosol concentration could be explained by an increase in the activity of the hydrolytic enzymes; in particular, in glycosidases which catalyse the hydrolysis of the oleuropein and secoiridoid derivatives, with the production of oleuropein aglycone and hydroxytyrosol [22, 37]. Esti et al. [38] reported data for phenolic content in four Italian olive fruits varieties (Gentile (Larino), Gentile (Colletorto), Coratina, and

Leccino) during maturation. Their results showed that as the olives ripen, hydroxytyrosol tenor increases, while oleuropein decreases.

Table 4. Oleuropein and Hydroxytyrosol contents in the six varieties

Cultivar	Hydroxytyrosol content ($g\ kg^{-1}$ of fruit FW)	Oleuropein content ($g\ kg^{-1}$ of fruit FW)
Chemlali	0.42 ± 0.05	3.79 ± 0.50
Chetoui	0.25 ± 0.02	5.86 ± 0.83
Zalmati	0.62 ± 0.08	1.92 ± 0.12
Chemchali	0.41 ± 0.03	3.15 ± 0.55
Dhokar	0.08 ± 0.01	0.25 ± 0.04
Gemri-Dhokar	0.12 ± 0.01	1.16 ±0.03

Antioxidant Potential of Olive Extracts

Antioxidants have become a topic of increasing interest recently to health and food science researchers or medical experts. The antioxidant activities of olive fruit extracts are showed in Table 5. Olive extracts showed similar trends of activity in DPPH and ABTS assays (Table 5). The comparison of IC_{50} and TEAC values of the six cultivars olive fruit extracts showed different antioxidant activity. Indeed, Chemchali and Chemlali extracts presented the highest antioxidant activity. While Gemri-Dhokar and Dhokar extracts exhibited the lowest one.

Table 5. Radical scavenging activity of olive extracts

Cultivar	IC_{50} (μgmL^{-1}) (DPPH assay)	TEAC (mM) (ABTS assay)
Chemlali	7.04 ± 0.92	1.91 ± 0.30
Chetoui	50.13 ± 3.20	0.82 ± 0.10
Zalmati	76.40 ± 3.85	0,76 ± 0.10
Chemchali	6.30 ± 0.85	2.04 ± 0.48
Dhokar	31.61 ± 2.14	0.13 ± 0.07
Gemri-Dhokar	26.30 ± 2.30	0.26 ± 0.08

The antioxidant activity and the level of the total phenols content in the extracts suggested that the radical scavenging effect in the extract could be attributed to the structure of phenolic compounds and their content [37]. The antioxidant activities of phenolic compounds are known for a long time [37] and they are principally based on the redox properties of their phenolic hydroxyl groups and the structural relationship between different parts of their chemical structure [38, 39]. For oleuropeosides and the other phenols it is mainly the dihydroxy (catechol) structure present in their moieties, which confers antioxidant properties to these compounds. However, their antioxidant capacity depends on mechanisms, which are influenced by structural factors, such as the presence or absence of glucosidic moieties in the polyphenol, the glycosylation site, the number and position of the free and esterified hydroxyls, etc [36, 37].

CONCLUSION

The data presented in this paper suggest that despite the limited number of cultivars examined, the study of the polyphenol composition might contribute to biochemical characterization of olive fruit varieties. Oleuropein was the main secoiridoid identified in the olive fruit extracts. However, hydroxytyrosol was its main degradation product and decreased during fruit maturation. The cultivar factor seemed to have an effect on the antioxidant activities. Indeed, the highest antioxidant potential was obtained for the cultivar with high total phenols and flavonoid contents. These findings could be used for controlling the production processes and correlating the oil sensorial characteristics to the polyphenolic pattern.

ACKNOWLEDGEMENTS

The authors would like to thank Prof. Monique Simmonds from Jodrell laboratory Kew Gardens UK for her help in LC/MS analysis.

REFERENCES

[1] P. Kris-Etherton, K. D. Hecher, A. Bonanome, S. M. Coval, A. E. Binkoski, K. F. Hilpert, A. E. Griel, T. D. Etherton. *Am. J. Med.* 113, 71 (2002).
[2] S. A. Wiseman, J. N. N. Mathod, N. J. De Fouw, Z. B. M. Tijnurg. *Atherosclerosis.* 120, 15 (1996).
[3] F. Gutiérrez-Rosales, J. J Ríos, L. Gómez-Rey. *J. Agric. Food. Chem.* 51, 6021 (2003).
[4] M. Servili, G. F. Montedoro. *Eur. J. Lipid Sci. Technol.* 104, 602 (2002).
[5] R. Mateos, M. Trujillo, C. Pérez-Camino, W. Moreda, J. A.Cert. *J.Agric. Food Chem.* 53, 5766, (2005).
[6] D. Boskou. Olive Oil: Chemistry and Technology. AOCS Press, Champaign, IL (USA) pp. 85, (1996).
[7] P. Pinelli, C. Galardi, N. Mulinacci, F. F. Vincieri, A. Cimato, A. Romani. *Food Chem.* 80, 331, (2003).
[8] N. Mulinacci, C. Giaccherini, M. Innocenti, A. Romani, F. F. Vincieri, F. Marotta, A. Mattei. *J. Sci. Food Agric.* 85, 662 (2005).
[9] R. W. Owen, A. Giacosa, W. E. Hull, B. Haubner, B. Spiegelhalder, H. Bartsch. *Eur J Cancer*, 36, 1235 (2000).
[10] G. Montedoro,M. Servili, M. Baldioli, E. Miniati. *J. Agric. Food Chem.* 40, 1571(1992).

[11] G. Pannelli, M. Servili, R. Selvaggini, M. Baldioli, G. F.Montedoro. *Acta Hortic* 356, 239 (1994).
[12] D. Ryan, K. Robards, S. Lavee. Evolution de la qualité de l'huile d'olive. Olivae 72, 23 (1998).
[13] International Olive Oil Council (IOOC) Olive oil exportations. http://www.internationaloliveoil.org, (2004).
[14] S. Ben Temime, E. Campeol, P. L. Cioni, D. Daoud, M. Zarrouk, *Food Chem.* 99, 315 (2006).
[15] L. Abaza, M. Msallem, D. Daoud, M. Zarrouk. *Oléagineux Corps Gras Lipides*, 9, 174 (2002).
[16] M. Uceda . M Hermoso, in El Cultivo del Olivo, ed. by Barranco D,Fernandez-Escobar R and Rallo L. Junta de Andalucya Ediciones Mundi-Prensa, Madrid, 547 (1998).
[17] M. Bouaziz, M. Chamkha, S. Sayadi, *J. Agric. Food Chem.* 52, 5476 (2004).
[18] M. Bouaziz, H. Jemai, W. Khabou, S. Sayadi, *J. Sci. Food Agric.* 90, 1750 (2010).
[19] J. Zhishen, T. Mengcheng, W, Jimming, *Food Chem.* 64, 555(1999).
[20] M. Bouaziz, S. Lassoued, Z. Bouallagui, S. Smaoui, A. Gargoubi, A. Dhouib, *Bioorg. Med. Chem.* 16, 9238 (2008).
[21] O. Laporta, L. P´erez-Fons, R. Mallavia, N. Caturla , V. Micol, *Food Chem.* 101, 1425 (2007).
[22] M. Bouaziz, M. Damak, S. Sayadi, *J. Soc. Chem. Tunisie*. 8, 11 (2006).
[23] M.Roca, M. I. Mınguez-Mosquera. *J. Agric. Food Chem.* 49, 832 (2001).
[24] H. Jemai, M. Bouaziz, S. Sayadi. *J. Agric. Food Chem.*, 57, 2961 (2009)
[25] MS, Cardoso, S. Guyot, N. Marne, AJ. Lopes-Da-Silva, MGCC. Renard, AM. Coimbra. *J. Sci. Food Agric.* 85, 21(2005).
[26] A. De Nino, N. Lombardo, E. Perril, A. Procopio, A. Raffaelli, G. Sindona. *J. Mass Spectrom.* 32, 533(1997).
[27] RT. La Londe, C. Wong, AI. Tsai, *J. Am. Chem. Soc.* 98, 3007 (1976).
[28] K. Machida, A. Kaneko, T. Hosogai, R. Kakuda, Y. Yaoita, M. Kikuchi. *Chem. Pharm. Bull.* 50, 493 (2002).
[29] S. Silva, L. Gomes, F. Leit˜ao, AV. Coelho and L. Vilas Boas. *Food Sci. Technol. Int.*12, 385 (2006).
[30] D. Caruso, R. Colombo, R. Patelli, F. Giavarini, G. Galli. *J. Agric. Food Chem.* 48, 1182 (2000).
[31] K. De la Torre-Carbort, O. Jauregui, E. Gimeno, AI. Castellote, RM. Lamuela-Ravent, MC. Lopez-Sabater. *J. Agric. Food Chem.* 53, 4331(2005).
[32] A. De Nino, F. Mazzotti, SP. Morrone, E. Perri, A. Raffaelli, G. Sindona: *J.Mass Spectrom.* 34, 10 (1999).
[33] S. Silva, L. Gomes, F. Leitão, M Bronze, A. V Coelho and Luís Vilas Boas. *Grasas y Aceites*, 61 (2), 157 (2010).
[34] S. Silva. M.Sc. Thesis, Faculty of Pharmacy, Lisbon University, Lisbon, Portugal.(2004)
[35] SM. Cardoso, S. Guyot, N. Marnet, JA. Lopes-da-silva, MS. Silva, CMGC. A. Renard. *J Sci. Food Agric.* 86, 1495 (2006).
[36] HK. Obied, Jr DR. Bedgood, D. Prenzler, K. Robards. *Anal. Chim. Acta.* 603, 176 (2007).

[37] M. Bouaziz, R.J. Grayer, M.S.J. Simmonds, M. Damak, S. Sayadi. *J. Agric.Food Chem.* 53, 236 (2005).
[38] M. Esti, L. Cinquanta, E. La Notte. *J. Agric. Food Chem.*, 46, 32 (1998).
[39] B. Le Tutour, D. Guedon. *Phytochem.* 31, 1173 (1992).
[40] O. Benavente-García, J. Castillo, J. Lorente, A. Ortuño, J.A.Del Rio. *Food Chem.* 68, 457 (2000).
[41] M. Bouaziz, S. Sayadi. *J. Lipid Sci. Technol.* 107, 118 (2005).

In: Cultivars
Editor: Katya Carbone ISBN: 978-1-62417-101-7

Chapter 8

STUDY OF THE OXIDATIVE STABILITY OF ALMONDS BASED ON DIFFERENT PARAMETERS AND TECHNIQUES: APPLICATION TO CULTIVAR AUTHENTICITY

A. Beltrán Sanahuja* and M. C. Garrigós Selva
Analytical Chemistry, Nutrition and Food Sciences Department,
University of Alicante, Alicante, Spain

ABSTRACT

In recent years, food authenticity and origin determination have become a crucial issue in food quality control and safety. Almonds (Prunus dulcis) are very appreciated for their nutritional and sensory characteristics. The consumption of almonds has slightly increased because they are great-tasting foods that greatly enrich the dishes and recipes, as they can be used all year round in different forms (natural, salty, dried, roasted or fried) or incorporated into a variety of manufactured food products, such as nougat, marzipan, pralines, chocolate and ice cream. Almonds are foods with low water content, while its fat and protein contents are high. Health benefits have been linked to their fatty acid composition and the presence of minor compounds with antioxidant activity (tocopherols, polyphenols). The lipid content of each cultivar is conditioned by itself and its geographical origin. This variability affects the fatty acid composition, especially oleic and linoleic acids content, as well as minor compounds profile with antioxidant character that subsequently leads to changes in the oxidative stability of almonds during processing.

In this chapter, the oxidative stability of different almond cultivars (Marcona, Guara, Garrigues and Butte) was evaluated based on different parameters and analytical techniques: Fourier transform infrared spectroscopy (FTIR); gas chromatography-flame ionization detector (GC-FID) for the determination of the fatty acids profile; headspace-solid-phase microextraction gas chromatography-mass spectrometry (HS-SPME-GC-MS) for volatiles content; and thermal analysis techniques, such as differential scanning calorimetry (DSC) and thermogravimetric analysis (TGA). Subsequent classification of

---

* Corresponding author: Email: ana.beltran@ua.es

cultivars by applying multivariate stepwise linear discriminant analysis (LDA) was also performed. The obtained results proved the suitability of the studied techniques combined with LDA for an easy and fast discrimination among different almond cultivars to protect the authenticity of almond samples in food processing against frauds, particularly for Protected Designation of Origin (PDO) products.

INTRODUCTION

In recent years, food and health aspects are receiving special attention, being food authenticity and origin determination a crucial issue in food quality control and safety [1]. Nuts and specially almonds are very important food products from a nutritional point of view, mainly due to their high content of numerous beneficial nutritive and bioactive compounds [2,3].

Almonds are grown as orchard crops, highly nutritious and with a high fat content, but they have a high cost. The fruit is highly valued for its dietetic, cosmetic and pharmaceutical properties [4]; being its health benefits linked to their fatty acid composition [5] in relation to risk of coronary heart disease, diabetes and cancer; and the presence of minor compounds with antioxidant activity (tocopherols, polyphenols) and cholesterol lowering effects (phytosterols) [6].

Alicante and Jijona nougats are typical Spanish confectionery food products manufactured in a traditional way made of toasted almonds, sugar and honey. They are elaborated using raw materials from Eastern Spain and are protected by the Regulation Council of the PDO Jijona and Alicante Nougats [7]. In recent years, production of almonds in Spain has decreased considerably compared with USA, where cultivation has increased due to the improvement of agricultural techniques and selection of new almond cultivars. To avoid adulteration practises, some studies have been focused on the characterization of different almond cultivars and criteria for almond quality evaluation [8]. In this sense, fatty acids profile, triglyceride and tocopherol composition have been used as discriminant parameters in order to differentiate between almond cultivars [9].

Since almonds are considered fatty foods, it is important to detail the processes occurring during lipid oxidation, which is influenced by factors including antioxidants [6], metals, photosensitizers and oxygen content, heat intensity and light energy and the number of double bonds present in fatty acids [10,11]. As a result, oxidized lipids can accelerate deterioration of sensory quality in edible oils and decrease consumer acceptances. In almond samples, lipid oxidation is the main cause of off-flavour development due to their high content of unsaturated fatty acids [5]. Since almonds are used as ingredients in food products, which are processed at rather high temperatures, the study of thermal stability of almond oils can be useful for identifying the most resistant almond cultivars to oxidative degradation.

The mechanism of oxidation processes of edible oils largely depends on the oxidative conditions. In this sense, the nature of the different compounds produced in the oxidation of edible oils at room temperature [12], in closed receptacles, could be different from those at higher temperatures [13]. Oxidative stability is an important indicator to determine oil quality and shelf life because low-molecular weight off-flavour compounds are produced during oxidation, which can make oil unacceptable to consumers or for industrial use as a food ingredient. In the oxidation process, the triglyceride structure undergoes changes with the consequence of the formation of a great number of compounds. It has been postulated that first of all, primary oxidation compounds, such as hydroperoxides, are generated. Among the secondary oxidation products, it is possible to find alkanals, (E)-2-alkenals and (E, E)-2,4-alkadienals, but also γ-oxygenated α,β-unsaturated aldehydes [14]. In previous studies, hexanal has been monitored for the determination of the degree of oxidation in the headspace of food samples [15]. The lack of double bonds of this compound makes it more oxidatively stable than unsaturated aldehydes. In food oxidized samples, the concentration of 2-alkenals increases with the oxidation level of the sample. Alkadienals detected appear with the highest proportions in the most oxidized food samples, being the most important those from 9 to 10 carbon atoms [12].

The majority of almond composition research has focused on the study of lipid and fatty acid fractions. A lipid content of about 44–61% has been reported in studies of commercial or locally important (sweet) almond cultivars [2]. Oxidation processes in almond oils have been monitored using different instrumental techniques, as FTIR. This technique has been increasingly used for studying the oxidative stability of oils and fats [5], as it has several advantages as low cost, simple sample preparation and reduced analysis time [16]. On the other hand, fatty acid profile has been widely investigated using GC techniques. In this sense, oleic and linoleic acids account for about 90% of total lipids, with very low levels of saturated fatty acids (<10%) found in all varieties from all regions. However, oleic/linoleic acid ratios vary widely among almond varieties [5].

Only a few studies about aroma profiles of almond cultivars are reported in the literature. Vázquez-Araújo et al. [17] used simultaneous steam distillation-extraction (SDE) followed by GC-MS for the analysis of volatile compounds in different toasted almonds and nougats. Volatile organic compounds (VOCs) emission of almonds and its relation with navel orangeworm have been also studied using a Tenax collection system [18] or HS-SPME [19] and identified via GC-MS analysis. Finally, HS-SPME-GC-MS has been applied to the determination of hexanal and other volatile compounds in γ-irradiated raw unpeeled almonds [20].

Regarding thermal properties, they have been evaluated on different nuts by using thermal analysis techniques, such as DSC and TGA. DSC was used for the characterization of different seed oils by obtaining their thermal profiles. The quality of oils was dependent on their composition and this may be influenced by agronomy factors, such as cultivar and industrial factors derived from the oil processing.

These modifications can be determined from DSC parameters such as melting and crystallization temperatures and enthalpies [21]. It has been demonstrated that the thermal profile is strongly dependent on temperature scanning rate and thermal history of samples [k22,23]. Oxidation kinetic profiles of different edible oils using DSC have also been reported [24]. TGA has been also used to study food composition under different thermal treatments [25] such as the effect of antioxidants on oxidative stability of edible fats and oils [26].

In this chapter, the oxidative stability of different almond varieties (Marcona, Guara, Garrigues and Butte) was evaluated based on different parameters and techniques, followed by subsequent classification of cultivars by applying LDA in order to differentiate the cultivar used and to detect adulteration practices in PDO products. For this purpose, fatty acid profiles determined by gas chromatography and infrared spectra were monitored at different stages of the oxidation process. On the other hand, almond cultivars were characterized based on volatiles content determined by HS-SPME-GC-MS in fresh and oxidized samples. Finally, thermal stability of cultivars was also evaluated by using thermal techniques (DSC and TGA).

EXPERIMENTAL

Samples

Fifty-five samples from 4 different almond cultivars were used in this work: Marcona, Guara, Garrigues and Butte. The first three cultivars were selected, as they are representative of the cultivars grown in Spain and were collected from different Spanish geographical areas i.e. Reus (Tarragona), Santomera (Murcia), Córdoba, Alicante, Agost (Alicante), Xixona (Alicante), Pinoso (Alicante), Zaragoza, Alcaníz (Teruel) and Velez Rubio (Almeria). Butte samples were ground in California and this cultivar was chosen because it is one of the most widely grown cultivar in the world [27]. Samples from the four different cultivars were obtained from two Spanish importers: Colefruse (San Juan, Alicante) and Almendras Llopis (San Vicente del Raspeig, Alicante). Almond shells were immediately removed by using a hammer and the seeds obtained were then stored at 7 ± 1 ºC to keep them fresh until oil extraction.

Extraction of Almond Oil

Almond seeds were ground in an electric grinder (Moulinex, Barcelona, Spain) just before extraction of the oil. The seed fragments passing through a 1.5-mm sieve were stored in a desiccator. For oil extraction, five grams of each ground sample were extracted using a commercial fat extractor (Selecta, Barcelona, Spain) with 40 mL of petroleum ether (analytical grade, Panreac, Barcelona, Spain) for 90 min.

The temperature of the extraction process was fixed at 60 ºC because of the boiling point of the petroleum ether. The obtained oil was a mixture of twelve independent fat extractions of each sample and it was dried under a nitrogen current and kept sealed in an amber vial at -21 ºC until required for analysis.

Oxidative Treatment

In order to carry out fatty acid analysis by GC and collecting FTIR spectra, two grams of each almond oil were put into a 3 cm high glass vial (25 mm outside diameter.) without a lid. The vial was then placed in an oven (Selecta, Barcelona, Spain) at 75 °C in order to accelerate the lipid oxidation process. The temperature was monitored using an external thermometer. A quantity of few drops of oil was sampled at the beginning of the oxidative treatment and every 5 days until secondary oxidation products appeared (i.e. around the 20th day of treatment). As oxidative processes depend on the amount of sample and the surface area in contact with air, different vials were used for each sample treatment day. Oil for a particular day of treatment was therefore sampled from a new vial, which always contained the same amount of oil (i.e. approximately 2 g). For volatiles content determination, one gram of almond oil was placed into a 20-mL dark glass vial sealed with an aluminium crimp cap provided with a polytetrafluoroethylene/ silicone septum. The sample vial was placed in an oven at 100 °C, in order to accelerate the lipid oxidation process. The temperature was monitored using an external thermometer. Oxidative stability of samples was evaluated by obtaining oxidation volatile profiles at 1, 3, 5, 7, 10, 15 and 20 days.

Fatty Acids Profile Determination by GC-FID

Methylation of fatty acids was carried out according to the American Oil Chemists' Society method CE-2-66 [28]. The fatty acids contained in 0.3 g of almond oil were transformed into their respective fatty acid methyl esters (FAMEs) by adding 6 mL of sodium methylate solution and heating the mixture under reflux until one phase was obtained. Sodium methylate solution was prepared by dissolving 3.4 g of sodium (Panreac, Barcelona, Spain) in 1 L of methanol (HPLC grade, Scharlau, Barcelona, Spain). A 5% solution of sulphuric acid (analytical grade, Panreac, Barcelona, Spain) in methanol was then added to the mixture.

After completion of the methylation reaction, FAMEs were extracted using 6 mL of hexane (96% GC grade, Scharlau, Barcelona, Spain). A saturated sodium chloride solution (25 mL) was added to the mixture to facilitate the separation of the two layers. 1 mL of the organic layer was mixed with 1 mL of the internal standard solution and the volume was made up to 10 mL with hexane in a volumetric flask. A 730 mg/L solution of methyl tridecanoate (98% GC, Sigma, Steingeim, Germany) was used as internal standard as this fatty acid was not present in almonds [29]. FAMEs were analyzed using a Carlo Erba Series 8000 gas chromatograph equipped with a split/splitless injector and a flame-ionization detector (FID). Calibration solution standards were prepared by mixing different amounts of these standards to cover the different concentration range of the almonds' FAME. An internal standard was added to all samples and standard solutions. Comparison of retention times and areas of standard solutions were performed for the identification and quantification of FAME´s with those obtained for the test samples. The chromatographic column selected was a BPX70 (70% cyanopropyl polysilphenylene-siloxane SGE, 30 m x 0.25 i.d). Experimental conditions were optimized to obtain a good separation of all fatty acid methyl esters. Samples (1 μL) were injected in the split mode (1:30 ratio) at 230 °C, in duplicate. Helium was used as the carrier gas at a flow rate of 2 mL/min. The FID temperature was set at 250 °C. The oven temperature was programmed from 110 °C (hold 2 min) to 170 °C (hold 4 min) at 3.0 °C/min and to 190 °C at 2.0 C/min and to 220 °C (hold 4 min) at 20.0 °C/min.

FTIR Analysis

FTIR spectra were collected using a Vector 22 FTIR spectrometer (Bruker Optic Gmbh, Ettlingen, Germany) with OPUS software (version 3.1). A film of a small amount of the oil (five drops) was placed on the ATR device, which was equipped with a ZnSe crystal [30]. Spectra were obtained in the 4000-400 cm^{-1} range, using 128 scans and 4 cm^{-1} resolution. A blank spectrum was obtained before each measurement to compensate for humidity effects and carbon dioxide in the air by spectra subtraction. After each measurement, the ATR plate was carefully cleaned with analytical grade acetone (Panreac, Barcelona, Spain) and dried with a soft tissue before analysis of the next sample. Three spectra replicates were obtained for each sample.

Volatiles Content Determination by HS-SPME-GC-MS

In order to evaluate suitable extraction conditions for the determination of volatile compounds in almond oils, a sample of Butte cultivar was selected.

Three variables were studied by a two-level factorial experimental design (2^3): extraction temperature, extraction time and stirring speed. This design consists of eight experiments, which were performed, in duplicate plus one central point which was replicated four times; and allows us to study the main effects and factor interactions. All experiments were run randomized. The type of fiber was not considered as a variable in the optimization process. A divinylbenzene/carboxen/polydimethylsiloxane (DVB/CAR/PDMS) fiber was selected based on previous studies where different solid-phase microextraction fibers were used and compared [31]. The best results for target compounds were obtained using a DVB/CAR/PDMS fiber, which can be explained not only by its nature, but by slightly larger capacity for the analytes. In addition, this fiber maintains its performance at well above 100 extractions with between-day precision below 10%.Table 1 lists the upper and lower values for each factor, which were selected considering instrumental limitations. For the maximum extraction temperature, this factor was set to 60 °C in order to avoid the oxidation of the almond oil.

Table 1. Factor levels and design matrix of experiments for optimization of HS-SPME factors

Factor			Levels			*Run*								
Variable	Key	Units	Low (-)	High (+)	Middle (0)	*1*	*2*	*3*	*4*	*5*	*6*	*7*	*8*	*9*
Stirring speed	A	rpm	0	110	55	–	+	–	+	–	+	–	+	*0*
Time	B	min	20	60	40	–	–	+	+	–	–	+	+	*0*
Temperature	C	°C	35	60	47.5	–	–	–	–	+	+	+	+	*0*

On the other hand, 60 min was considered as the maximum time in order not to have extremely long experimental extraction times.HS-SPME analyses were carried out by weighing 1 g of almond oil into a 20 mL dark glass vial sealed with an aluminium crimp cap provided with a polytetrafluoroethylene/silicone septum. Then, the vial was placed into a water bath where the sample was maintained under magnetic stirring.

After 5 min of sample conditioning, the needle of the SPME device was inserted into the vial through the septum, and the fiber was exposed to the vial headspace for different time periods (20, 40 or 60 min). After that, the fiber was retracted into the needle assembly, removed from the vial, transferred to the injection port of the GC unit and immediately desorbed.

Analysis of volatile compounds was performed by using a Perkin Elmer TurboMass Gold GC-MS equipped with a split/splitless injector, a SPB-5 capillary column (30 m x 0.25 mm i.d. x 0.25 μm film thickness) (Supelco, Bellefonte, PA, USA) and a quadrupole mass spectrometer operating in electronic impact (EI) ionisation mode (70 eV). Helium was used as the carrier gas (1 mL min^{-1}). The column temperature was programmed from 50 °C (hold 10 min) to 280 °C (hold 5 min) at 4 °C min^{-1} heating rate. Ion source and detector temperatures were 180 and 300 °C, respectively. The injector temperature was 270 °C and the time for fiber desorption into the injection port was fixed at 10 min in splitless mode (1.5 min splitless-period). After every run, the SPME fiber was conditioned for 30 min at 270 °C in the injector of the gas chromatograph followed by a blank analysis to avoid carryover of the fiber. Identification of the volatile compounds in almond oil samples was performed in full scan mode (40–550 m/z), by the combination of NIST mass spectra and gas chromatographic retention times of standard compounds. Chromatographic responses of detected volatile compounds (peak area counts) were monitored for comparative measurements of each compound in the studied samples.

Thermal Analysis by DSC and TGA

Calorimetric tests were performed with a TA Instruments DSC Q2000 V23.12 Build 103 (New Castle, DE, USA). Nitrogen was used as an inert atmosphere at a flow rate of 50 mL min^{-1}. The DSC instrument was calibrated using indium (melting point 156.6 °C; melting enthalpy 28.45 J g^{-1}), according to standard procedure described in the user's manual issued by the manufacturer. Then, 6–7 mg of oil was weighed to the nearest 0.1 mg into hermetically sealable aluminum pans (40 μL) and placed in the sample chamber. An empty sealed pan was used as a reference. The following thermal program was used: (1) sample loading at 50 °C (5 min hold); (2) cooling to -80 °C (5 min hold); and (3) heating to 50 °C. Analyses were performed in duplicate obtaining a good reproducibility (relative standard deviation between 1 and 5%). Due to the strong dependence of results on the scanning rate, calorimetric tests were performed at 2, 5 and 10 °C min^{-1}. Details of methods and criteria for selecting the time–temperature program are reported elsewhere [32].

Dynamic TGA tests were carried out by using a TGA/SDTA 851 Mettler Toledo (Schwarzenbach, Switzerland) thermobalance. 6–7 mg of oil samples were placed in alumina pans (70 μL) and heated from 30 to 700 °C at 2, 5 and 10 °C min^{-1} under nitrogen atmosphere (50 mL min^{-1}).

Statistical Analysis

Experimental data were processed with the aid of the SPSS statistical package Version 15.0 using analysis of variance (ANOVA) in order to determine significant differences among samples at $p < 0.05$ according to Tukey's post-hoc test. The presence of different categories within almond samples was investigated using stepwise linear discriminant analysis (LDA). This is a supervised pattern recognition technique used not only to recognize different classes in a set of data but also to obtain classification functions (i.e., vectors) that make it possible to predict to which group a sample belongs using the appropriate latent variables [33].These latent variables are linear combinations of the initial selected variables that maximize the resolution among the groups. The method used for variable selection was to minimize the Wilks' lambda statistics (k_w). In order to validate the classification rule developed for this study, a test set was constructed on the basis of the leave-one-out algorithm.

The validation of the methodology was carried out by leaving randomly 1/3 of our data "out", using the remaining 2/3 to build an LDA model and testing the "stability" of our model with the 1/3 of the samples left out as unknowns.

On the other hand, the results of the two-level factorial experimental design for optimization of the HS-SPME conditions were analyzed using the statistical package Statgraphics Plus for Windows Version 5.1.

RESULTS

Major Fatty Acid Composition Study

As it can be seen in Table 2, the most abundant fatty acid present in almond oils was oleic acid (57–74%), followed by linoleic acid (16– 28%). There were also smaller amounts of palmitic (6–7%), stearic (1.7–2.7%) and palmitoleic (0.37–0.54%) acids.

Table 2. Major fatty acid content (g/100 g oil) of the four almond cultivars at different days of the oxidative treatment [a]

Almond Cultivar	Fatty acid methyl esters	0 days	11 days	20 days
Garrigues	Palmitic (C16.0)	6.935±0.256b	6.380±0.088c	6.110±0.201c
Guara	Palmitic (C16.0)	6.055±0.439b	5.719±0.303bc	5.469±0.369c
Butte	Palmitic (C16.0)	6.368±0.590b	5.916±0.438bc	5.272±0.536c
Marcona	Palmitoleic (C16.1)	0.541±0.058b	0.498±0.051bc	0.457±0.032c
Garrigues	Palmitoleic (C16.1)	0.494±0.018b	0.450±0.015c	0.369±0.015d
Guara	Palmitoleic (C16.1)	0.371±0.023b	0.348±0.025b	0.313±0.026c
Butte	Palmitoleic (C16.1)	0.447±0.029b	0.402±0.032c	0.332±0.030d
Marcona	Stearic (C18.0)	2.029±0.224b	1.861±0.235bc	1.724±0.220c
Garrigues	Stearic (C18.0)	1.948±0.103b	1.759±0.275b	1.669±0.208b
Guara	Stearic (C18.0)	2.690±0.162b	2.449±0.143c	2.335±0.092c
Butte	Stearic (C18.0)	1.711±0.144b	1.554±0.104bc	1.454±0.046c
Marcona	Oleic (C18.1)	65.876±3.569b	59.987±4.399c	54.698±4.083d
Garrigues	Oleic (C18.1)	74.090±1.045b	69.387±2.276c	56.933±1.107d
Guara	Oleic (C18.1)	67.427±2.124b	59.811±2.479c	53.590±4.252d
Butte	Oleic (C18.1)	57.258±1.111b	50.610±2.631c	43.360±1.689d
Marcona	Linoleic (C18.2)	16.728±1.581b	13.082±1.826c	3.685±0.937d
Garrigues	Linoleic (C18.2)	20.626±1.933b	18.105±1.461b	4.716±0.249c
Guara	Linoleic (C18.2)	16.090±1.026b	13.634±1.196c	4.102±1.657d
Butte	Linoleic (C18.2)	28.317±2.066b	14.953±3.277c	2.914±1.077d

[a] Mean ± SDm, n = 12 (Marcona), *n* = 4 (Garrigues), *n* = 9 (Guara) and *n* = 6 (Butte). Values followed by different letters in the same row are significantly different at the 5% level.

Regarding the comparison between the different cultivars at day 0 of the oxidative treatment, Butte cultivar showed the lowest concentrations of oleic and stearic acids and the highest linoleic acid content. On the other hand, Guara cultivar had the highest stearic acid content and the lowest palmitoleic acid content. The concentration of all the major fatty acids monitored over the course of oxidation decreased from those obtained on day 0. In relative terms, the reduction of the content followed the order linoleic > oleic ~ palmitoleic acids. This order falls in line with the data published in the literature which states that linoleic acid reacts with oxygen 10 times faster than oleic acid does [34]. Finally, a small reduction in the amount of saturated fatty acids was observed in all cultivars.

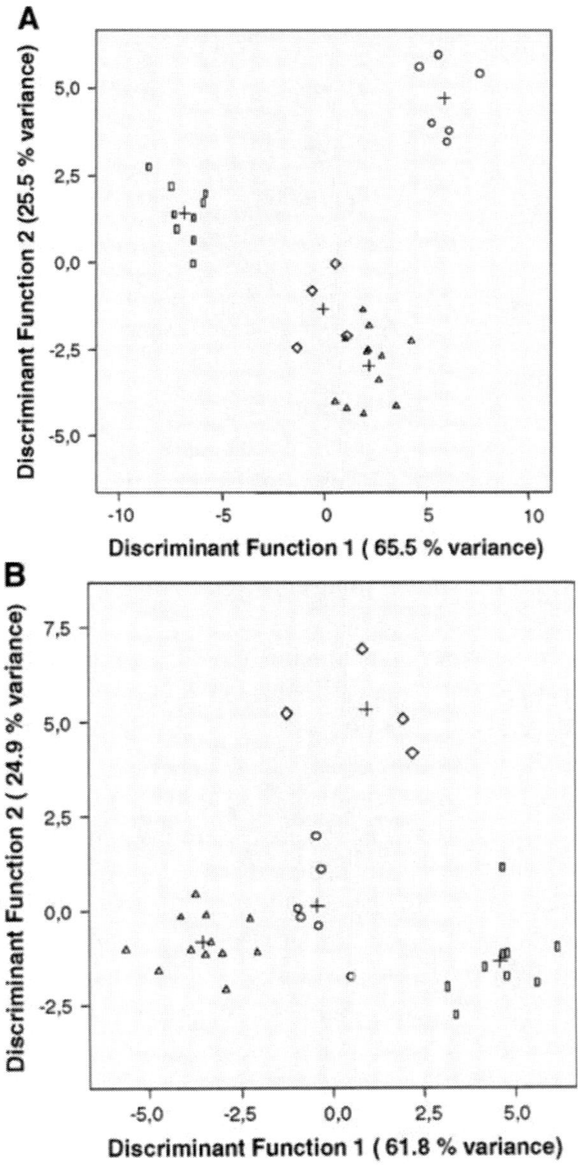

Figure 1. (Continued).

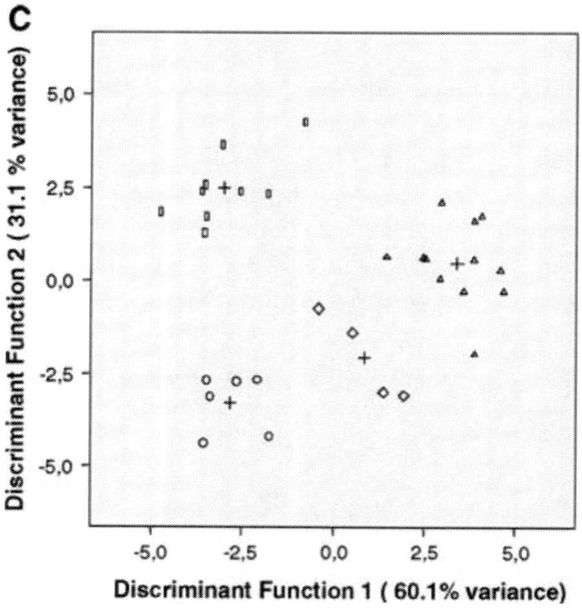

Figure 1. Mean scores of almond cultivars for the two discriminant functions at different stages of the oxidative process: 0 (A), 11 (B) and 20 (C) days of treatment.+ signs Centroid, Δ Marcona, □ Guara, ◊ Garrigues, ○ Butte.

When comparing the fatty acid profile for the four almond cultivars, the different behaviour of Butte samples is worth highlighting, as there was a greater drop from the initial concentration of linoleic acid when compared to the other three cultivars. On day 20 of the process, 90% of the linoleic acid in the Butte cultivar had degraded.

This percentage was lower in samples of Marcona (78%), Garrigues (77%) and Guara (74%) cultivars. This feature could be used to help differentiate this cultivar from the other three evaluated.

In order to obtain a separate classification for each of the four almond cultivars, a discriminant analysis was applied to the data collected on days 0, 11 and 20 of treatment. This analysis was conducted stepwise using the Wilk's lambda statistic for variable selection. The discriminant analysis led to three discriminant functions for all the treatment days selected. In all cases, with the first two discriminant functions, 85% of the total variance was retained.

Figure 1 shows the mean scores for the almond cultivars projected on the reduced space of the first two discriminant functions at different stages of the oxidation process. By using the calculated discriminant functions, samples were correctly classified in 100% of the cases using the data from days 0 or 20. Samples from day 11 were correctly classified 97% of the time. In this case, one sample was misclassified, causing the lower correct classification value. As the oxidation progressed; the differences between the four almond cultivars became more obvious. As a result, at the end of the oxidative treatment, the four almond cultivars appeared totally separated in the reduced space formed by the first two discriminant functions.

The first discriminant function was predominantly influenced by palmitoleic, oleic and linoleic acids; being the second discriminant function predominantly determined by stearic acid. The third discriminant function was mainly influenced by oleic and palmitic acids. Butte cultivar could be differentiated using the initial linoleic acid content and the drop in the concentration of this fatty acid as the oxidative process progresses. However, the other three cultivars could not be differentiated based on this parameter alone. By using the discriminant analysis, based on the major fatty acid content, all four cultivars could be differentiated.

ATR-FTIR Study

FTIR spectra of the analyzed almond oils showed typical characteristic absorption bands of common vegetable oils. Figure 2 shows, as an example, the FTIR spectrum obtained for Butte almond oil at the beginning of the oxidative treatment (day 0). Similar spectra were found for all the studied cultivars as all almond oils had similar composition. As the oxidation process progresses, several changes took place in the FTIR spectrum of almond oil. Figure 2 also shows spectra recorded on days 5, 11, 15 and 20, which were manually shifted by adding a constant value to the actual absorbance value (i.e. by adding 0.2, 0.4, 0.6 and 0.8, respectively) of the spectra. The main changes observed appeared at the following regions of spectra:

a) 3700 – 3150 cm^{-1}: this region remained unaltered for the first 5 days of exposure of the sample under the degradation conditions. After this period, the profile of this region began to change. Oxidation of almond oil led to a wider and more intense absorption band in this region compared with the raw almond oil spectra. In addition, the frequency of this absorption band decreased slowly as the process progressed (from 3470 to 3466 cm^{-1}). This decrease in the frequency value was due to the overlap with the original band (3470 cm^{-1}) caused by the overtone of the glyceride ester carbonyl absorption and new absorptions due to the hydroperoxides or primary oxidation products generated in the oxidation process. In advanced stages of oxidation, the frequency of this band increased towards values similar to those of the original band in the raw samples (3467 – 3468 cm^{-1}). This can be attributed to the decrease in hydroperoxide concentration and the appearance of new bands due to secondary oxidation products such as alcohols or acids, which overlap with those of the hydroperoxide groups. Secondary oxidation products also contributed to the morphology of this band. As such, changes in this band throughout the oxidative treatment were only able to provide us qualitative information.
b) 3010 – 2999 cm^{-1}: the absorbance of the band at approximately 3006 cm^{-1} (due to the stretching vibration of CH cis-olefinic groups) remained unaltered for approximately 5 days while the sample was under oxidative conditions. Oils with absorbance values lower than the original ones for this band were in advanced stages of oxidation. While the oxidative treatment was taking place, double bonds of unsaturated fatty acids underwent isomerization from cis to trans. This explained the gradual disappearance of this band as shown in Figure 2.
c) 1800 – 1700 cm^{-1}: the band, which appeared at approximately 1743 cm^{-1}, was due to the ester carbonyl functional group of the triacylglycerides. This band underwent an asymmetric modification throughout the oxidative treatment. Having remained practically unchanged for the first 5 days of the oxidative treatment, its frequency value began to decrease (from 1743 to 1741 cm^{-1} approximately) at different rates

depending on the oil. This change could be associated with the appearance of saturated aldehyde functional groups of secondary oxidation products, causing an absorption band at 1728 cm^{-1}, which overlaps the band for the ester functional group. As a result of this overlap, the width of this band increased as the oxidative treatment progressed.

d) 1000 – 900 cm^{-1}: the absorbance of the band originally at 967 cm^{-1}, associated with bending vibrations of CH functional groups of isolated trans-olefins, increased at different rates depending on the oil. This band provided us information about the gradual cis to trans isomerization throughout the oxidative treatment.

The extent of the changes in these regions was related to differences in almond oil composition. As such, some specific features could be selected to compare the behaviour of the cultivars throughout the oxidation process: the absorbance values of the bands appearing at 3468, 3006 and 967 cm^{-1} and the width and wavenumber shift of the 1743 cm^{-1} band.

In Figure 3, the shift in the 1743 cm^{-1} band and changes in the absorbance at 3006 cm^{-1} during the oxidative treatment are presented. At the beginning of the oxidative treatment (data for days 0 and 5), no significant differences were found between the almond cultivars in the position of the 1743 cm^{-1} band (Figure. 3A).

However, when the oxidative treatment progressed (data for days 11, 15 and 20) differences between Butte cultivar and the other ones began to appear.

The shift in this band was related to the production of some oxidation products (i.e. the more the oxidation process is advanced, the lower wavenumber value of this band will be). This different behaviour in Butte oil is an indication of its lower resistance to oxidative deterioration.

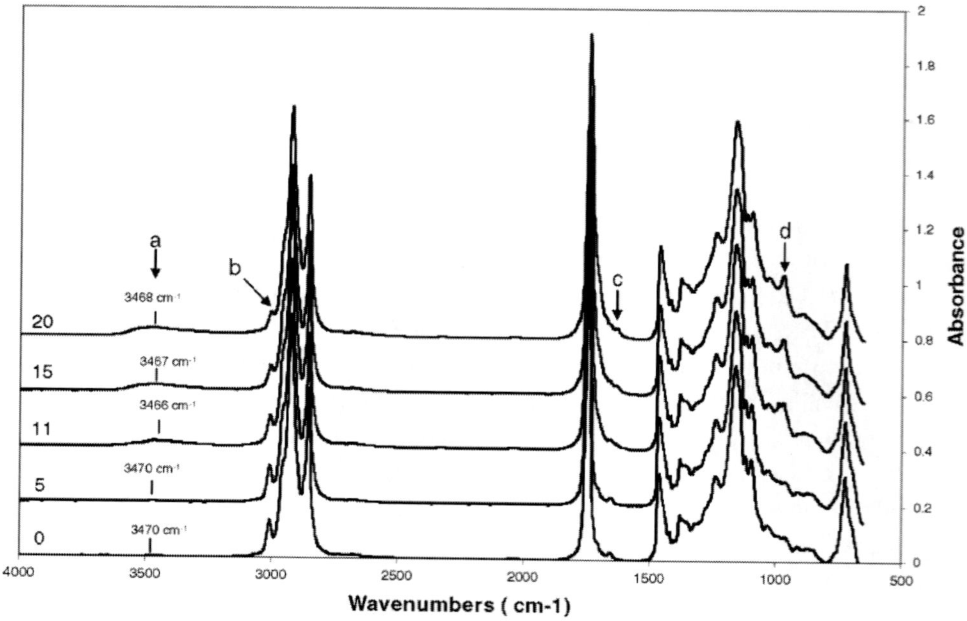

Figure 2. Fourier transformed infrared spectra of Butte almond oil after 0, 5, 11, 15 and 20 days under oxidative conditions.

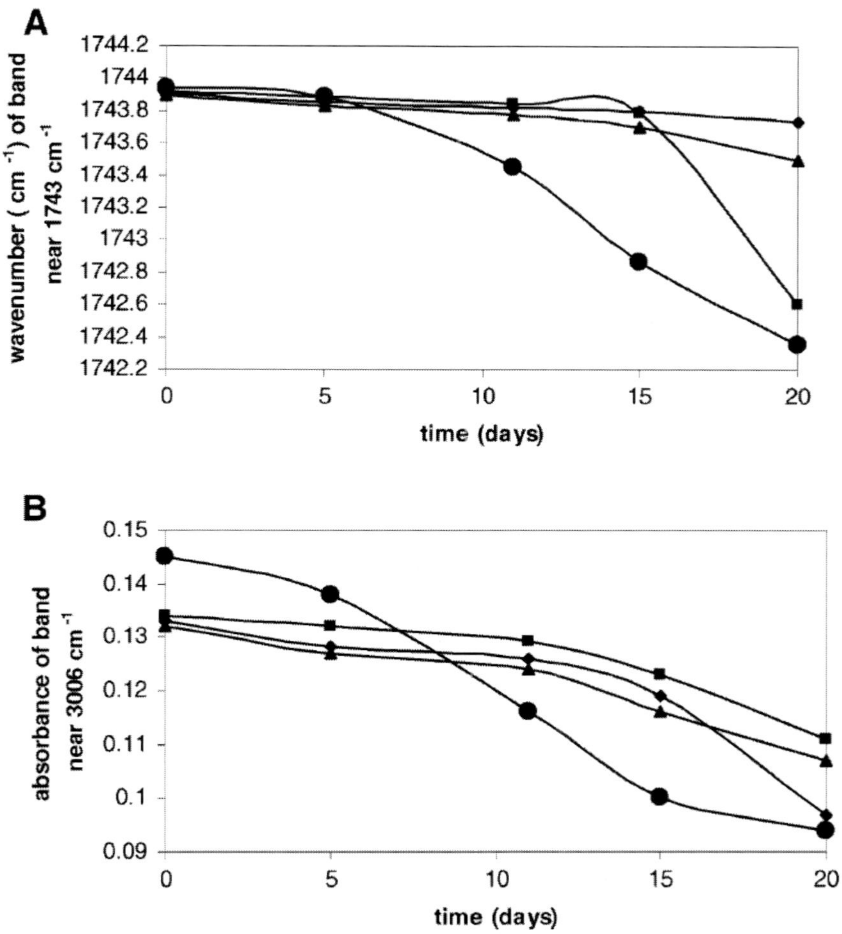

Figure 3. Changes in the selected features of the FTIR spectra of the four almond cultivars at different days of the oxidation process. ▲Marcona, ■ Guara, ♦ Garrigues, and ● Butte. A) Shift of 1743 cm^{-1} band. B) Absorbance of 3006 cm^{-1} band.

Previous studies had showed that the initial absorbance value of the 3006 cm^{-1} band can be related to the composition of the oil. Almonds with a large proportion of polyunsaturated acyl groups have higher absorbance values than those with a lower amount. Since Butte almond oil had the highest linoleic acid content at the beginning of the oxidative treatment, a higher initial absorbance value of this band was expected [5]. This fact was experimentally confirmed by data shown in Figure 3B.

As the oxidation progressed, a decrease in the absorbance value was registered for all cultivars, although Butte cultivar showed the greatest variation of this parameter. This trend can be attributed to the higher oxidation rate of Butte oil. It is interesting to note that there were also some differences in the absorbance at 3006 cm^{-1} between Guara, Garrigues and Marcona on day 20 of treatment.

As in the case of GC data, a linear discriminant analysis was carried out in order to classify the studied samples in different groups according to the different almond cultivars. Again, three discriminant functions were obtained using the variable selection rule for minimizing Wilk's lambda.

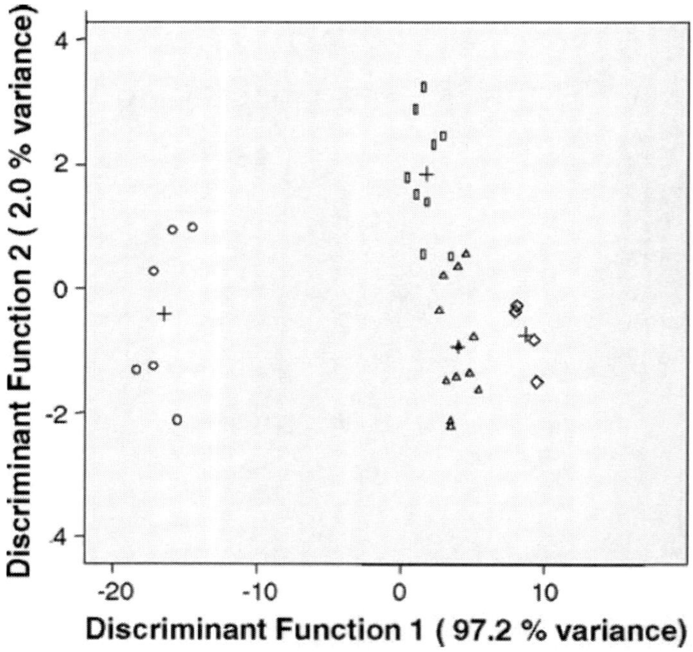

Figure 4. Mean scores for almond cultivars for the first two discriminant functions after 20 days under oxidative conditions. + signs Centroid, Δ Marcona, □ Guara, ◊ Garrigues, ○ Butte.

For day 20, the three discriminant functions account for 97.2, 2.0 and 0.8% of the total variance, respectively. The scores for the variables in the first discriminant function were -0.651, 0.533, 1.188 and -0.857 for the 3006 cm^{-1} region maximum absorbance; 1743 cm^{-1} band width, 1743 cm^{-1} band wavenumber and 967 cm^{-1} region maximum absorbance, respectively.

It is important to note that the data for the maximum absorbance at 3468 cm^{-1} is not included in the discriminant functions after the stepwise selection of variables. Figure 4 shows the scores for the almond cultivars projected on the reduced space of the first two discriminant functions obtained after 20 days under oxidative conditions. It is very interesting to note that Spanish cultivars showed great differences compared with the American one.

However, there were not a clear separation between Marcona, Guara and Garrigues samples. By using the calculated classification functions, all the samples were correctly classified into their groups except for one Guara, which was misclassified in the Marcona group.

Optimization of HS-SPME Process

The effect of main variables affecting the HS-SPME process in Butte almond oil was studied by performing the experiments shown in Table 3. Different volatile compounds were detected depending on the experiment performed.

The influence of each experimental factor on the extraction of each volatile compound was evaluated by means of standardized Pareto diagrams.

Table 3. Influence of factors and interactions on the response of nine main volatile compounds from Butte almond oil

Volatile compound	Standardized Effects [a]						
	A	B	C	AB	BC	AC	ABC
1,3-dimethylbenzene	ns	ns	ns	+	ns	ns	ns
Nonane	ns	ns	ns	ns	ns	ns	ns
Octanal	ns	ns	ns	+	-	-	ns
Nonanal	ns	+	ns	+	ns	ns	ns
Dodecane	ns	+	+	+	+	ns	+
Nonanoic acid	ns	+	+	ns	+	ns	ns
2-undecenal	ns	+	ns	ns	ns	ns	ns
Dodecanal	ns	ns	ns	ns	ns	ns	ns
Tetradecanal	ns	ns	ns	ns	ns	ns	ns

[a] ns: no significant, + positive effect, - negative effect. A, B, C are the factors shown in table 1.

Table 3 shows the influence of each experiment factor and interactions on the response (peak areas) of nine main volatile compounds detected: 1,3-dimethylbenzene, nonane, octanal, nonanal, dodecane, nonanoic acid, 2-undecenal, dodecanal and tetradecanal.

For the nine volatile compounds presented in Table 3, it was observed that the effect of stirring speed (A) was not significant. The extraction time (B) showed to be significant with a positive influence on the chromatographic responses of nonanal, dodecane, nonanoic acid and 2-undecenal. As expected, the positive effect of extraction temperature (C) was higher for less volatile compounds as dodecane and nonanoic acid, probably due to an increase of their transfer to the headspace. It is important to note that negative significant effects have been observed only in the interactions BC and AC of octanal. This fact can be attributed to the extraction temperature, a factor present in both interactions BC and AC.

In HS extraction mode, an increase in the sample temperature increases the volatility of the analytes and finally the headspace sensitivity but with SPME this effect is counteracted by simultaneously lowering the concentration in the fibre coating due to the fact that adsorption is an exothermic process [35].

On the basis of these results, an extraction temperature of 60 °C and an extraction time of 60 min were selected as optimal conditions. This value of extraction time was selected as being significant for four of the nine studied compounds (nonanal, dodecane, nonanoic acid and 2-undecenal) and the presence of positive significant interactions between the extraction time and other factors (as seen in Table 3).

In spite of no significant positive effects were observed for stirring speed on any of the volatile compounds evaluated, 110 rpm was chosen as optimal stirring speed due to the positive significant interactions observed between this factor and the rest of variables for some of the compounds studied like dodecane.

Analysis of Major Volatile Compounds in Almond Oil Samples

The same main volatile compounds were detected in all almond oil samples after analysis by HS-SPME-GC-MS as expected, since differences among cultivars are lower than those presented from different nuts. The presence of straight chain alkanals was detected in all oils (octanal, nonanal, decanal, dodecanal and tetradecanal). In general terms, it can be said that volatile content was higher for Spanish almond cultivars (Marcona and Guara) in comparison with Butte samples, as it can be seen in Figure 5.

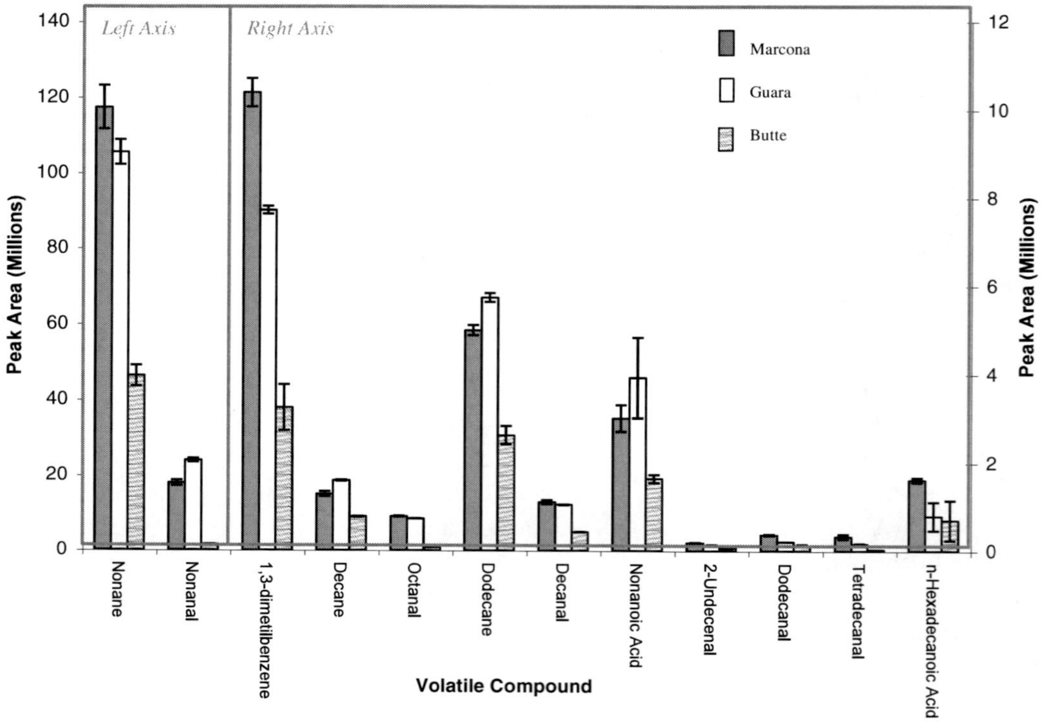

Figure 5. Comparison of peak areas obtained for twelve volatile compounds found in almond oils from different almond cultivars: Marcona, Guara and Butte (60 ºC, 60 min, 110 rpm).

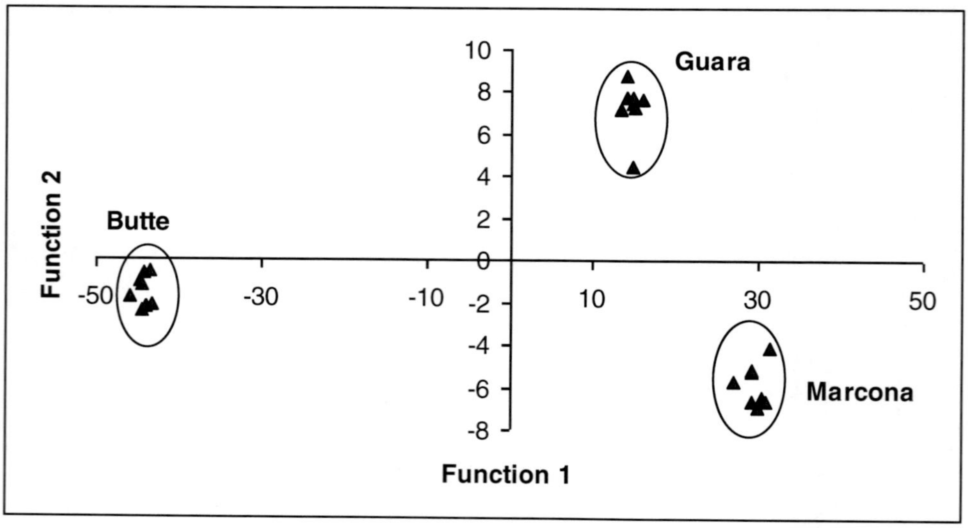

Figure 6. Projections of cultivar scores on the space determined by the two discriminant functions.

Based on this information, a discriminant analysis was applied to individual values of volatile compounds according to a stepwise method with the following three groups consisting each of a single cultivar: Marcona (group 1), Guara (group 2) and Butte (group 3). Two discriminant functions were obtained using the variable selection rule for minimizing Wilks´ lambda. The variance explained by the two discriminant functions was 92.2 and 7.8 %, respectively. Projections of cultivars scores on the two determined discriminant functions are shown in Figure 6. The variables octanal, nonanal, tetradecanal, nonanoic acid, decane and nonane were included in the analysis. The first discriminant function was influenced predominantly by nonanal, tetradecanal and nonanoic acid content. By using the calculated discriminant functions, samples were correctly classified in 100 % of the cases.

Analysis of Major Volatile Compounds in Oxidized Almond Oil Samples

Volatile compounds of oxidized almond oil samples were extracted and analyzed by HS-SPME-GC-MS. Among the detected compounds there were alkanals, 2-alkenals, 2,4-alkadienals, ketones, acids, esters, alcohols, as well as aliphatic and aromatic hydrocarbons. Seven main aldehydes were selected for oxidative stability evaluation: hexanal, (*E*)-2-heptenal, (*E*)-2-octenal, nonanal, (*E*)-2-nonenal, (*E,E*)-2,4-nonadienal and (*E,E*)-2,4-decadienal. These compounds are associated with rancidity of almond oils and have been usually reported in previous studies [8]. Hexanal is directly related to the development of oxidative off-flavours, it has a low odour threshold and is considered an indicator of oil quality [11].

This compound is derived from 13-hydroperoxide, which is one of the major hydroperoxides formed by autoxidation of linoleic acid whereas nonanal is derived from oleic acid [36]. From the unsaturated aldehydes, (*E*)-2-heptenal and (*E*)-2-octenal are the major volatiles formed during oxidation and they are produced by decomposition of the most abundant linoleic acid hydroperoxides [37], (*E,E*)-2,4-decadienal and (*E*)-2-nonenal content derive from oxidation of linoleic acid [38].

Figure 7 shows oxidative volatile profiles obtained for samples by considering the seven studied aldehydes content vs time of thermal treatment. As it can be seen, between the first and the fifth day of thermal treatment, (*E*)-2-heptenal was the main volatile compound formed in Butte sample. On the other hand, (*E, E*)-2,4-decadienal was the major compound found for Marcona sample at the same period of time.

From the seventh to tenth day of thermal treatment, the differences highlighted above between Spanish and American almond oils were found in seven volatile compounds selected. After fifteen days of oxidative treatment, it is important to note that oil samples from Butte and Marcona cultivars presented very similar behaviour related to hexanal, (*E*)-2-nonenal, (*E, E*)-2,4-nonadienal, (*E*)-2-octenal and nonanal content.

In contrast, significant differences between the two studied almond cultivars were observed for (*E*)-2-heptenal and (*E, E*)-2,4-decadienal content. Finally, at the end of the thermal treatment (20 days), significant differences could be observed for hexanal (*E*)-2-nonenal, (*E*)-2-octenal, nonanal and (*E, E*)-2,4-decadienal content between the two studied almond cultivars. On the other hand, the content of (*E*)-2-heptenal and (*E, E*)-2,4-nonadienal were very similar for both samples. Based on the data and conclusions obtained for the selected aldehydes in Figure 7, seven days were selected as the optimum time to carry out oxidative treatment for the studied samples since significant differences were obtained relative to aldehydes content and it was a reasonable short period of analysis time.

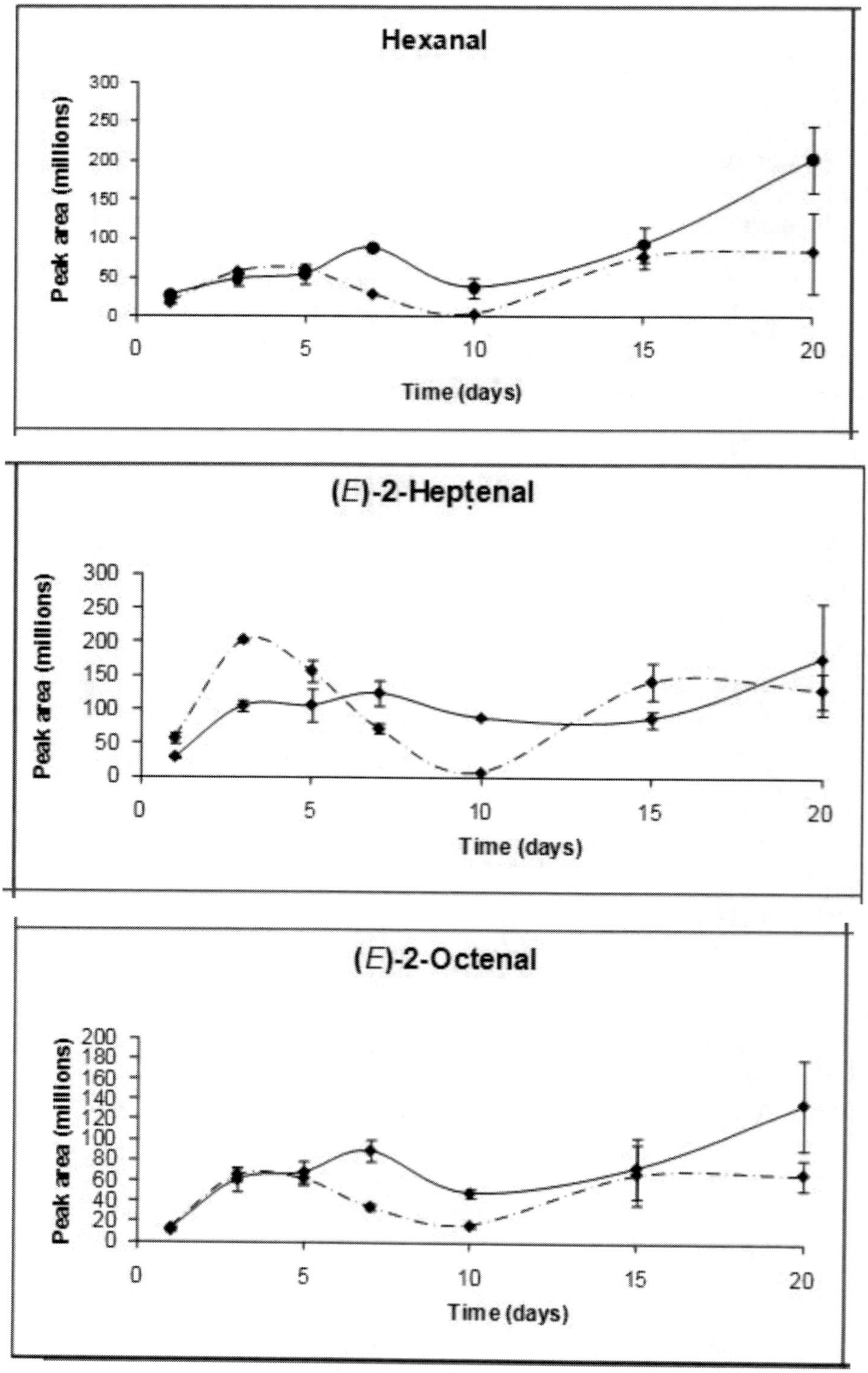

Figure 7. (Continued).

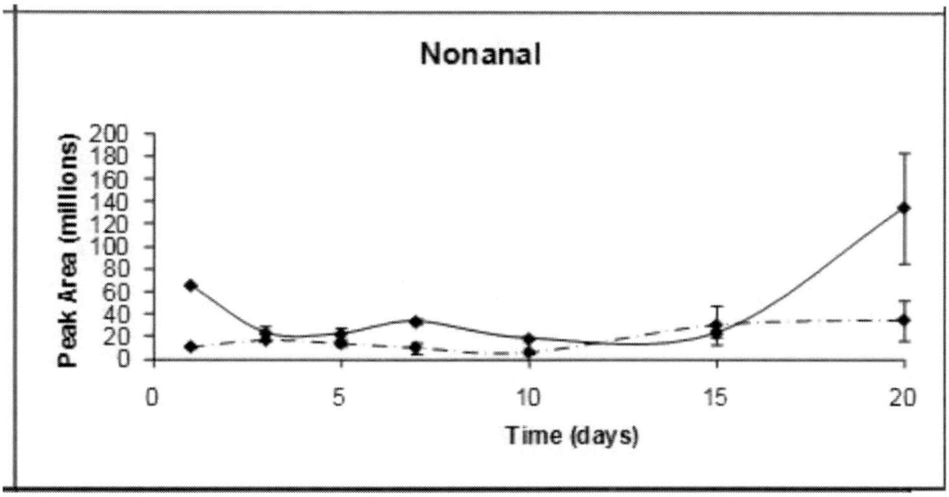

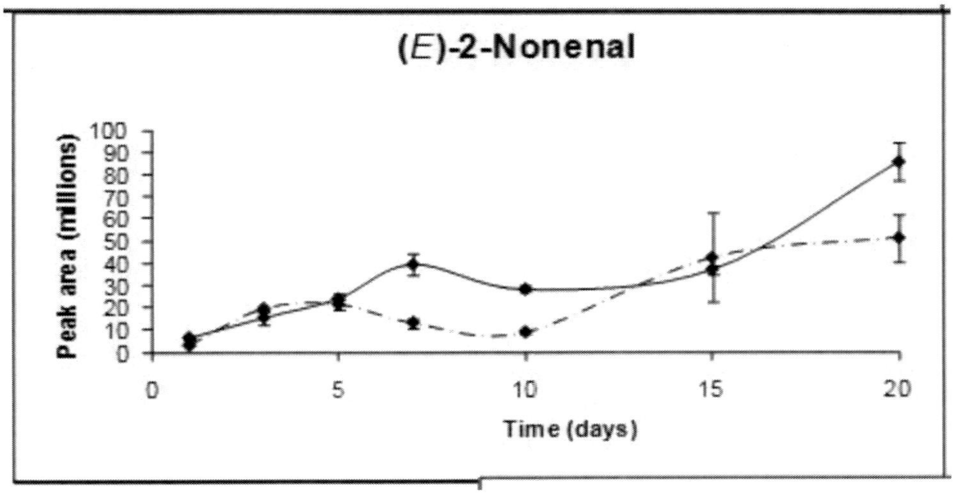

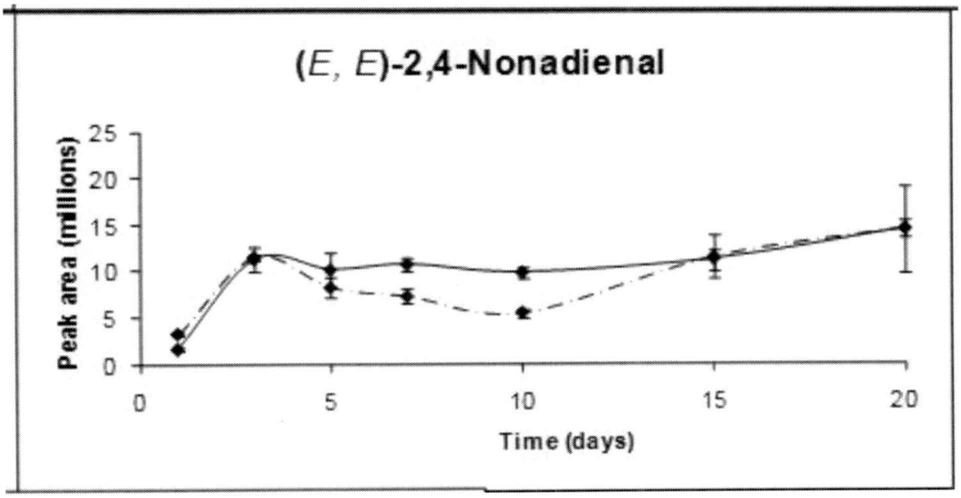

Figure 7. (Continued).

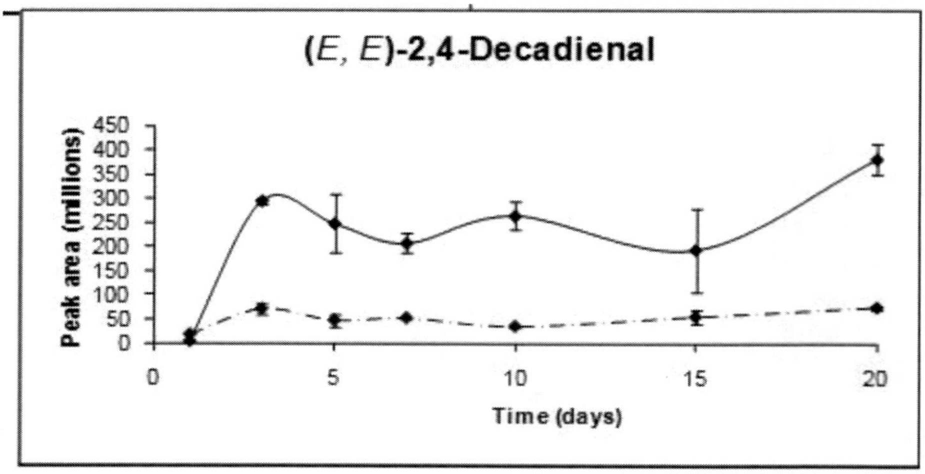

Figure 7. Oxidative volatile profiles obtained for Marcona and Butte samples by plotting the content of seven volatile compounds (hexanal, (*E*)-2-heptenal, (*E*)-2-octenal, nonanal, (*E*)--2-nonenal, (*E, E*)-2,4-nonadienal and (*E, E*)-2,4-decadienal) vs time of thermal treatment at 100 ºC. _._._.. Butte, ______ Marcona.

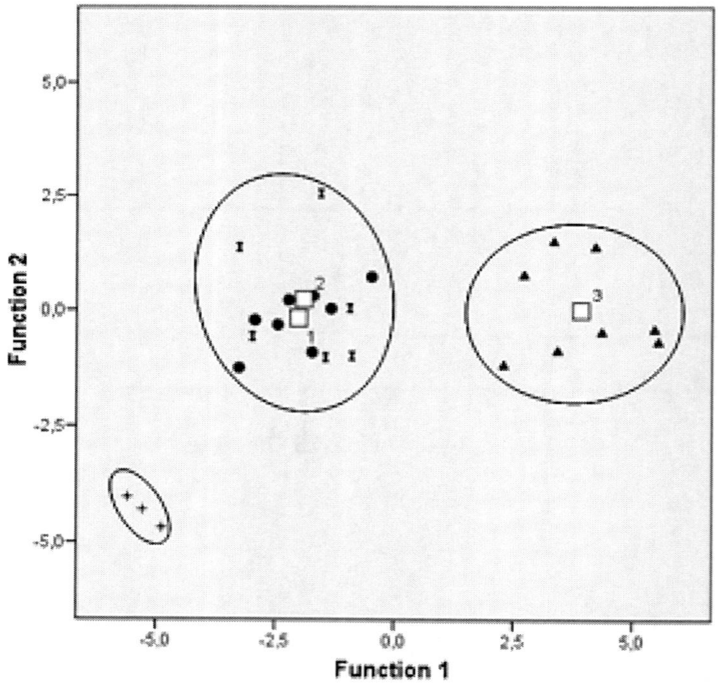

Figure 8. Projections of almond cultivar scores on the space determined by the two discriminant functions (● signs Marcona,⧗Guara, ▲Butte, + Not grouped Guara samples, □ Centroid).

The decrease in headspace volatile compound content observed from 5 to 10 days of thermal treatment may be due to the depletion of the headspace oxygen in the closed dark vial where the almond oil sample is contained. This depletion of the headspace oxygen is due to the reaction between unsaturated fatty acids and oxygen.

As the oxidative treatment progresses, an increase in volatiles content is observed. This fact can be attributed to an increase in the headspace oxygen content generated by chemical degradation of the oil components containing the peroxyl moiety.

In order to obtain discriminant parameters to differentiate Spanish and American almond cultivars, HS-SPME-GC-MS and LDA techniques were applied. In this sense, the next part of the study was the analysis of selected main volatile compounds in all twenty-four oxidized almond oils (Marcona, Guara and Butte) after seven days of thermal treatment, for determining the oxidized volatile profile of samples belonging to different almond cultivars. Comparing the studied almond cultivars, it was found that Butte cultivar had the highest (*E*)-2-heptenal, (*E*)-2-octenal, (*E, E*)-2,4-decadienal and (*E, E*)-2,4-nonadienal content.

These results agreed with data obtained in previous studies that confirm high linoleic acid content in samples belonging to Butte cultivar. This fact could explain the higher (*E*)-2-heptenal and (*E*)-2-octenal content as these volatile compounds are produced by decomposition of the most abundant linoleic acid hydroperoxides. The high (*E, E*)-2,4-decadienal content could be due to the formation of this compound during the oxidation of linoleic acid [5]. On the other hand, no statistical differences between the Spanish cultivars Marcona and Guara were found, as these almond cultivars present very similar composition related to the fatty acid content. Discriminant analysis of all individual data within every previously established group lead to two discriminant functions which account for 99.6 and 0.4 % of the total variance, respectively. The variables included in the stepwise linear discriminant analysis were (*E*)-2-heptenal and (*E*)-2-nonenal content. The variables hexanal, (*E*)-2-octenal, nonanal, (*E, E*)-2,4-nonadienal and (*E, E*)-2,4-decadienal were not included in the statistical analysis. Figure 8 shows a plot of individual data on the plane determined by the two discriminant functions whereas cultivars appear associated suggesting three groups. Clearly, the largest group included many of the Spanish almond samples belonging to Marcona and Guara cultivars. On the other hand, the American cultivar Butte was classified in a separated group. This differentiation between Spanish and American almond cultivars can be useful to avoid adulteration practices.

DSC Analysis of Almond Oil Samples

The same thermal behavior was observed for all almond oil samples by DSC analysis as expected, since differences among cultivars are lower than those presented from different nuts. As an example, DSC curves obtained for Guara Pinoso almond oil at different cooling and heating rates are shown in Figure 9. As expected, the DSC profile of this oil showed an exothermic peak at temperatures between -50 and -30 ºC (depending on the program rate) corresponding to the oil crystallization and an endothermic peak at around -10 ºC related to the melting of the previously formed crystals. It was also observed that as the cooling rate increased, the crystallization temperature shifted to lower values. This result was expectable since high cooling rates lead to crystallization of a certain amount of the fat in a less stable form at a lower temperature [22].

The use of low cooling rates gave more time to allow interactions between triacylglycerols, leading to complete crystallization at narrow temperature ranges.

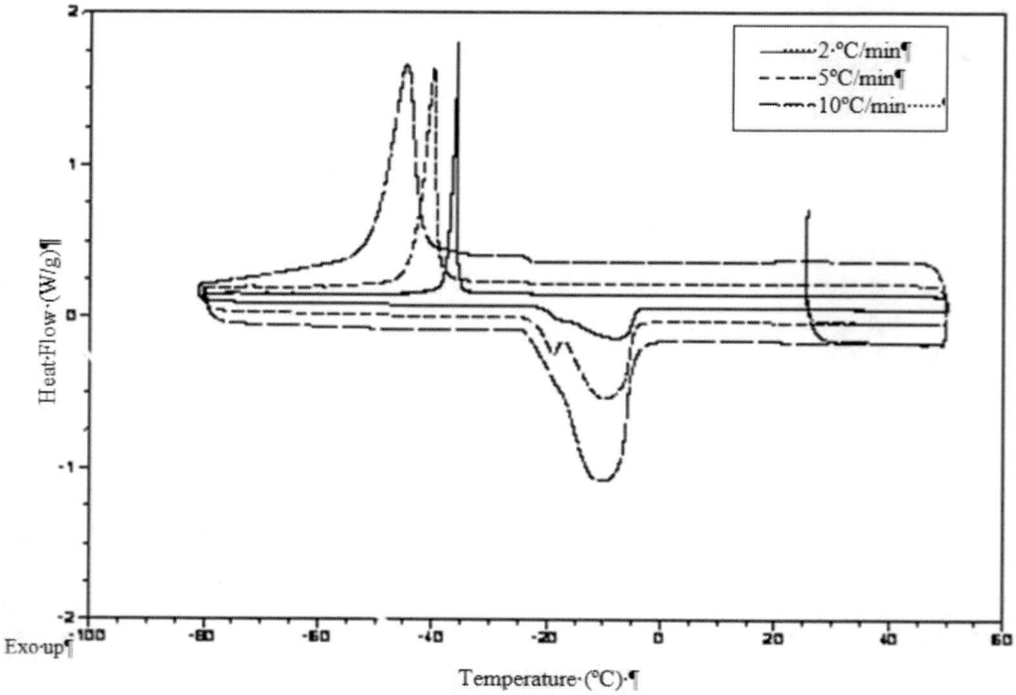

Figure 9. DSC curves obtained at different program rates (2, 5 and 10 ºC min^{-1}) for Guara Pinoso oil sample in nitrogen atmosphere (50 mL min^{-1}).

Table 4. Crystallization and melting parameters obtained from different almond oils at three program rates (2, 5 and 10 °C min^{-1})

Almond Cultivar	Analysis rate (ºC min^{-1})	DSC Parameters[a,b]			
		T_c (ºC)	ΔH_c (J/g)	T_m (ºC)	ΔH_m (J/g)
Marcona	10	-46.89	57.08	-11.61	65.45
Butte	10	-50.67	52.64	-12.56	61.21
Guara	10	-45.06	56.69	-10.59	64.59
Marcona	5	-42.32	58.01	-10.88	65.33
Butte	5	-45.67	52.06	-12.63	59.53
Guara	5	-40.67	56.53	-9.92	63.74
Marcona	2	-37.68	59.79	-8.94	58.95
Butte	2	-40.76	58.91	-11.50	58.29
Guara	2	-35.91	63.96	-7.75	62.28

[a]T_c (ºC), crystallization temperature; ΔH_c (J g^{-1}), crystallization enthalpy; T_m (ºC), melting temperature; ΔH_m (J g^{-1}), melting enthalpy. [b]Mean value (n=8).

The well-known effect of the poor thermal conductivity of oils was also observed in DSC results, since heating rate also influenced the sharpness of endothermic peaks [39]. In this sense, the endothermic melting peaks were more pronounced in curves with high heating rates, as it has been reported elsewhere [23].

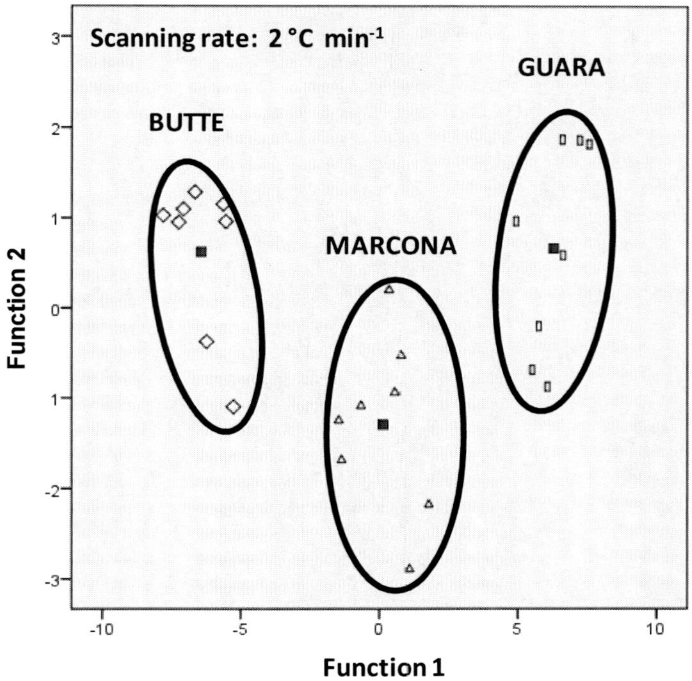

Figure 10. DSC mean values for the two discriminant functions obtained from the analysis of oil samples.

As it can be seen in Table 4, Butte samples showed lower values for all parameters, i.e. crystallization temperature (T_c), crystallization enthalpy (ΔH_c), melting temperature (T_m) and melting enthalpy (ΔH_m), at all testing rates. These results can be attributed to the different fatty acid profiles between Butte and the Spanish almond cultivars. In this sense, it was observed that Butte cultivar shows higher linoleic acid content making it more prone to lipid oxidation. On the other hand, all Spanish almonds showed similar DSC patterns and no conclusions for tendencies in their thermal parameters could be obtained. Using a stepwise LDA with the Wilk's lambda statistics for variable selection can carry out a deeper study for classification of almond cultivars from DSC results. In this case, the application of this statistical analysis led to two discriminant functions where 100 % of the total variance was retained. Melting temperature was the only variable not included in the analysis since differences in results for different oil samples were not high enough to get clear results. Figure 10 shows the mean scores obtained for DSC results at 2 ° C min^{-1}, projected on the reduced space of the two discriminant functions. Similar results were obtained for each testing rate in all DSC runs. In conclusion, samples were correctly classified in 100% of the cases by using the calculated discriminant functions obtained from DSC results.

TGA Analysis of Almond Oil Samples

Thermal stability behaviour of all almond oils used in this work was studied by TGA. Figure 11 shows an example of TGA curves obtained for Butte oil at 10 °C min^{-1}. In this case, the almond oil was stable up to a temperature of 289 °C where the mass loss reached 5 %.

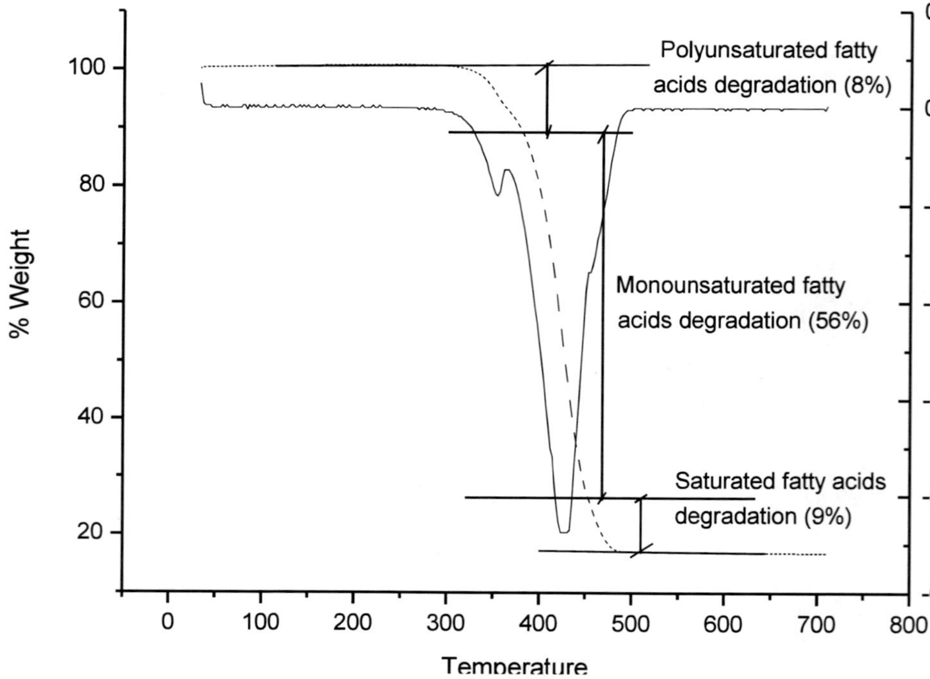

Figure 11. Thermogravimetric profile (———) and first derivative curve (_ . _ . _ ..) of Butte almond oil at 10 ºC min^{-1} in nitrogen atmosphere (50 mL min^{-1}).

The relatively high stability of almond oil when compared to other vegetable oils [40] could be attributed to the high amount of saturated fatty acids in its composition (ranging from 7 to 9%) and to the presence of α-tocopherol as a major component [6]. A typical profile of sample weight loss with temperature occurring at different stages was observed. The first step, starting at 289 ºC and reaching its maximum degradation rate at 352 ºC, represents the initial phase of triglycerides degradation where the oxidation of the poly-unsaturated fatty acids takes place.

The second and third stages, i.e. the main degradation with maximum rate at around 400 ºC and a small shoulder observed in the DTGA curve at around 470 ºC, represent the decomposition of monounsaturated and saturated fatty acids, respectively.

Significant differences at the 5% level were observed in TGA results for almond oil samples from different cultivars, being more important when the analyses were carried out at the highest heating rate (10 ºC min^{-1}). In this case, linear statistical analysis led to two discriminant functions. The percentage of polyunsaturated, monounsaturated and saturated fatty acids obtained from the relative weight loss in TGA runs and T_i for the polyunsaturated fatty acids peak were included in this analysis.

Figure 12 shows the mean scores for all almond cultivars, projected on the reduced space of the two discriminant functions. 100% of the total variance was retained and the samples were correctly classified in 91.7 % of the cases.

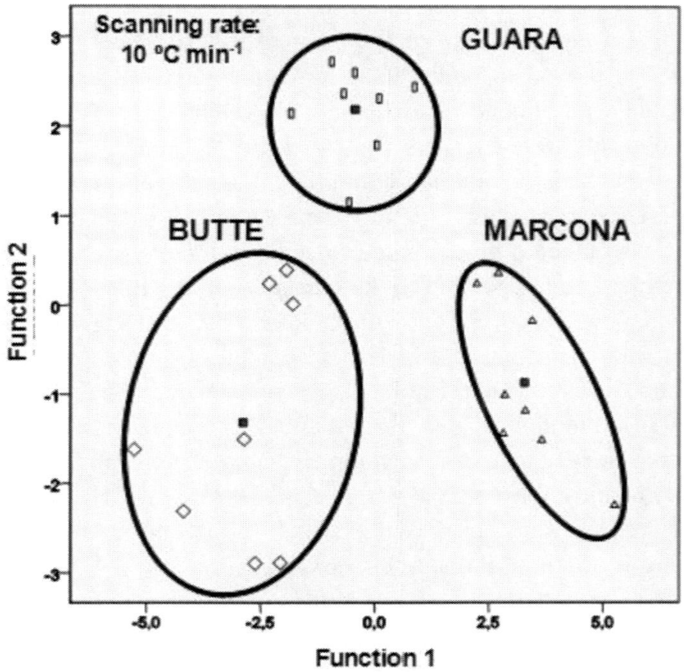

Figure 12. TGA mean values for the two discriminant functions obtained from the analysis of oil samples.

CONCLUSIONS

In conclusion, the results presented in the present chapter have shown that FAME content determined by GC-FID and variation in ATR-FTIR spectra of almond oils can be used to monitor the progress of the oxidation process. By applying multivariate data treatment to FAME and ATR-FTIR obtained results, differences between the cultivars were highlighted. In this way, Butte cultivar showed a lower oxidative stability, which could, in part, be attributed to the fact that it had the highest linoleic acid content at the beginning of the process. On the other hand, HS-SPME–GC–MS has proved to be a valuable method for the extraction and detection of a large range of volatile compounds in almond oils.

It has shown itself to be a rapid, repeatable and sensitive technique. DVB/CAR/PDMS was employed as the extracting fiber and a 2^3 factorial design demonstrated that temperature and extraction time were the parameters having significant influences on the efficiency of the extraction process. A total of 22 volatile compounds were detected in almond oils being the main chemical group present the straight chain alkanal (octanal, nonanal, decanal, dodecanal, tetradecanal). From the cultivars studied, American Butte showed a lower nonanal content, which could be attributed to its lowest oleic acid content in comparison with Spanish almond cultivars.

Finally, the use of multivariate techniques applied to the obtained volatile compounds profile has been successful for the classification and discrimination of different almond cultivars.

The results obtained related to the oxidised volatile compounds determined by HS-SPME/GC-MS can be used to monitor the progress of the oxidation process of almond oils. HS-SPME/GC-MS has proved to be a valuable methodology for determination of volatile compounds associated with almond oil oxidation. Highlighted differences between Spanish and American almond oils were found in samples kept under thermal treatment for 7 days, especially for the seven aldehydes studied. From the three cultivars analysed, American Butte showed the highest (E)-2-heptenal, (E)-2-octenal, (E,E)-2,4-decadienal and (E,E)-2,4-nonadienal content, which could be attributed to its high linoleic acid content in comparison with the Spanish almond cultivars. Due to the content and low odour thresholds of some of the aldehydes found in oxidised almond oil samples (such as (E,E)-2,4-decadienal), it is possible that these compounds play an important role in the aroma of oxidised almonds.

Finally, DSC and TGA have proved to be valuable and reliable techniques for the characterization of different almond oils. Both thermal techniques have shown their inherent advantages, including small sample size, minimal sample preparation, short experimental times and no use of chemical reagents.

Significant differences were observed in crystallization and melting parameters for oil samples extracted from different almond cultivars, in particular for DSC tests carried out at low rates (2 °C min^{-1}). Thermal degradation profiles were obtained, with higher differences in TGA curves among almond cultivars observed at a 10°C min^{-1} heating rate. In conclusion, the use of multivariate LDA applied to DSC and TGA results has been successful for the classification and discrimination of different almond cultivars. The results reported in this chapter constitute a useful basis for the control of adulteration in food processing, particularly for PDO products.

ACKNOWLEDGMENTS

Authors would like to thank ''Almendras Llopis S.A'' and ''Colefruse S.A'' for kindly providing the almond samples.

REFERENCES

[1] O. Kodad, R. Socias, *J. Agric. Food Chem.*, 56, 4096 (2008).
[2] S.Yada, K. Lapsley, G. Huang, *J. Food Compos. Anal.,* 24, 469 (2011).
[3] A. Esfahlan, R. Jamei, R. Esfahlan, *Food Chem.*, 120; 349 (2010).
[4] M. Sulek, K. Pytlas, *J. Appl. Cosmetol.,* 28, 1 (2010).
[5] A. Beltrán Sanahuja, M. S. Prats Moya, S. E. Maestre Pérez, N. Grané Teruel, M. L. Martín Carratalá, *J. Am. Oil Chem.,* 86, 51 (2009).
[6] C. M. López Ortiz, M. S. Prats Moya, A. Beltrán Sanahuja, S.E.Maestre Pérez, N. Grané Teruel, M. L. Martín Carratalá, *J. Food Comp. Anal.,* 21, 144 (2008).
[7] Boletín Oficial del Estado (BOE). Orden de 22 de Marzo de 1996 por la que se ratifica el Reglamento de las Denominaciones Específicas "Jijona" y "Turrón de Alicante" y su Consejo Regulador Común (BOE núm. 98 de 23 de Abril). Madrid (Spain).

[8] L. Vázquez Araujo, D. Chambers, A. Carbonell Borrachina, *J. Sens. Stud.*, in press (2011).
[9] O. Beyhan, M. Aktas, N. Yilmaz,N. Simsek ,R. Gercekcioglu, *J. Med. Plant Res.*, 5, 4907(2011).
[10] L. Merrill, O. Pike, L. Ogden, M. Dunn, *J. Am. Oil Chem. Soc.*, 85, 771 (2008).
[11] S. F. Mexis, A. V. Badeka, M. G. Kontominas, *Innov. Food Sci. Emerg. Technol.,* 10, 580 (2009).
[12] M. D. Guillén, E. Goicoechea, G. Palencia, N. Cosmes, *J. Agric. Food Chem.*, 56, 2028 (2008).
[13] L. Sanchez Prado, S. Risticevic, J. Pawliszyn, E. Psillakis, *J. Photochem. and Photobiol. A: Chemistry* 206, 227 (2009).
[14] M. S. Guillen and E. Goicoechea, *Food Chem.*, 116, 183 (2009).
[15] G. Pignoli, R. Bou, M. T. Rodríguez, E. A. Decker, *Meat Sci.,* 83, 412 (2009).
[16] A. Beltrán Sanahuja, M. Ramos Santonja, N. Grané Teruel, M. L. Martín Carratalá, M. C. Garrigós Selva, *Food Chem.*, 126, 603 (2011).
[17] E. Narbona, E. García, L.Vázquez, A. Carbonell, *Int. J. Food Sci. Tech.*, 45, 569 (2010).
[18] J. J. Beck, B. S. Higbee, G. B. Merrill, J. N. Roitman, *J. Sci. Food Agric.,* 88, 1363 (2008).
[19] J. J. Beck, G. B. Merrill, B. S. Higbee, D. M. Light, W. S.Gee, *J. Agric. Food Chem.,* 57, 3749 (2009).
[20] S. F. Mexis, A. V. Badeka, E. Chouliara, K. A. Riganakos, M. G Kontominas, *Innovat. Food Sci. Emerg. Tech.*, 10, 87 (2009).
[21] P. Samyn, G. Schoukens, L. Vonck, D. Staussens, H. Van den Abbeele, *J. Therm. Anal. Calorim.,* in press (2011).
[22] Y. B. Che Man and C. P. Tan, *Phytochem. Anal.*, 13, 142 (2002).
[23] C. P. Tan and Y. B. Che Man, *Phytochem. Anal.,* 13, 129 (2002).
[24] S. Arain, S. T. H. Sherazi, M. I. Bhanger, F. N. Talpur, S. A. Mahesar, *Thermo. Acta.,* 484, 1 (2009).
[25] J. A. Solís Fuentes, G. Camey Ortíz, M. R. Hernández Medel, F. Pérez Mendoza, M. C. Durán de Bazua, *Bior. Tech.,* 101, 799 (2010).
[26] M. Van Aardt, S.E. Duncan, T. E. Long, S. F. O' Keefe, J. E. Marcy, S. R. Sims. *J. Agric. Food Chem.*, 52, 587 (2004).
[27] Almond Board of California, http://www.almondboard.com/.
[28] Official methods and recommended practices of the American Oil Chemists' Society, 4th edn. (1992) American Oil Chemists' Society, Champaign, Method Ce-2-66.
[29] N. Grané Teruel, M. S. Prats Moya,V. Berenguer Navarro, M. L. Martín Carratalá, *J. Am. Oil Chem. Soc.*, 78, 617 (2001).
[30] F. R. Van de Voort, A. A. Ismail, J. Sedman, G. Emo, *J. Am. Oil Chem. Soc.*, 71, 243 (1994).
[31] M. Adam, M. Juklova, T. Bajer, A. Eisner, K. Ventura, *J. Chromatogr., A* 1084, 2 (2005).
[32] A. Jiménez Márquez, G. Beltrán Maza, M. P. Aguilera Herrera, M. Uceda Ojeda, *Grasas y aceites,* 58, 122 (2007).
[33] B. Tabachnik and L. Fidell, Using multivariate statistics, 12th edn. Harper Collins Publishers, New York (1992).

[34] T. P. Coultate In: Food. The chemistry of its components, 4th edn. RSC, Cambridge (2007).
[35] D. A. Lambropoulou, I. K. Konstantinou, T. A. Albanis, *J. Chrom., A.* 1152, 70 (2007).
[36] A. Fullana, A. Carbonell Barrachina, S. Sidhu, *J. Sci. Food Agric.,* 84, 2015 (2004).
[37] M. C. García Martínez, G. Márquez Ruiz, J. Fontecha, M. H. Gordon. *Food Chem.,* 113, 926 (2009).
[38] J. Lee, D. H. Kim, P. S. Chang, *Food Chem.*, 105, 414 (2007).
[39] C. P. Tan and Y. B. Che Man, *Food Chem.*, 76, 89 (2002).
[40] S. P. Kochhar, C. J. K. Henry, *Int. J. Food Sci. Nutr.*, 60, 289 (2009).

In: Cultivars
Editor: Katya Carbone

ISBN: 978-1-62417-101-7

Chapter 9

DIFFERENT CULTIVARS OF THE SAME SPECIES OF PLANT MAY PRODUCE PROTEINS THAT DIFFER IN STRUCTURE AND FUNCTION

Tzi Bun Ng[1,*], Evandro Fei Fang[1,2] and Jack Ho Wong[1]
[1]School of Biomedical Sciences, Faculty of Medicine,
The Chinese University of Hong Kong, Shatin New Territories, Hong Kong, China
[2]Laboratory of Molecular Gerontology, National Institute on Aging,
National Institutes of Health, Baltimore, MD, US

ABSTRACT

The intent of this article is to review reports in the literature that lead to the observation that proteins with different structures and functions may be produced by different cultivars of the same species.

Examples include production of antifungal proteins with different N-terminal amino acid sequences and sometimes even specificity of antifungal action by green and gold kiwi fruits, different *Phaseolus vulgaris* (bean) cultivars, different *Allium sativum* (garlic) cultivars, and different *Pisum sativum* (pea) cultivars.

Serine protease inhibitors with different molecular masses, N-terminal amino acid sequences and trypsin/chymotrypsin specificity may be produced by different *Vicia faba* (broad bean) and *Glycine max* (soybean) cultivars.

Lectins/hemagglutinins with variations in N-terminal amino acid sequence, sugar specificity and presence/absence of biological activities such as mitogenic activity toward spleen cells, antiproliferative activity toward tumor cells, and inhibitory activity toward HIV-1 reverse transcriptase are found in different cultivars of *P. vulgaris*.

* Correspondence: Professor Tzi Bun NG, Choh Ming Li Basic Medical Sciences Building, The Chinese University of Hong Kong, Hong Kong. Fax: 852-26035123; Tel: 852-26098031; Email: b021770@mailserv.cuhk.edu.hk

INTRODUCTION

A cultivar is a plant or group of plants selected for desirable characteristics that can be maintained by propagation (1, 2). Different cultivars of the same plant species may have different appearances. For instance, red kidney beans, black kidney beans, brown kidney beans, pinto beans, and French beans are different cultivars of *Phaseolus vulgaris.* They differ in color, size and shape. Knowledge on the proteins produced by different cultivars of the same species has been accumulating. This chapter aims to review such information.

Actinidia Delicisa (Green Kiwi Fruit) and *A. Chinensis* (Gold Kiwi Fruit) Antifungal Proteins

The genus *Actinidia* Lindl. is composed of about 60 species, and two of them are common in the fruit market including *Actinidia delicisa* (green kiwi fruit) *and A. chinensis* (gold kiwi fruit) [3, 4]. *A. delicisa* is a Chinese-only species, originally named yang tao, and some commercial names including kiwi fruit, green kiwi fruit have come up. The wild *A. delicisa* is grown in the mountainous area of the Yangtze River, such as Huangshan city and Xuancheng city in Anhui Province. *A. delicisa* was carried to New Zealand in the early 20th century and then spread to become a fruit of worldwide significance. *A. chinensis* is a new species developed in New Zealand which has a less hairy peel and a sweeter taste than the green one [5]. Kiwifruit is of high health benefits (including rich in vitamin-C and other vitamins, soluble dietary fiber, and antioxidants) and has been extensively investigated [6].

A single-chain 21 kDa thaumatin-like protein (TLP) exhibiting antifungal activity against *Botrytis cinerea, Mycosphaerella arachidicola* and *Coprinus comatus* was isolated from green kiwi fruits. The sequence of the first 20 N-terminal acid residues are: ATFNFINNCPFTVWAAAVPG. A comparison of N-terminal sequence analysis among TLP and other medicinal proteins is shown in Table 1. Besides antifungal activity, TLP manifested anti-human immunodeficiency virus (HIV) potential as evidenced by a 30.6% inhibition of HIV-1 reverse transcriptase at a TLP concentration of 27 μM [7].

An antifungal protein designated actinchinin, with an N-terminal sequence different from that of TLP from green kiwi fruit. Unlike its counterpart was isolated from the gold kiwi fruit. It was active against *Fusarium oxysporum* which did not respond to TLP from green kiwi fruit-Unlike its counterpart from green kiwi fruit, actinchinin did not exert antifungal activity against *Botrytis cinerea* and lacked HIV-1 reverse transcriptase inhibiting activity [5]. This is explainable since TLP and actinchinin are two different proteins based on the following evidence. Firstly, the molecular weight of actinchinin (25 kDa) is larger than that of TLP which is 21 kDa. Secondly, the N-terminal sequences are different (see Table 1 for details). Thirdly, they exhibit different biochemical characteristics as supported by distinct chromatographic affinities and different antifungal as well as anti-HIV reverse transcriptase activities [5, 7].

Though kiwi fruit is of high nutritional value, there are some people who are allergic to this berry and the first report was described in 1981 [8]. It is estimated that there are around 12 allergens in kiwifruit with a characteristic of IgE-binding activity (9). For example, Pastorello and colleagues purified some Kiwi proteins, including proteins of 30, 28, 24, and

17 kDa, and found that the 30-kDa protein, with the name actinidin, is a major allergen of kiwi fruit [9]. Lately, another research group reported the isolation and characterization of another kiwi allergen, named thaumatin-like protein (10). This purified protein consisted of 2 isoforms with a molecular weight of 24 kD under reducing conditions. All 7 poly-sensitized patients with oral allergy syndrome (OAS) recognized this protein as an allergen. Furthermore, it expressed antifungal activity toward *Saccharomyces carlsbergensis* and *Candida albicans*. Though it also owns antifungal activity, it is distinct from actinchinin. Its first 8 N-terminal amN-terminal amino acid sequence is ATFNIINN which was different from that of actinchinin (GAKRAYDEVEAQN) (5, 10). Further in-depth studies on kiwifruit allergens may enable the allergic people to enjoy the fruit.

Allium Sativum (Garlic) Antifungal Proteins

Garlic (binominal name *Allium sativum*) is a very common product in the food market with both culinary and medicinal applications [11]. It is among the oldest of all cultivated plants and it has been speculated that there are over 600 cultivated sub-varieties of garlic all over the world [11, 12]. Garlic is classified into two subspecies, including hard-necked garlics (Ophioscorodon) and soft-necked garlics (Sativum), and comprises around ten distinct varieties (12). A majority of investigations have been carried out on the biomedical applications of garlic [11, 13, 14].

A 13-kDa protein designated as alliumin and isolated from multiple-cloved garlic bulbs, demonstrated an N-terminal sequence (DDFLCAGGCL) similar to a partial sequence of glucanase and antifungal activity against *Mycosphaerella arachidicola*, but not against *Fusarium oxysporum* [14]. Alliumin inhibited the bacterium *Pseudomonas fluorescens* and expressed antiproliferative activity toward leukemia L1210 cells. However, it did not have ribonuclease activity, protease activity, mitogenic activity toward mouse splenocytes, and antiproliferative activity toward hepatoma Hep G2 cells [14].

Another 13-kDa antifungal protein, designated allivin, was isolated from bulbs of the round-cloved garlic. Its N-terminal sequence is DTFSDAGSFLDRGL, which showed very little similarity to sequences of *Allium sativum* chitinases and ribosome inactivating proteins [13]. It exhibited antifungal activity against *Botrytis cinerea, Mycosphaerella arachidicola* and *Physalospora piricola* [13].

Glycine Max (Soybean) Trypsin Inhibitors

Cultivated soybean (*Glycin max*) belongs to the Fabaceae family and falls under the genus Glycine. The species is a legume native to East Asia, and now it is widely planted all over the world due to its numerous applications and high commercial value [15]. Soybean was cultivated in China about 3000 to 5000 years ago and then introduced to United States in 1765 [15]. In China, both cultivated soybean (*Glycine max*) and wild soybean (*Glycine soja*) are found [15]. An investigation conducted by Wang and coworkers revealed that the soybeans (264 soybean landraces) cultivated in China occupy at least 12 out of the 13 soybean maturity groups existing in the United States [16]. Soybeans are health staple as they manifest antitumor and antioxidant capacities, reduce the risk of cardiovascular diseases and obesity, and even relieve menopause-related symptoms [17]. The abovementioned activities can be attributed to different bioactive components including protease inhibitors and others.

A 8-kDa Bowman-Birk trypsin inhibitor from the seeds of Hokkaido large black soybeans inhibited proliferation in breast cancer (MCF-7) cells and hepatoma (Hep G2) cells with an IC_{50} (concentration that brings about 50% inhibition) of 35 and 140 μM, respectively [18]. The trypsin inhibitory activity of the inhibitor was completely preserved after exposure to temperatures up to 100 °C for 30 min and to the pH range 2-13 for the same duration. The trypsin inhibitor inhibited HIV-1 reverse transcriptase with an IC_{50} of 38 μM, but was devoid of antifungal activity toward *Fusarium oxysporum* and *Mycosphaerella arachidicola* [18]. No Kunitz-type trypsin inhibitor was found in this cultivar.

A 19-kDa Kunitz-type trypsin inhibitor from seeds of Chinese black soybean *Glycine max* cv. 'Small Glossy Black' was reported recently [19]. It inhibited trypsin and chymotrypsin with an IC_{50} of 19 μM and 14.3 μM, respectively. Its trypsin inhibitory activity was stable in the pH range of pH 3-pH 13 and in the temperature range of 0 °C to 60 °C. The trypsin inhibitor was inhibited by dithiothreitol (from 5 to 25 mM) in a dose-dependent manner. It inhibited the activity of HIV-1 reverse transcriptase with an IC_{50} of 0.16 μM, and proliferation of MCF-7 breast cancer cells with an IC_{50} of 4.3 μM and HepG2 hepatoma cells with an IC_{50} higher than 25 μM. The trypsin inhibitor lacked antifungal activity and mitogenic activity towards mouse splenocytes. No Bowman Birk-type trypsin inhibitor was found in Chinese black soybeans (19).

A 20-kDa Kunitz type trypsin inhibitor from Korean large black soybeans displayed remarkable thermal stability (0-100 °C) and pH stability (pH 3-11). It weakly inhibited alpha-chymotrypsin activity, and inhibited HIV-1 reverse transcriptase activity with an IC_{50} value of 0.71μM and induced the release of pro-inflammatory cytokines including TNF-alpha, IL-1beta, IL-2 and interferon-gamma at the mRNA level. It weakly inhibited proliferation of CNE-2 and HNE-2 nasopharyngeal cancer cells, Hep G2 hepatoma cells, and MCF-7 breast cancer cells. It lacked antifungal, mitogenic, and ribonuclease activities. No Kunitz-type trypsin inhibitor was found in Korean large black soybeans [17]. In addition to the trypsin inhibitor, a 48-kDa lectin, designated as KBL, was also isolated from the same seeds (20). The hemagglutinating activity of KBL could be inhibited by some sugars, such as D-(+)-galactose, D-(+)-raffinose, L-(+)-arabinose, α-D-(+)-melibiose, and α-lactose. KBL inhibited HIV-1 reverse transcriptase activity with an IC_{50} of 1.38 μM. However, it was destitute of cytokine releasing, mitogenic, ribonuclease and antifungal activities. In addition, inhibitory activity toward nasopharyngeal cell lines was undetectable in KBL at concentrations up to 20 μM [20].

Phaseolus Vulgaris (Common Bean) Antifungal Proteins

Common bean (*Phaseolus vulgaris*) is included in the same family of Fabaceae as soybean. The common bean is a common annual plant cultivated worldwide, and it represents around 50% of legumes for human consumption (21). There are over 200 landraces of common bean in China with a cultivation period of over 400 years [21]. Many well-known bean varieties are classified into this species, such as black turtle bean, spotted bean, pinto beans, cloud bean, red kidney bean, Hokkaido red bean, among others. The seeds of common bean are rich in different medicinal components, such as antifungal defensins, tannins, protease inhibitors, and lectins [22, 23].

A 7.3-kDa antifungal peptide from *P. vulgaris* cv. 'Spotted Bean' exhibited N-terminal sequence homology to other leguminous defensins. It exerted an antifungal action on *Fusarium oxysporum* (IC_{50}1.8 μM) and *Mycosphaerella arachidicola.* It inhibited [methyl-(3)H]-thymidine incorporation by leukemia L1210 cells and MBL2 cells with an IC_{50} value of 4.0 and 9.0 μM, respectively. It was devoid of HIV-1 reverse transcriptase inhibitory activity when tested up to 100 μM [24].

Another 7.3-kDa antifungal peptide with a defensin-like sequence was isolated from seeds of *P. vulgaris* 'Cloud Bean'. It inhibited mycelial growth in *Fusarium oxysporum* and *Mycosphaerella arachidicola* with an IC_{50} value of 2.2 μM and 1.8 μM, respectively. Proliferation of L1210 mouse leukemia cells and MBL2 lymphoma cells was inhibited by the antifungal peptide with an IC_{50} of 10 μM and 40 μM, respectively. It had no HIV-1 reverse transcriptase inhibitory activity when tested up to 100 μM [25].

A 5443-Da defensin-like peptide from purple pole beans (*P. vulgaris* cv. 'Extra-long Purple Pole bean') inhibited mycelial growth in *Fusarium oxysporum, Mycosphaerella arachidicola, Helminthosporium maydis, Verticillium dahliae, Rhizoctonia solani, Candida albicans* and *Setosphaeria turcica* with an IC_{50} of 2.3, 0.8, 0.9, 3.2, 4.3, 4.8 and 9.8 μM, respectively. It hindered growth in *M. arachidicola* by inducing chitin accumulation at hyphal tips as disclosed by Congo Red staining. It induced membrane permeabilization in *C. albicans* as evidenced by SYTOX Green uptake, but did not alter erythrocyte membrane permeability. The antifungal activity was pH stable and thermostable. The peptide inhibited the proliferation of hepatoma (HepG2), breast cancer (MCF7), colon cancer (HT29) and cervical cancer (SiHa) cells, but not that of human embryonic liver (WRL68) cells. Its anti-HepG2 activity (IC_{50} = 4.1± 0.8 μM, n=3) was higher than that of another plant defensin, gymnin (IC_{50} > 50 μM). Its anti-MCF7 activity (IC_{50} = 8.3 ± 0.3μM, n=3) was similar to that of other plant defensins. It manifested higher HIV-1 reverse transcriptase inhibitory activity (IC_{50} = 0.5± 0.1 μM, n=3) than other plant defensins (IC_{50} > 40μM) (26).

A 7.3-kDa antifungal peptide from dried red kidney beans displayed the N-terminal sequence DGVCFGGLANGDRT. The peptide exerted an antifungal action on *Fusarium oxysporum* and *Mycosphaerella arachidicola.* It suppressed growth of lymphoma MBL2 cells and leukemia L1210 cells with an IC_{50} of 5.2 ± 0.4 μM and 7.6 ± 0.6 μM, respectively. HIV-1 reverse transcriptase was inhibited with an IC_{50} of 40 ± 3.2 μM (27).

Phaseolus Vulgaris (Common Bean) Hemagglutinins/Lectins

Besides antifungal proteins, common bean is rich in lectins. For more than a century ago, there was evidence for the presence of a class of proteins with erythrocyte-agglutinating and carbohydrate-specific binding activities, which were then known as lectins/hemagglutinins (28). The most well-known lectin is ricin, which is a 60-kDa, heterodimeric glycoprotein consisting of an A chain (with ribosome inactivating activity) and a B chain exhibiting lectin activity (29). A large number of lectins were reported in recent years. Lectins are in the spotlight of some investigations since this class of components exhibit potential medicinal activities, such as antitumor activity, cytokine-inducing activity, nitric oxide (NO)-inducing activity, and anti-HIV potential (28, 29).

A 60-kDa galactose-specific lectin from *P. vulgaris* L. cv. 'Extralong Autumn Purple Bean' (EAPL) manifested hemagglutinating activity toward mouse, rat, rabbit, and human erythrocytes. It manifested anti-HIV-1 reverse transcriptase activity, and inhibited the proliferation of human tumor cells by inducing the production of apoptotic bodies. Its NO-inducing activity may find application in tumor therapy [22].

A 60-kDa dimeric polygalacturonic acid-specific lectin from seeds of *P. vulgaris* cv. 'Blue tiger king' displayed broad-spectrum of hemagglutinating activity toward human, rabbit, rat, and mouse erythrocytes. It exhibited potent anti-Hep G2 activity which is associated with [1] induction of DNA fragmentation, (2) production of apoptotic bodies and chromatin condensation, (3) triggering of cell apoptosis and necrosis, and [4] depolarization of mitochondrial membrane. Furthermore, the lectin could induce inducible NO synthase (iNOS) expression and subsequent NO production in vitro in mouse macrophages, which may contribute to its antitumor activity. In addition, the lectin could bring about a significant dose-dependent increase in the production of mRNAs of proinflammatory cytokines including interleukin-1 beta, interleukin-2, tumor necrosis factor alpha, and interferon-gamma. It lacks significant toxicity on normal liver WRL 68 cells [23].

A dimeric lectin with specificity for glucuronic acid and galacturonic acid has been isolated from seeds of the Indian cultivar of the French bean [30]. Its hemagglutinating activity was stable within the pH range of 1-13 and the temperature range of 10-60 °C. The lectin neither exhibited any antiproliferative activity against tumor cells nor stimulated NO production by murine peritoneal macrophages. The lectin did not evoke any mitogenic response from murine splenocytes. It was devoid of HIV-1 reverse transcriptase inhibitory and antifungal activities [30].

A dimeric 64-kDa hemagglutinin, with an N-terminal sequence resembling those of other *Phaseolus* hemagglutinins, was isolated with a high yield from dried *P. vulgaris* cv. 'French bean number 35' seeds. Its hemagglutinating activity was stable in pH 6-8 and at 0-50 °C. It inhibited HIV-1 reverse transcriptase with an IC_{50} of 2μM. It inhibited proliferation of hepatoma HepG2 cells and breast cancer MCF-7 cells with an IC_{50} of 100 and 2μM, respectively but there was no antiproliferative effect on normal embryonic liver WRL68 cells. It suppressed mycelial growth in *Valsa mali* with an IC_{50} of 10 μM [31].

A glucuronic acid-specific dimeric 64-kDa lectin was purified from seeds of French bean (*P. vulgaris*) cultivar number 1. Full hemagglutinating activity was observed when the lectin was exposed to a pH ranging from 3 to 11 and temperatures within the range 20-60 °C. Maximum mitogenic activity of the lectin toward murine splenocytes occurred at a lectin concentration of 0.488 μM. The mitogenic activity was almost abolished in the presence of 250 mM glucuronic acid.

The lectin had no effect on proliferation of hepatoma (HepG2) cells, breast cancer (MCF-7) cells, and nasopharynegeal carcinoma (NPC) CNE-1 and CNE-2 cells. It was also devoid of significant anti-HIV reverse transcriptase activity [32].

A lectin from seeds of the *P. vulgaris* cv. 'Anasazi beans ' was comprised of two 30-kDa subunits with N-terminal sequence similarity to other Phaseolus lectins [33].

Its hemagglutinating activity was stable within the pH range of 1-14 and the temperature range of 0-80 °C. The lectin potently suppressed proliferation of MCF-7 (breast cancer) cells with an IC_{50} of 1.3μM, and inhibited the activity of HIV-1 reverse transcriptase with an IC_{50}

of 7.6 μM. The lectin did not inhibit fungal growth and failed to stimulate NO production by murine peritoneal macrophages. Tryptophan was crucial for its hemagglutinating activity [33].

A 67-kDa homodimeric lectin from *P. vulgaris* cv. 'red kidney beans' manifested inhibitory activity on HIV-1 reverse transcriptase and alpha-glucosidase. Its hemagglutinating activity was inhibited by glycoproteins but not by simple saccharides. It inhibited growth of the fungal species *Fusarium oxysporum, Coprinus comatus,* and *Rhizoctonia solani* [34].

The hemagglutinating activity of a 67-kDa dimeric lectin from *P. vulgaris* cv. 'Dark Red Kidney Bean' was stable between 25 °C and 70 °C, and between pH 4 and pH 11, and in the presence of a variety of divalent metal chlorides at 500 mM concentration. A variety of saccharides tested failed to inhibit its hemagglutinating activity. Although the lectin exhibited mitogenic activity toward murine splenocytes, it had no HIV-1 reverse transcriptase inhibitory or antifungal activity. It inhibited proliferation of leukemia L1210 cells [35].

A galactose-specific dimeric 162-kDa lectin from pinto beans evoked a mitogenic response from murine splenocytes, inhibited HIV-1 reverse transcriptase, but had no inhibitory effects on proliferation of L1210 leukemia cells. Its hemagglutinating activity was stable within the pH range of 3-12 and the temperature range of 0-70 °C [36].

The hemagglutinating activity of a dimeric 60-kDa lectin from *P. vulgaris* cv. 'Hokkaido red bean' could not be inhibited by a various simple saccharides at 200 mM concentration including alpha-L-fucose, D(+)-galactose, D(+)-glucose, D(+)-glucosamine, D(-)galactosamine, galacturonic acid, (+)-lactose, D(+)-melibose, L(-)-mannose, D(+)-mannose, D-mannosamine, D(+)-raffinose, L-rhamnose, (+)-xylose and galacturonic acid. The hemagglutinating activity was fully preserved at pH 4-11 and at 0-80 °C, but completely disappeared at extreme pH values (0-2 and 13-14) and at very high temperatures (90 °C and 100 °C). The lectin weakly stimulated the mitogenic response of murine splenocytes, exhibited a more potent anti-proliferative activity than Con A toward HepG2 (human hepatoma) cells, and potently inhibited HIV-1 reverse transcriptase inhibitory activity. However, it was devoid of anti-fungal activity [37].

Pisum Sativum (Pea) Antifungal Proteins and Ribosome Inactivating Proteins

Pisum sativum L. also vernacularly called pea, dry pea, Chinese pea pod, sugar snap pea, and others. *P. sativum* is an annual plant and cultivated worldwide in warm climates. The pea seeds and pea sprouts are nutritious staples in many countries for they are scored high for nutrition value. Nutritional analysis unveils that peas are high in soluble dietary fiber, vitamins, minerals, proteins, sugars, antioxidants, and others [38, 39].

From the legumes of the sugar snap pea *Pisum sativum* var. 'macrocarpon' (also known as honey pea), two antifungal proteins were isolated. A 38-kDa antifungal protein designated as sativin with an N-terminal amino acid sequence which showed similarity to those of miraculin (a sweet protein) and pisavin (a ribosome-inactivating protein from *Pisum sativum* var 'arvense Poir' manifesting similarity to miraculin) [40]. The second one is a 31-kDa antifungal protein designated as pisumin with a novel N-terminal sequence GVGAAYGCFG [41]. Sativin exerted antifungal activity against *Fusarium oxysporum, Coprinus comatus* and *Pleurotus ostreatus* but not against *Rhizoctonia solani* [40]. Pisumin displayed antifungal activity against *Coprinus comatus* and *Pleurotus ostreatus* and much weaker activity against *Fusarium oxysporum* and *Rhizoctonia solani* [41].

An 11-kDa antifungal protein with a lysine-rich N-terminal sequence from seeds of the garden pea *Pisum sativum* var. 'arvense Poir' inhibited mycelial growth in *Physalospora piricola* (IC_{50} = 0.62 μM), *Fusarium oxysporum* and *Mycosphaerella arachidicola*. It also inhibited HIV-1 reverse transcriptase (IC_{50} = 4.7μM) [42].

Ribosome inactivating proteins/RIPs also existed in *Pisum sativum* var. 'arvense Poir'. RIPs are a class of RNA glycosylases that cleave an adenine-ribose glycosidic bond at position A-4324 within the conserved α-sarcin/ricin loop in the eukaryotic 28S ribosomal RNA bound by elongation factors or adenine-2660 in *Escherichia coli* 23S rRNA [29]. Two ribosome inactivating proteins designated as alpha- and beta-pisavins, which were non-glycoproteins with a molecular mass of 20.5 kDa and 18.7 kDa respectively, were purified from the variety arvense Poir. The sequences of the first sixty N-terminal amino acids of alpha- and beta-pisavins were identical and exhibited similarity to that of miraculin. They inhibited protein synthesis by a rabbit reticulocyte lysate system with an IC_{50} of approximately 0.5 nM. The proteins acted on ribosomal RNA through its RNA N-glycosidase activity to release an Endo's fragment, and converted the conformation of DNA from supercoiled and circular forms into a linear form [43].

Solanum Tuberosum (Potato) Serine Protease Inhibitors

The scientific name of potato is *Solanum tuberosum*. It belongs to the Solanaceae family. Potato originated in the Andes of South America, and is now cultivated in 160 countries. Same as soybean, common bean and pea, potato is also a very common staple with high commercial value. Potato tuber, rice and wheat are the world's top 3 food crops for human consumption [44]. This starchy and tuberous crop is best known for its fruitful carbohydrates, such as starch and dietary fiber.It is a good energy provider, with little fat. Potato also contains vitamins (vitamin C and some B vitamins), minerals, phytochemicals, and different bioactive proteins [44].

Potide-G, a 5578.9-Da thermostable antimicrobial peptide from potato tubers (*S. tuberosum* L. cv. 'Golden Valley') potently inhibited growth of a variety of fungi (*Candida albicans* and *Rhizoctonia solani*) and bacteria (*Staphylococcus aureus*, *Listeria monocytogenes*, *Escherichia coli*, and *Clavibacter michiganense* subsp. michiganinse). It almost completely inhibited the activities of trypsin, chymotrypsin and papain. In addition, the first 11 N-terminal sequence (QICTMCCAGRK) of potide-G is identical to that of potato proteinase inhibitor, a member of the Kunitz superfamily [45].

Besides Potide-G, Potamin-1 is a different antimicrobial peptide from another cultivar of potato reported by the same research group from Republic of Korea (46). Potamin-1 is a 5.6-kDa thermostable trypsin-chymotrypsin protease inhibitor from tubers of the potato *(S. tuberosum* L cv. 'Gogu') strongly inhibited pathogenic microbial strains, including *Candida albicans*, *Clavibacter michiganense* subsp. michiganinse and *Rhizoctonia solani*. Its N-terminal sequence was NH2-DICTC CAGTK GCNTT SANGA FICEG QSDPK KPKAC PLNCD PHIAYA showing 62% homology with a serine protease inhibitor belonging to the Kunitz-type family, and the peptide inhibited chymotrypsin, trypsin, and papain. The disulfide bridge is crucial to its protease inhibitory and antifungal activities [46].

This research group also reported a 13.5-kDa Kunitz serine protease inhibitor, AFP-J, from potato (*S. tuberosum* L cv. 'Jopung') tubers [47]. AFP-J strongly inhibited the yeasts *Candida albicans, Saccharomyces cerevisiae*, and *Trichosporon beigelii,* but there was no activity against crop fungal pathogens. Its partial N-terminal sequence was NH2-Leu-Pro-Ser-Asp-Ala-Thr-Leu-Val-Leu-Asp-Gln-Thr-Gly-Lys-Glu-Leu-Asp-Ala-Arg-Leu. It inhibited chymotrypsin, pepsin, and trypsin [47].

Vicia Faba (Broad Bean) Trypsin Inhibitors

Broad bean (binominal name *Vicia faba* in the family Fabaceae) is also commonly known as fava bean, broad bean, and field bean. Broad bean is mainly categorized into spring and winter ecotypes. It is a leguminous crop cultivated in many countries for thousands of years and serves as a common staple in our everyday lives. It has been reported that broad bean is full of bioactive and nutritional components, such as trypsin inhibitors, lectins, galactosidase, procyanidins, and tannins (48, 49). A 7.5-kD trypsin-chymotrypsin inhibitor from Chinese broad beans with HIV-1 reverse transcriptase inhibitory activity manifested antifungal activity toward *Mycosphaerella arachidicola, Fusarium oxysporum,* and *Botrytis cinerea.* It stimulated proliferation of murine splenocytes (Ye and Ng,2001).Another trypsin-chymotrypsin inhibitor from Chinese broad beans had a molecular mass of 13 kDa. It exerted antifungal activity toward *Mycosphaerella arachidicola* and *Physalospora piricola*, elicited a mitogenic response from mouse splenocytes, and inhibited HIV-1 reverse transcriptase [50]. A 15-kDa Bowman-Birk type trypsin inhibitor (termed VFTI-E1) from Egyptian fava beans (*Vi. faba* cv. 'Egypt 1') displayed antiproteolytic activity against trypsin (K(i) 11.9×10^{-9} M) but no activity against chymotrypsin [49]. It demonstrated antifungal activity toward *Valsa mali* with an IC_{50} of 20 μM. The mechanism of its antifungal action toward *V. mali* included [1] induction of changes in hyphal morphology, [2] growth inhibition by chitin deposition at hyphal tips, and [3] permeabilization of fungal membrane. The antifungal activity was dependent on the ambient ionic strength as increasing concentrations of NaCl, $CaCl_2$, and $MgCl_2$ diminished the activity. The membranolytic action of VFTI-E1 was confined to fungus, but not exerted on human and rabbit erythrocytes (49).Another 15-kDa Bowman-Birk type trypsin inhibitor (termed VFTI-G1) from *Vicia faba* cv. Giza 843 demonstrated significant antiproteolytic activity against trypsin (5761 BAEE units/mg, K(i) 20.4×10^{-9} M), but only very weak chymotrypsin inhibitory activity (< 10 BTEE units/mg). It inhibited HIV-1 reverse transcriptase activity with an IC_{50} of 0.76 μM, and exhibited antiproliferative activity toward HepG2 hepatoma cells by inducing chromatin condensation and cell apoptosis [48].

Concluding Remarks

Soybean, common bean, pea, potato, and broad bean are main crops/vegetables in the world, while garlic is always a popular flavor and kiwifruit serves as a widely welcomed fruit.

It is confirmed that some bioactive components in the plant can help to protect the plant from external stress/attacks, such as the invasion of fungi and pests.

Table 1. Comparison of N-terminal amino acid sequences of proteins/peptides from different cultivars of the same species

Plant species and its medicinal component(s)	N-terminal sequence
Actinidia chinensis (cv. gold kiwi fruit) antifungal protein	GAKRAYDEVEAQN
Actinidia delicisa (cv. green kiwi fruit) antifungal protein	ATFNINNCPTVWAA
Allium sativum (cv. round clove) antifungal protein	DTFSDAGSFLDRGLA
Allium sativum (cv. multiple clove) antifungal protein	DDFLCAGGCL
Glycine max (cv. Hokkaido large black) trypsin inhibitor	DDESSKPCCDQCACT
Glycine max (cv. Korean black) trypsin inhibitor	DFLDNEHNMLENGG
Phaseolus vulgaris (cv. white cloud bean) antifungal protein	KTCENLADTFRGPCF
Phaseolus vulgaris (cv. cloud bean) antifungal protein	KTYENLADTYKGPYF
Phaseolus vulgaris (cv. red kidney bean)antifungal protein	DGVCFGGLANGDRT
Phaseolus vulgaris (cv. pinto bean) lectin	SETSFSFQRFVETNL
Phaseolus vulgaris (cv. red kidney bean)lectin	ANQTSFNFQRFDETN
Phaseolus vulgaris (cv. dark red kidney bean)lectin	SNDIYFNFQR
Phaseolus vulgaris (cv. extralong autumn purple)lectin	ANEIYFSFQRFNETN
Phaseolus vulgaris (cv. Chinese French bean number 35) lectin	ATETYSAFQRFCETN
Phaseolus vulgaris (cv. Chinese French bean number 12) lectin	ATETYFNFQRFCETN
Pisum sativum (cv. arvense Poir) antifungal protein: pisumin	GVGAAYGCPN
Pisum sativum (cv. acrocarpon) antifungal protein: sativin	APGVAPAVGHADLR
Pisum sativum (cv. arvense Poir) ribosome inactivating protein	APEPVLDVSGKKLIT
Solanum tuberosum (cv. Golden Valley)serine protease inhibitor	QICTMCCAGRK
Solanum tuberosum (cv. Jopung) serine protease inhibitor	LPSDATLVLDQTGK
Solanum tuberosum (cv. Gogu) serine protease inhibitor	DICTCCAGTKGCNT
Vicia faba (Chinese) serine protease inhibitor	GDDVKSACCDTCL
Vicia faba (cv. Egypt 1)serine protease inhibitor	GDKVKSACCDTTLLK

The antifungal mechanisms are different in distinct types of components, including induction of alteration of hyphal morphology, inhibition of the growth of hyphal tips, damage of fungal membrane, membrane channels and pore formation [49]. On the other hand, these bioactive components own some potential clinical applications. For example, the incidence of human fungal infections has increased significantly during the past two decades, but the efficacy of current drugs is limited [51]. Encouragingly, a number of antifungal components are in preclinical trials and the results are anticipated [51]. Besides the well-investigated antifungal activities, some bioactive components in these plants also exhibited other medicinal applications as exemplified by anti-HIV, and antitumor effects. Among them, trypsin inhibitors, lectins, RIPs, and ribonucleases own distinguished clinical setting [52 – 54].

From the foregoing account it can be seen that different cultivars of the same plant species may produce dissimilar proteins as judged by different N-terminal amino acid sequences, molecular masses, and other properties such as biological activities. Hopefully more cultivars with desirable characteristics can be developed in the future to the benefit of mankind. Studies on isolation and characterization of antifungal peptides/proteins and other bioactive/medicinal components from plants have a good potential to lead to very interesting clinical tools.

REFERENCES

[1] J.B. Bowne, T.A. Erwin, J. Juttner, T. Schnurbusch, P. Langridge, A. Bacic, et al., *Mol. Plant.,* 18, 4410 (2011).

[2] K. Sujatha, P. Natarajkumar, G.S. Laha, B. Mishra, K.S. Rao, B.C. Viraktamath, et al., *Genet. Res. (Camb)*, 93, 397 (2011).

[3] J. S. Lucas, S. A. Lewis, J. B. Trewin, K. E. Grimshaw, J. O. Warner, J. O Hourihane, *Pediatr. Allergy Immunol.,* 16, 647 (2005).

[4] A. R. Ferguson, New temperate fruits: Actinidia chinensis and Actinidia deliciosa. In: Janick J, editor. Perspectives on new crops and new uses. Alexandria, VA.: ASHS Press (1999).

[5] L. Xia, T. B. Ng, *Peptides,* 25, 1093 (2004).

[6] A. Karlsen, M. Svendsen, I. Seljeflot, M. A. Sommernes, J. Sexton, A. Brevik, et al., *Br. J. Nutr.*, 106, 557 (2011).

[7] H. Wang, T. B. Ng, *Phytochemistry* 61, 1 (2002).

[8] J. S. Lucas, S. A. Lewis, J. O. Hourihane, *Pediatr. Allergy Immunol.,* 14, 420 (2003).

[9] E. A. Pastorello, A. Conti, V. Pravettoni, L. Farioli, F. Rivolta, R. Ansaloni, et al., *J. Allergy Clin. Immunol.,* 101, 531 (1998).

[10] M. Gavrovic-Jankulovic, T. cIrkovic, O. Vuckovic, M. Atanaskovic-Markovic, A. Petersen, g. Gojgic, et al., *J. Allergy Clin. Immunol.,* 110, 805 (2002).

[11] M. Thomson, M. Ali, *Curr. Cancer Drug Targets.,* 3, 67 (2003).

[12] R. C. Anderson, Gourmet garlic gardens. 1997.

[13] H. X. Wang, T. B. Ng, *Life Sci.,* 70, 357 (2001).

[14] L. Xia, T. B. Ng, *Peptides.,* 26, 177 (2005).

[15] H. M. Lam, X. Xu, X. Liu, W. Chen, G. Yang, F.L. Wong, et al., *Nat. Genet.,* 42, 1053 (2010).

[16] Y. Wang, J. Qin, J. Gai, G. He, *Genetic Resources and Crop Evolution.,* 53, 803 (2006).

[17] E. F. Fang, J. H. Wong, T. B. Ng, *J. Biosci. Bioeng.,* 109, 211 (2010).

[18] V. S. Ho, T. B. Ng, *J. Pept. Sci.,* 14, 278 (2008).

[19] X. Ye, T. B. Ng, *Planta Med.,* 75, 550 (2009).

[20] E. F. Fang, J. H. Wong, P. Lin, T.B. Ng, *Protein Pept. Lett.,* 17, 690 (2010).

[21] X. Zhang, M. W. Blair, S. Wang. *Theor. Appl. Genet.,* 117, 629 (2008).

[22] E. F. Fang, P. Lin, J. H. Wong, S. W. Tsao, T. B. Ng, *J. Agric. Food Chem.,* 58, 2221 (2010).

[23] E. F. Fang, W. L. Pan, J. H. Wong, Y. S. Chan, X. J. Ye, T. B. Ng, *Arch. Toxicol.*, 85, 1551 (2011).

[24] H. X. Wang, T. B. Ng, *Appl. Microbiol. Biotechnol.,* 74, 125 (2007).

[25] X. Wu, J. Sun, G. Zhang, H. Wang, T. B. Ng, *Phytomedicine,* 18, 104 (2011).

[26] P. Lin, J. H. Wong, T. B. Ng, *Biosci. Rep.,* 30, 101 (2010).

[27] M. Li, H. Wang, T. B. Ng, *Protein Pept. Lett.,* 18, 594 (2011).

[28] N. Sharon, H. Lis. *Glycobiology.,* 14, 53 (2004).

[29] E. F. Fang, C. Z. Zhang, T. B. Ng, J. H. Wong, W. L. Pan, X. J. Ye, et al., *Cancer Prev. Res. (Phila),* 5, 109 (2012).

[30] A. Sharma, J. H. Wong, P. Lin, Y. S. Chan, T. B. Ng, *Protein Pept. Lett.,* 17, 221 (2010).

[31] S. K. Lam, T. B. Ng, *Phytomedicine*, 17, 457 (2010).

[32] Y. S. Chan, J. H. Wong, T. B. Ng, *Protein Pept. Lett.,* 18, 194 (2011).

[33] A. Sharma, T. B. Ng, J. H. Wong, P. Lin, *J. Biomed. Biotechnol.,* 2009, 929568 (2009).

[34] X. Y. Ye, T. B. Ng, P. W. Tsang, J. Wang, *J. Protein Chem.,* 20, 367 (2001).

[35] L. Xia, T. B. Ng, *J. Chromatogr. B Analyt. Technol. Biomed. Life Sci.,* 844, 213 (2006).

[36] J. H. Wong, C. C. Wong, T. B. Ng, *Biochim. Biophys. Acta.*, 1760, 808 (2006).

[37] J. H. Wong, C. T. Wan, T.B. Ng, *J. Sci. Food Agric.,* 90, 70 (2010).

[38] J. S. Chapman, L. K. Jefferies, O. A. Pike, *J. Food Sci.,* 75, 162 (2010).

[39] T. L. Pownall, C. C. Udenigwe, R. E. Aluko, *J. Agric. Food Chem.,* 58, 4712 (2010).

[40] X. Y. Ye, H. X. Wang, T. B. Ng, *Life Sci.,* 67, 775 (2000).

[41] X. Y. Ye, T. B. Ng, *Comp. Biochem. Physiol. C Toxicol. Pharmacol.,* 134, 235 (2003).

[42] H. X. Wang, T. B. Ng, *Peptides.,* 27, 1732 (2006).

[43] S. S. Lam, H. Wang, T. B. Ng, *Biochem. Biophys. Res. Commun.,* 253, 135 (1998).

[44] M. E. Camire, S. Kubow, D. J. Donnelly, *Crit. Rev. Food Sci. Nutr.,* 49, 823 (2009).

[45] M. H. Kim, S. C. Park, J. Y. Kim, S. Y. Lee, H. T. Lim, H. Cheong, et al., *Biochem. Biophys. Res. Commun.,* 346, 681 (2006).

[46] J. Y. Kim, S. C. Park, M. H. Kim, H. T. Lim, Y. Park, K. S. Hahm, *Biochem. Biophys. Res. Commun.,* 330, 921 (2005).

[47] Y. Park, B. H. Choi, J. S. Kwak, C. W. Kang, H. T. Lim, H. S. Cheong, et al., *J. Agric. Food Chem.,* 53, 6491 (2005).

[48] E. F. Fang, A. A. Hassanien, J. H. Wong, C. S. Bah, S. S. Soliman, T. B. Ng, *Protein Pept. Lett.,* 18, 64 (2011).

[49] E. F. Fang, A. A. Hassanien, J. H. Wong, C. S. Bah, S. S. Soliman, T. B. Ng, *J. Agric. Food Chem.,* 58, 10729 (2010).

[50] X. Y. Ye, T. B. Ng, *J. Pept. Sci.,* 8, 656 (2002).

[51] C. P. Selitrennikoff, *Appl. Environ. Microbiol.,* 67, 2883 (2001).

[52] E. F. Fang, T. B. Ng, *Biochim. Biophys. Acta.,* 1815, 65 (2011).

[53] E. F. Fang, T. B. Ng, *Curr. Mol. Med.,* 11, 417(2011).

[54] E. F. Fang, T. B. Ng, P. C. Shaw, R. N. Wong, *Curr. Med. Chem.*, 18, 4410 (2011).

In: Cultivars
Editor: Katya Carbone
ISBN: 978-1-62417-101-7

Chapter 10

CORN BIOTECHNOLOGY

Pil Son Choi[1*], Setyo Dwi Utomo[2], Min Kyoung Shin[3], Han Sang Yoo[3], Raza Ahmad[4] and Suk Yoon Kwon[4]

[1]Medical Plant Transformation Center, Nambu University, Korea,
[2] Department Crop Science, Faculty of Agriculture, University of Lampung, Indonesia
[3] Department of Infectious Diseases, College of Veterinary Science, Seoul National University, Seoul, Korea
[4] Korea Research Institute of Bioscience and Biotechnology, Daejeon, South Korea

ABSTRACT

Among genetically modified (GM) crops, corn is the second most widely cultivated crop in the world. GM corn cultivars tolerant to herbicide and/or insect pests were cultivated on about 45 million hectares worldwide. Other GM corn cultivars having novel traits such as drought tolerant, efficient nitrogen utilization, and vaccine production, resistant to other abiotic stresses and potentially high yielding are in a pipeline. Like these, genetic transformation has been regarded as a potential approach for validation of target genes and crop improvement. However, one of the limitations to *Agrobacterium*-mediated transformation of corn is the scarcity of immature embryos being used as explants. To ensure uninterrupted availability of immature embryos requires well-equipped greenhouse and laborious artificial pollination. To overcome these limitations, stable *Agrobacterium*-mediated transformation using a type II embryogenic callus is needed. Additionally, before being commercialized, GM corn should undergo mandatory safety testing and regulatory assessment. To ensure international consistency, uniform standards for both human health and environmental safety assessment should be applied. This chapter describes the development of herbicide resistant transgenic corn via *Agrobacterium*-mediated transformation by using a type II embryogenic callus and potential of transgenic corn-derived ApxIIA and CTB-ApxIIA#5 fusion proteins as candidates for animal vaccines.

[*] Corresponding author Tel: 82-62-970-0161, e-mail: cps6546@hanmail.net.

INTRODUCTION

Corn is one of the most important crops as a source of food, feed, and energy. The global demand for corn utilization for food and feed has been estimated 852 million metric tons for the year 2020 [1]; while its production in 2009 was 818,823,434 tons based on FAO data. The demand is expected to be higher due to the utilization of corn for bio-energy production. It is, therefore, imperative to increase world corn production to meet the estimated demands in future. The yield or production can be increased through the utilization of superior cultivars developed either by classical plant breeding or modern genetic engineering approaches. Besides increasing the yield, breeding via genetic transformation can develop cultivars having improvedagronomic and other novel characters, i.e., tolerant to herbicides and producing high-value products such as edible vaccines.

Three important steps to develop a new cultivar are the creation or the enhancement of genetic variability, selection, and yield trials. The genetic transformation contributes to the creation or to the enhancement of genetic variability by introducing selected gene or genes of interest into the plant genome. The advantages of using biotechnology compared to the conventional breeding are, but not limited to: 1) the source of genes of interest can be from unrelated or distantly-related organisms (the concept of border less genome); 2) relatively faster due to simple-Mendelian inheritance; and 3) minimizing linkage drag or carrying unwanted characters [2]. Moreover, adaptability of Biotech crops is increasing exponentially and these crops (including corn) have been cultivated by 16.7 million farmers in 29 countries (160 million hectares) in 2011. More countries like Indonesia, Vietnam, Bangladesh, Mali, Togo, Nigeria, Kenya, Uganda, Tanzania, and Malawi are expected to adopt biotech crops in 2015.

The procedure of genetic transformation of corn starts with the construction of gene cassettes containing gene/s of interest and selectable marker for transformation and followed by, gene transfer into explants, *in vitro* regeneration of putative transgenic plants from transformed tissue, confirmation of integration of transgenes into the genome of plants, transcriptional and post transcriptional evaluation of chimeric genes and agronomic evaluation. Transgenic corn plants have been successfully developed using particle bombardment [3, 4], electroporation [5], and *Agrobacterium tumefaciens*-mediated gene transformations [6 - 10]. Generally, genetic transformation protocols of corn have been developed using immature zygotic embryo's explants of Hi-II [11,12], and reproducible protocols for *Agrobacterium*-mediated transformation have been developed either using super or standard binary vectors [6 - 9, 13 - 16]. Although the protocol of *Agrobacterium*-mediated transformation using explants of immature zygotic embryos is reproducible, but it requires well-equipped greenhouse and laborious artificial pollination to ensure continuous supply of immature embryos round the year. To overcome these limitations, development of another *Agrobacterium*-mediated transformation system by using explants of a type II embryogenic callus is needed. After successful production of primary type II embryogenic calli, itcan be sub-cultured and proliferated conveniently and maintained easily all-year-long [10]. Efforts have been made to use embryogenic callus for corn transformation by electroporation method [17] and production of transgenic cornby particle bombardment using type II embryogenic callus has been reported [18]. *Agrobacterium*-mediated transformation was reported by using explants of embryogenic callus [19] and type I embryogenic callus [15]. Recently, stable

Agrobacterium-mediated transformation of corn by using explants of a type II embryogenic callus was developed by Kim et al. [10].

Several vaccines have been developed and commercialized to control animal disease, which is a major economic loss in livestock industry, worldwide. During the development of a vaccine, several delivery systems have been applied to overcome adverse effects that could be induced by parental administration of conventional vaccines. Recently, oral delivery systems including plants are focusing in the animal vaccines since the system has several advantages in compared with parental administration. Therefore, needs on the development of new safe vaccines, using important virulence factors and/or new delivery system had been increasing. Those requirements made us give an eye to corn as a delivery system. *A. pleuropneumoniae*, a causative agent of porcine pleuropneumoniae, produces several virulence factors such as exotoxins, lipopolysaccharide, capsular polysaccharide, pili, etc. Of the virulence factors, cytotoxin from the bacteria (Apx) has been recognized as the most important virulence factor [20-22]. Four different Apx toxins with different cytotoxic activities, ApxI, II, III and IVhave been identified. Among the Apx toxins, ApxIIA is thought to be a common antigen for the development of a vaccine against the bacterial infection. Several delivery systems to induce effective efficacy of the vaccines have been developed. In addition to injectable vaccines, new delivery system expressing ApxIIA using either *Saccharomyces cerevisiae* or *Nicotiana tabacum*, has proved to be capable of inducing protective immune responses against *A. pleuropneumoniae* in mouse model [23-25]. Transgenic plants have recently been focused as an effective agent for delivery of vaccines due to several advantages: decreased manufacturing costs and potential for alternate route like oral delivery [26]. The plant-derived antigens are bio-encapsulated by the plant cell wall and cell membrane, thus increasing resistance to degradation in the gut [26]. In addition, plant tissues can be edible materials for oral delivery without any extra processing [27]. The development of a subunit vaccine candidate directed against porcine pleuropneumonia is thought to be very important both clinically and economically in swine industry worldwide.

This chapter describes the biotechnological aspects of corn with an emphasis atthe status and advancements made to develop transgenic corn tolerant to herbicide and/or producing edible vaccines along with a focus on the regulations for commercialization of the its products.

Herbicide Resistant Corn Using Efficient Transformation Protocols

Since transgenic sterile corn was, firstly, developed from protoplast cultures via electroporation, an effective direct gene transfer system using suspension cells and immature embryos of corn has been reported by particle bombardment, which resulted in high copy numbers and extensive rearrangement of the inserted DNA [28]. For the last two decades, *Agrobacterium tumefaciens*–mediated gene transfer system has only been used in dicotyledonous plants. A major advantage of *Agrobacterium*-mediated transformation is that a small number of copies of relatively large segments of T-DNA are integrated into the plant genome with minimal rearrangement, resulting in transgenic plants of high-quality [28]. However, its transformation scope has been extended to monocotyledonous plants as well, and efficient transformation protocols were successfully developed for several monocotyledonous plants such as rice [29], corn [6], wheat [30] and barley [31]. The important factors in these achievements include the optimization of types of plant material (explants), the choice of vectors and *Agrobacterium* strains, and the optimization regeneration through tissue culture [28]. Numerous studies reported the efficient protocols of *Agrobacterium*-mediated transformation using embryogenic callus [36], type I embryogenic callus [32], immature embryo [14, 16, 28, 33-35], and the characteristics of herbicide resistance in the transgenic corns have been confirmed (Table 1).

Table 1. Summary of various studies on the development of herbicide-resistant corn using different explants

Observation	Explant/method	Genes	References
Herbicide resistant	Embryogenic callus /microprojectile bombardment	*Bar, GUS*	Gordon-Kamm et al. (1990) [3]
	Type II callus /bombardment	*GUS, bar*	Dennehey et al. (1994) [40]
	Type I callus /microprojectile bombardment	*Bar, GUS*	Wan et al. (1995) [32]
	Immature embryos /particle bombardment	*GUS, Pat*	Brettschneider et al. (1997) [33]
	Immature embryo /agrobacterium co-cultivation	*GUS, bar*	Zhao et al. 2001 [14]
	Immature embryo /agrobacterium co-cultivation	*GUS, bar*	Vega et al. (2008) [16]
	Immature embryo /agrobacterium co-cultivation	*GUS, bar*	Takavar et al. (2010) [35]
	Immature embryo /particle bombardment	*GUS, bar*	O'Kennedy et al. (2001) [34]
	Immature embryo /agrobacterium co-cultivation	*Bar*	Ishida et al. (2007) [28]
	Embryogenic callus /bombardment	*EPSPs*	Howe et al. (2002) [36]
	Immature embryo /agrobacterium co-cultivation	*GUS, bar*	Frame et al. (2002) [8]

Kim et al. [10] had recently developed a stable *Agrobacterium*-mediated transformation of corn by using explants of type II embryogenic callus. The Type II callus obtained from immature embryos [11] is characterized by rapid proliferation, friability and lack of differentiation. The protocol can be routinely used for the production of large numbers of transgenic plants. The protocol consisted of the preparation of embryogenic callus, preparation of *Agrobacterium* suspensionand inoculation of the explants with *Agrobacterium*, co-cultivation, delay step, selection on selection medium, and germination (Fig. 1). Here we describe its details step by step for better understanding of readers. **Step I**, immature embryos are aseptically dissected from greenhouse-grown ears, and then were cultured on callus initiation medium (MS salts, Eriksson's vitamins, Thiamine-HCl, 2,4-D, MES, L-proline, sucrose, myo-inositol, casamino acid vitamin, BBL agar, $AgNO_3$, pH 5.8) in the dark at 24°C. The primary Type II embryogenic callus was produced in 2 months. **Step II**, the clumps of type II embryonic callus were chopped into small pieces, and then were submerged into a suspension of *Agrobacterium tumefaciens* strain C58C1 [38] carrying the binary vector. After 5 min of infection *Agrobacterium solution* was pipette out, and the callus clones were cultured on solid co-cultivation medium for 2 days (24°C, dark). **Step III**, after co-cultivation callus clones were cultured on delay medium for 7 days in the absence of selective agents to promote preferential growth of transgenic versus non-transgenic plant cells and, subsequently calli were transferred on selection medium for 4 x 14 days (24°C, dark). The dead or brown tissues were removed during sub-culturing on selection medium. **Step IV**, the callus clones that grew and proliferated on selection medium were sub-cultured on the primary regeneration medium for 7 days (24°C, dark), and then on secondary regeneration medium for 7 days (24 °C, light), and lastly on the germination medium for 7 days. The plantlets were acclimatized and grown in greenhouse to maturity. The R_0 plants were artificially self-pollinated, and the R_1 seeds were harvested. R_1 plants were employed for leaf painting assay, northern-blot analysis, and the production of R_2 seeds. By using this protocol, 7 independent phosphinothricin (selectable marker) resistant callus clones were obtained in 6 different transformation cycles having 1.3% highest transformation efficiency with an overall average of 0.60% (Table 2). Total 36 plantlets (R_0) were acclimatized and grown to maturity in the greenhouse. Out of 30 R_0 plants, 20 R_0 plants were artificially self-pollinated and R_1 seeds were harvested; the other 16 R_0 plants were out-crossed or failed to produce seeds. Five R_1 seeds per R_0 line were randomly selected and grown in the greenhouse for leaf painting assay, Northern-blot analysis, and to obtain of R_2 seeds.

An efficient protocol of genetic transformation is always required in order to study *in planta* gene function and to develop crop plant cultivars carrying superior traits, i.e., resistance to herbicide. *Agrobacterium*-mediated transformation is rare event and dependent on the different factors, therefore, its efficiency varied among protocols and plant species. Frame et al. [8] and Utomo [9] reported transformation efficiency of maize (5.5 and 2.2%, respectively), based upon the number of independent transformation events. Hinchee et al.[37], Olhoft et al. [39], and Utomo [41] also reported transformation efficiency (0.3 – 2.2%, 16.4%, and 2.6 – 6.5%, respectively) in soybean. Although the transformation efficiency is low (0.6%) by using described protocol, yet it is useful to generate a sufficient number of transgenic corn plants. Mammoth number of transformation cycles can be done because of the type II embryogenic callus as the explants, which can be proliferated conveniently [10] around the year.

The expression analysis of integrated gene into corn genome by Northern blot analysis and leaf-painting assay was conducted at the R_1 generation (Fig. 2, 3, [10]). In this experiment, total RNA was prepared using Tri-Reagent, according to the manufacturer's instructions (MRC Inc. no. tr118) from randomly selected transgenic plants and non-transgenic control of Hi II genotype. Approximately, 30 μg total RNA was electrophoresed on 1% agarose gel containing 5.1% (v/v) formaldehyde and blotted onto nylon membrane (Zeta-Probe GT genomic tested blotting membranes; Bio-Rad) in 20X SSC (saline sodium citrate buffer). The 622-bp *bar* polymerase chain reaction (PCR) product was used and subsequentlylabeled with [^{32}P] dCTP using the Random Primed DNA Labeling kit (Boehringer Manheim). The mRNA from transcription of *bar* gene in transgenic R_1 maize was strongly detected in 5 out of 9 R_1 plants tested (Lines 24, 29, 30, 31, and 33). On the other hand, it was weakly detected Lines 14 and 19, while no signals were detected in Lines 2 and 12. Among all, Line 31 showed the highest expression of *bar* gene (Fig. 2) [10].

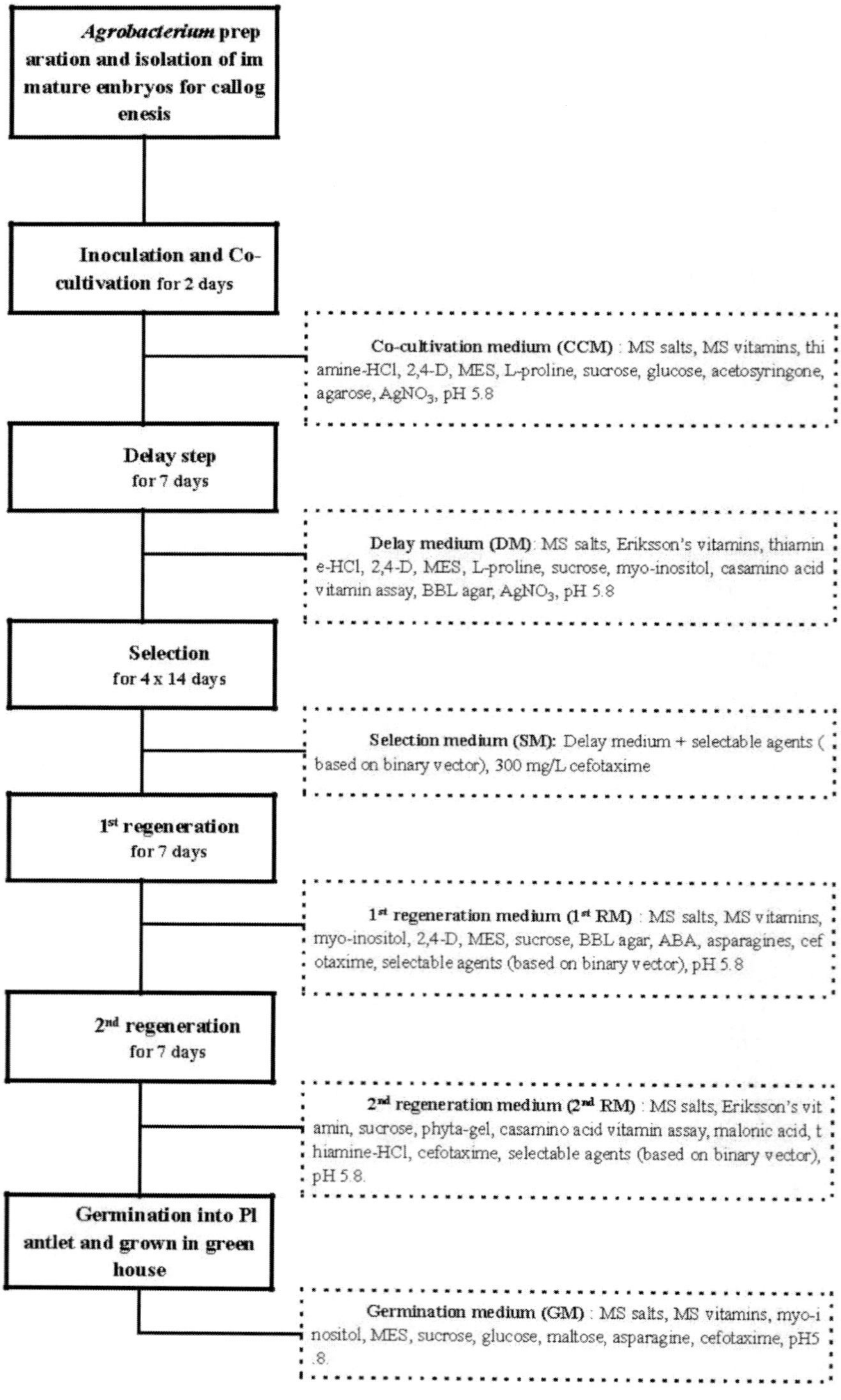

Figure 1. Astep-by-step description of *Agrobacterium*-mediated transformation protocol of corn by using type II embryogenic callus.

Table 2. ***Agrobacterium*****-mediated transformation efficiency using type II embryogenic calli explants**

Experiment No.	The number of callus clones inoculated with *Agrobacterium*	Callus clones resistant to phosphinothricin		Number of putative R_0 plants
		Number	Transformation efficiency (%)[a]	
(A)	*(B)*	*(C)*	*(C/B)*	*(D)*
I	156	2	1.3	10
II	170	1	0.6	5
III	367	3	0.8	16
IV	87	0	0.0	0
V	178	1	0.6	5
VI	200	0	0.0	0
Total	1,158	7	0.60	36

[a]Transformation efficiency (%) was the number of callus clones resistant to phosphinothricin (column C) divided by the number of callus clones inoculated with *Agrobacterium* (column B – see ref. [10]).

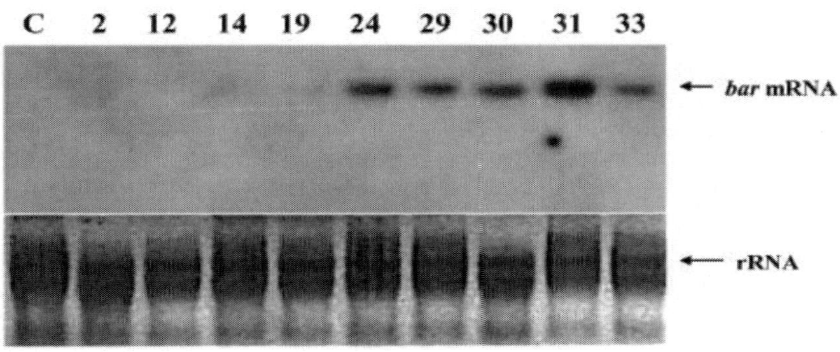

Figure 2. Northern blot analysis of 9 lines of R_1 generation of corn transformed with *bar* gene. Transformation was mediated by *Agrobacterium*. The gel was 1% agarose (w/v) containing 5.1% formaldehyde in 1x MOPS buffer. Line C: non-transformed control, Hi II genotype. Transgenic line 24, 29, 30, 31, and 33: messenger RNA (mRNA)after transcription of *bar* gene in transgenic R_1 maize. Transgenic line 14 and 19: mRNA was weakly detected. Transgenic line 2 and 12: mRNA was not detected. (*Kim et al. 2009,* [10])

Glufosinate leaf-painting assay [40] was used to show and confirm the expression of the herbicide (*bar*) gene in R_1 progenies. A line across of the upper leaves was drawn using a black permanent marker and then herbicide solution (Basta solution contained 0.01% Tween 20) was painted. The leaves were scored for herbicide resistance after 7 days of herbicide application. Based on the glufosinate leaf painting assay, three R_1 plants of line 31 with the highest expression in mRNA level (Fig. 3b, c, and d, [10]) showed resistant to 200 mg l^{-1} glufosinate; on the other hand, the negative control plant was susceptible to herbicide (Fig. 3a). This result indicated the stable insertion, successful transcription and posttranscriptional processes of herbicide resistant gene in transgenic corn plants.

Glufosinate is a potent inhibitor of glutamine synthetase and its application results in impaired assimilation of ammonia, which results in tissue death. The *bar* gene encodes phosphinothricin acetyl transferase (PAT) enzyme, which negates the lethal effects of glufosinate by acetylation of the amino group. Resultantly, plants expressing *bar* gene can tolerate toxic levels of glufosinate applications. [10]. In conclusion, based on Northern-blot analysis and leaf painting assay, line 31 was selected as elite line resistant to herbicide. This line can be utilized as herbicide-resistant parent in breeding programs for crossing with commercial corn cultivars. The line can also be advanced to be an inbred line as a herbicide-resistant parent in development of herbicide tolerant hybrid corn cultivars.

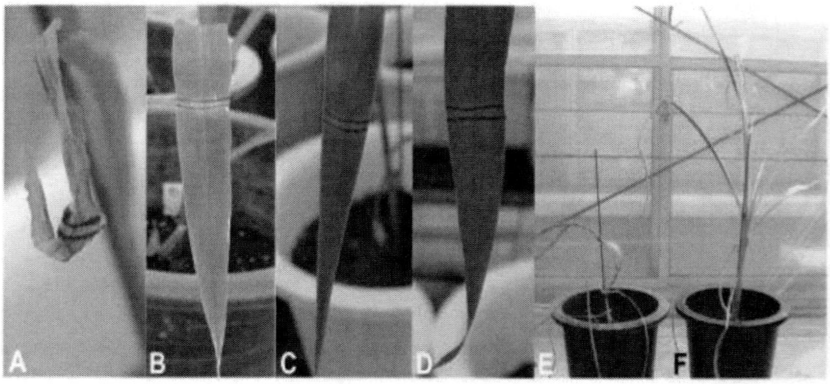

Figure 3. Glufosinate leaf-painting assay using Basta at concentration of 200 mg/ml glufosinate. A, E: Non-transformed control of Hi II painted with 200 mg l^{-1} glufosinate showing the dead tissue on the painted area, B, C, and D: Transgenic R1 plants (line 31) of Hi II painted with 200 mg l^{-1} glufosinate showing resistant to glufosinate, F: Transgenic R1 plants (line 31) of Hi II (right) painted with 1000 mg/ml glufosinate showing resistant to glufosinate. (*Kim et al. 2009,* [10]).

TRANSGENIC CORN AS A SOURCE OF VACCINE

Porcine pleuropneumonia caused by *Actinobacillus pleuropneumoniae* has been known as the most destructive respiratory disease in the swine farming-industry, worldwide. Therefore, several attempts have been made to control the disease and vaccination is recognized as an effective method for its prevention. During the development of vaccines, exotoxins known as Apx toxin from the causative bacteria have been recognized as the most important virulence factor. Among other Apx toxins, ApxIIA is thought to be a common antigen for the development of a vaccine against the bacterial infection. A number of delivery systems to induce impressiveefficacy of the vaccines have been developed. Transgenic plants have recently been focused for efficient delivery of vaccines due to several advantages: increased safety, anticipated low cost of mass vaccination programs and wider use of vaccination [42]. Several plant-derived vaccine candidates have been developed, and evaluated for their immunogenicity and protection against microbial infection [43]. Corn, one of the major forage crops available all over the world, is particularly attractive for its intensively studied genetics and the availability of established transformation procedures. In addition, grain seeds are appropriate systems for the oral delivery of subunit vaccines because of their low-water contents and long periods of storage [44].

Many antigen have been expressed in plants using integration techniques of genes encoding antigen into plant's genome; hetero-protein in tobacco plant [45], hepatitis B surface antigen (HBsAg) in tobacco plants and lettuce[49], Norwalk virus capsid protein in tobacco plants and potato tubers [47]. Especially, potato tubers transformed with synthetic

LT-B gene showed higher expression than bacteria-derived LT-B gene [48]. Like these, the expression of antigen-protein in plants could determine the suitability of the plants as a host and the availability of plant derived-antigen protein by injecting more visible. Recently, Kim et al. [52] have developed a transgenic corn with ApxIIA gene by using an efficient transformation protocol.

The CTB-ApxIIA#5 fusion protein and full-size ApxIIA expressed in corn seed, as a subunit vaccine candidate, have been observed to induce Apx-specific immune responses in a murine model. These results support the speculation that the transgenic corn-derived ApxIIA and CTB-ApxIIA#5 fusion protein has a potential to be used as candidate vaccine against *A. pleuropneumoniae* infection. The immunogenicity of CTB-ApxIIA#5 fusion protein (0.93kb) and full-size ApxIIA (2.8kb) expressed in corn was investigated in mice [49]. The soluble corn extract was subcutaneously injected into mice to evaluate the transgenic corn-derived ApxIIA and CTB-ApxIIA#5 fusion protein as an oral vaccine candidate. The results demonstrated that the higher serum antigen-specific IgG levels were obtained in the mice boosted with the transgenic corn expressing CTB-ApxIIA#5 fusion protein and full-size ApxIIA two weeks after the final immunization ($p<0.01$) (Fig. 4A). Moreover, the production of ApxIIA-specific IgG memory B cells significantly increased in the mice boosted with the transgenic corn expressing CTB-ApxIIA#5 fusion protein and full-size ApxIIA by ELISPOT ($P<0.01$) (Fig. 4B). Especially, ApxIIA-specific IgA memory B cells were observed in the mice injected with transgenic corn expressing CTB-ApxIIA#5 fusion protein ($p<0.05$).

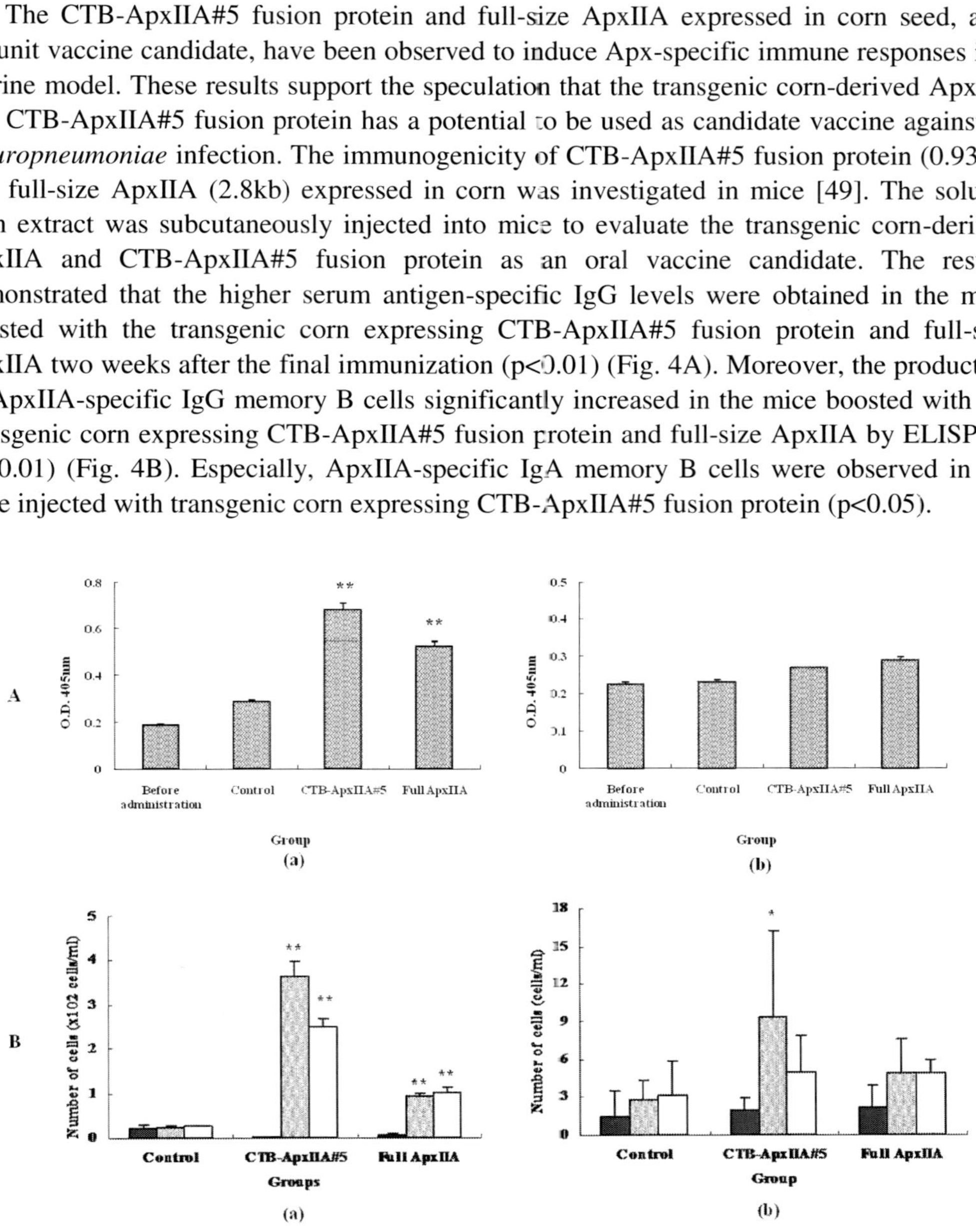

Figure 4. Induction of antigen-specific immune responses in mice administrated with transgenic corns. **A:** ApxIIA-specific IgG (a) and IgA (b) levels showed in serum of mice, immunized with transgenic corns expressing CTB-ApxIIA#5 and full-ApxIIA (**, $p<0.01$) and **B :** Administration of corn-derived ApxIIA resulted in predominant numbers of ApxII-specific memory B cell for IgG (a) and IgA (b), compared with control group (■, Non-treated control; ▒, ApxIIA; □, LPS; *, $p<0.05$; **, $p<0.01$). (*Shin et al. 2011,* [49]).

In addition, the secretion of interferon-gamma (IFN-γ), interleukin (IL)-4 and nitric oxide (NO) were examined as an indicator for the induction of immune response in the spleen cells from the experimental mice. High levels of IFN-γ, T helper type 1 (Th1) cytokine, and NO were detectable in the mice injected with transgenic corn expressing full-size ApxIIA ($p<0.05$) (Fig. 5), while IL-4, T helper type 2 cytokines were not detectable in any group. Based on the results of the induced antigen-specific immune responses, we reported that transgenic corn-derived ApxIIA is capable of effectively producing antigen-specific antibodies and inducing the cytokines related with Th1 immune response [49].

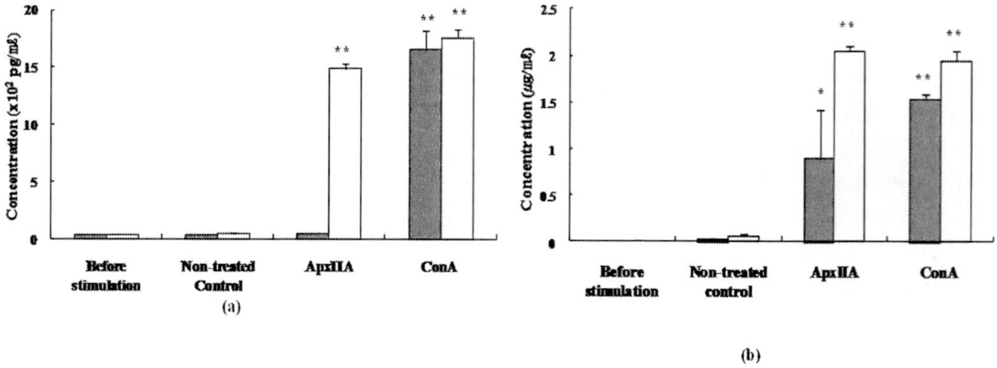

Figure 5. Secretion of IFN-γ (a) and iNOs (b) levels from splenocytes of mice administrated with transgenic corn expressing CTB-ApxIIA (▒) and full ApxIIA(□) after stimulation with non-treated control, ApxIIA, and ConA (*, $p<0.05$; **, $p<0.01$). (*Shin et al. 2011,* [49]).

Regulation for Release and Commercialization of Transgenic Crops

Biotech corn was first commercialized in 1996. Due to the positive contributions towards the income of farmers, the biotech corn has been successfully cultivated in 29 countries in 2011 [59]. Biotech crops (including corn) have been cultivated by 16.7 million farmers (160 million hectares) in 2011 [59]. Biotech corn occupied 51 million hectars or 32% of total area of biotech crops in 2011. Ten countries (Mali, Togo, Nigeria, Kenya, Uganda, Tanzania, Malawi, Indonesia, Vietnam, and Bangladesh) are expected to adopt biotech crops in 2015. Transgenic corn for food or feed can be commercialized if the products comply with the regulation of biosafety and food/feed safety. Although the main objectives of release of biotech crops including corn are to develop cultivars with beneficial traits, still regulations are required to avoid or minimize unintentionally hazardous effects on humans and the environment. Therefore, before being commercialized, biotech corn must undergo mandatory safety testing and regulatory assessment. Biosafety regulation is enforced to avoid biotech crops or products disrupt or jeopardize humans, other organisms, and environment. The risk assessment consists of identification and characterization of hazards, exposure assessment, and risk characterization [55].

To ensure international consistency, uniform standards for both human health and environmental safety assessment shall be applied. Up to now, biotech crops are regulated differently in the USA and Europe. In the USA, the regulation is vertical – a product-based system; regulation depends on the characteristics of biotech crops. On the other hand, the regulation in Europe is horizontal – a process-based system applied to all plants produced by genetic transformation methods. The responsibility for regulating biotech crops in the USA is shared by three federal agencies—the Environmental Protection Agency (EPA), the United States Department of Agriculture (USDA), and Food and Drug Administration (FDA) [60]. EPA controls or regulates plant properties and registers herbicides applied on herbicide-tolerant plants. USDA regulates permits for inter-state movement and assesses the pest characteristic of transgenic plants. FDA determines whether biotech crops have been adequately evaluated in accordance with policy of food or feed biotechnology. In Europe, the deliberate release of biotech crops is regulated by European Union Directive 2001/18/EC. The regulation contains a step-by-step approval process of assessment of the risk of biotech crops to human health and the environment [54]. In Indonesia, the biosafety and food safety of biotech crops are regulated based on Joint Decree of 4 ministers concerning Biosafety and Food Safety of Genetically Engineered Agricultural Product [60], Government Regulation No. 21issued in year 2005 on biosafety of genetically engineered product [58], and President Regulation No. 39 issued in year 2010 on Biosafety Committee of genetically modified product [56]. The Ministry of Environment coordinates biosafety evaluation, whereas the Ministry of Agriculture coordinates food safety. The regulation mentions that the biosafety evaluation of biotech crops should be conducted in the laboratory, controlled greenhouse facility, and controlled field facility, respectively. Biosafety of biotech crops is based on evaluation of weediness of crops or on the probability of biotech crops to become weeds. The weediness traits include an ability to grow wildly without human intervention, rapid growth, highly resistance to biotic and abiotic stress, and seed dormancy [53]. The degree of weediness is evaluated based on the agronomic similarity between a biotech cultivar and its counterpart.

CONCLUSION

Biotech crops are being cultivated at a steady pace since the commercial release of first biotech crop in 1996. Their popularity among the farmers of various countries resulted in an exponential increase of biotech crop cultivation during the last couple of years. It is believed that area under cultivation of these crops will touch new records in coming couple of years because many countries of Asia and Africa are in a queue to adopt them. Genetic transformation of plants was initiated by using gene gun and electroporation, but a natural genetic engineering tool called as *Agrobacterium tumefaciens* now has overcome it.

Agrobacterium originally was utilized to transform dicot plants, but now it is being employed to transform monocots as well with satisfactory results. There are various advantages of using *Agrobacterium* over other mechanistic ways of plant transformation, which have been documented in previous sections.

Genetic transformation requires a robust and reliable plant regeneration protocol to successfully get the transgenic plants. In this connection, optimizing regeneration protocol acquires more importance when plant species is considered to be recalcitrant.

Monocots (Corn, wheat, rice, etc.) are considered to be hard to regenerate as compared to dicots (tobacco, potato), and require special tissue for regeneration like immature embryos, which are difficult to obtain throughout the year. The study, which is summarized in this chapter, states an easy and reproducible protocol to regenerate transgenic corn plants. In this protocol type II embryogenic calli instead of immature embryos were used to transform and regenerate corn, which makes it a protocol of choice because embryogenic calli can be proliferated throughout the year.

Biotechnology provides the researchers an opportunity to introduce novel characters via genes of any source in crop plants, which are naturally impossible like transformation of bacterial genes into crop plants to obtain herbicide or insect resistance. Transgenic corn plants resistant to a non-selective herbicide phosphinothricin were obtained as summarized above. Herbicide tolerance helps farmers obtain more yields per unit area by eradicating unwanted plants/weeds with a single application of non-selective herbicide like phosphinothricin. Transgenic crop plants with such characteristics can be utilized for commercial purposes or they may serve as a new germplasm material to be used in classical breeding programs as novel parents.

Recently, safety of animal products and animal welfare has been major concerned issues in livestock industry.

Therefore, new technologies, especially in the delivery system have been applied to develop veterinary vaccines to overcome the problems induced by conventional vaccination methods [23, 50, 51]. Oral vaccination in veterinary medicine could be a potential application owing not only to induction of local and systemic immune responses but also a safe and convenient way [23]. Several subunit vaccines expressed in plants are mainly targeted to virus infection since the complex of pathogenesis of bacterial infection [26]. However, the transgenic corn-derived ApxIIA has been proven to be effective in inducing immune responses, showing its potential as a promising candidate for an effective vaccine against *A. pleuropneumoniae*.

Although various countries are cultivating biotech crops since first release but the growth rate of accepting these crops is lower than estimated. Before being commercialized, biotech crop must undergo mandatory safety testing and regulatory assessment to avoid or minimize unintentionally hazardous effects to human and environment.

At present, in USA, the regulation system is vertical, which is a product-based system; regulation depends on the characteristics of biotech crops. On the other hand, the regulation is horizontal – a process-based system applied to all plants developed by genetic transformation in Europe.

In USA, the responsibility for regulating biotech crops is shared by EPA, USDA and FDA. While in Europe, the release of biotech crops is regulated by European Union Directive. The regulation contains a step-by-step approval process of assessment of the risk of biotech crops to human health and environment.

However, in Indonesia, it is controlled by the Ministry of Environment and Ministry of Agriculture, while the biosafety evaluation based on evaluation of weediness of crops is being conducted in the laboratory, controlled greenhouse facility, and controlled field facility, respectively. We urge that GM crops including corn must undergo mandatory safety testing and regulatory assessment to minimize hazardous effects on human and environment by country's own regulations, however, we consider that there should be worldwide consistency, and minimum extra mandatory regulations should be applied by every nation for biosafety evaluation of any biotech crop.

ACKNOWLEDGMENT

This study was supported by a grant (PJ008103) from the Biogreen 21 program, the Rural Development Administration (RDA).

REFERENCES

[1] M. Rosegrant, M.S. Paisner, S. Meijer, and J. Witcover, at 2001 http://www.ifpri.org/sites/default/files/publications/gfp.pdf
[2] P. Gepts, *Crop Sci.* 42, 1780 (2002)
[3] W.J, Gordon-Kamm, T. Spencer, M.L. Mangano, T.R.Adams, R.J. Daines, W.G. Start, J.V. O'Brien, S.A. Chambers, W. Jr. Adams, N.G. Willets, T.B. Rice, C.J. Mackey, R.W. Kreuger, A.P. Kausch, and P.G. Lemaux, *Plant Cell* 2: 603-618 (1990)
[4] M. Sutrino, M. Herman, S.J. Pardal, E. Listanto, A. Sisharmini, S.G. Budiarti, D. Damayanti, T.J. Santoso, and R. Wu, , S., T. Purwadaria, M. Herman, A. Rukyani, Sutrisno, H. Kasim, dan I. N. Orbani (eds.). Prosiding Ekspose Hasil Penelitian Bioteknologi Pertanian dalam rangka 25 Tahun Badan Penelitian dan Pengembangan Pertanian. Balitbangtan, Deptan (2000)
[5] K. D'Halluin, E. Bonne, M. Bossut, M.D. Beuckeleer, and J. Leemans, *Plant Cell* 4: 1495-1505 (1992)
[6] Y. Ishida, H. Saito, S. Ohta, Y. Hiei, T. Komari, and T. Kumashiro, *Nat Biotechnol* 14:745-750 (1996)
[7] D. Negrotto, M.J.S. Beer, A.R. Wench, and G. Hansen, *Plant Cell Rep* 19:798-803(2000)
[8] B.R. Frame, H. Shou, R.K. Chikwamba, Z. Zhang, C. Xiang, T.M. Fonger, S. Ellen, K. Pegg, B. Li, D.S.Nettleton, D. Pei, and K. Wang, *Plant Physiol* 129:13-22 (2002)
[9] S.D. Utomo, Pengaruh, *Buletin Agronomi* XXXIII:7-16 (2005)
[10] H.A. Kim, S.D. Utomo, S.Y. Kwon, S.R. Min, J.S. Kim, H.S. Yoo, and P.S. Choi, *Plant Biotech Rep* 3: 277-283 (2009)
[11] C.L. Armstrong, C.E. Green, and R.L. Phillips, *Maize Genet Coop Newslett* 65:92–93 (1991)
[12] C.L. Armstrong, and C.E. Green, *Planta* 164:207-214 (1985)
[13] Z.Y. Zhao, W. Gu, T. Cai, L.A. Tagliani, D.A. Hondred, D. Bond, S. Krell,M.L. Rudert, W.B. Bruce, and D.A. Pierce, *Maize Genet Coop Newslett* 72:34-37 (1998)
[14] Z.Y. Zhao, W. Gu, T. Cai T, L.A. Tagliani, D.A. Hondred, D. Bond, S. Krell, M.L. Rudert, and D.A. Pierce, *Mol Breed* 8:323-333 (2001)
[15] V. Sidorov, L. Gilbertson, P. Addae, and D. Duncan, *Plant Cell Rep* 25:320-328 (2006)
[16] J.M. Vega, W. Yu, A.R. Kennon, X. Chen, and Z.J. Zhang, *Plant Cell Rep* 27:297-305 (2008)
[17] S.M. Pescitelli, and K. Sukhapinda, *Plant Cell Rep* 14 : 712-71 (1995)

[18] B.R. Frame, H. Zhang, S. Cocciolone, L. Sidorenko, C. Dietrich, S. Pegg, S. Zhen, P. Schnable, and K. Wang, *In Vitro Cell Dev Biol-Plant* 36:21-29 (2000)

[19] S.A. Danilova, and Y.I. Dolgikh, *Russian J Plant Phys* 52:535-541 (2005)

[20] J. Frey, *Trends Microbiol* 3: 257-61 (1995)

[21] R.I. Tascon, J.A. Vazquez-Boland, C.B. Gutierrez-Martin, I. Rodriguez-Barbosa, and E.F. Rodriguez-Ferri, *Mol Microbiol* 14: 207-16 (1994)

[22] D. Reimer, J. Frey, R. Jansen, H.P. Veit, and T.J. Inzana, *Microb Pathog* 18: 197-209 (1995)

[23] S.J. Shin, J.L. Bae, Y.W. Cho, D.Y. Lee, D.H. Kim, M.S. Yang, Y.S. Jang, and H.S. Yoo, *FEMS Immunol Med Microbiol* 43: 155-64 (2005)

[24] S.J. Shin, S.W. Shin, M.L. Kang, D.Y. Lee, M.S. Yang, Y.S. Jang, and H.S. Yoo, *J Vet Sci* 8: 383-92 (2007)

[25] K.Y. Lee, D.H. Kim, T.J. Kang, J. Kim, G.H. Chung, H.S. Yoo, C.J. Arntzen, M.S. Yang, and Y.S. Jang, *FEMS Immunol Med Microbiol* 48: 381-9 (2006)

[26] J. Rice, W.M. Ainley, and P. Shewen, *Anim Health Res* Rev 6: 199-209 (2005)

[27] S.J. Streatfield, J.R. Lane, C.A. Brooks, D.K. Barker, M.L. Poage, J.M. Mayor, B.J. Lamphear, C.F. Drees, J.M. Jilka, E.E. Hood, and J.A. Howard, *Vaccine* 21: 812-5 (2003)

[28] Y. Ishida, Y. Hiei, and T. Komari, *Nature Protocols* 2: 1614-1621 (2007)

[29] Y. Hiei, S. Ohta, T. Komari, and T. Kumashiro, *Plant J* 6: 271-282 (1994)

[30] M. Cheng, T. Hu, J. Layton, C.N. Liu, and J.E. Fry, *In Vitro Cell Dev Biol Plant* 39: 595-604 (2003)

[31] S. Tingay, D. McElroy, R. Calla, S. Fieg, M. Wang, S. Thornton, and R. Bretteli, *Plant J* 11: 1369-1376 (1997)

[32] Y. Wan, J.M. Widholm, and P.G. Lemaux, *Planta* 196: 7-14 (1996)

[33] R. Brettschneider, D. Becker, and H. Lorz, *Theor Appl Genet* 94: 737-748 (1997)

[34] M.M. O'Kennedy, J.T. Berger, and T.G. Watson, *S Afr J Sci* 94: 188-192 (1998)

[35] S. Takavar, H. Rahnama, H. Rahimian, and K. Kazemitabar, *J Sci Islamic Rep Iran* 21: 21-29 (2010)

[36] A.R. Howe, C.S. Gasser, S.M. Brown, S.R. Padgette, J. Hart, G.B. parker, M.E. Fromm, and C.L. *Mol Breed* 10: 153-164 (2002)

[37] M.A.W. Hinchee, D.W. Connor-Ward, C.A. Newell, E.E. McDonnell, S.J. Sato, C.S. Gasser, D.A. Fischhoff, D.B. Re, R.T. Fraley, and R.B. Horsch, *Bio/Technol* 6:915-922 (1998)

[38] C. Koncz, and J. Schell, *Mol Gen Genet* 204:383-39(1986)

[39] P.M. Olhoft, L.E. Flagel, C.M. Donovan, and D.A. Somers, *Planta* 216:723-735(2003)

[40] B.K. Dennehey, W.L. Peterson, C. Ford-Santino M. Pajeau, and C.L. Armstrong CL, *Plant Cell Tiss Org Cult* 36:1–7(1994)

[41] S.D. Utomo, *Jurnal Agrotropika* IX :95-101(2004)

[42] H.S. Mason, H. Warzecha, T. Mor, and C.J. Arntzen, *Trends Mol Med* 8: 324-9(2002)

[43] H. Daniell, S.J. Streatfield, and K. Wycoff, *Trends Plant Sci* 6: 219-26(2001)

[44] B.J. Lamphear, J.M. Jilka, L. Kesl, M. Welter, J.A. Howard, and S.J. Streatfield, *Vaccine* 22: 2420-2424 (2004)

[45] R.I. Curtiss, and C.A. Cardineau, *World Patent Application*. WO 90/02484 (1990)

[46] H.S. Mason, L. Dominic Mam-Kit, and C.J. Arntzen, *Proc Natl Acad Sci USA* 89: 11445-11749(1992)

[47] H.S. Mason, J.M. Ball, J.J. Shi, X. Jiang, M.K. Estes, and C.J. Arntzen, *Proc Natl Acad Sci USA* 93: 5335-5340(1996)

[48] H.S. Mason, T.A. Haq, J.D. Clements, and C.J. Arntzen, *Vaccine* 16: 1336-1343(1998)

[49] M.K. Shin, M.H. Jung, W.J. Lee, P.S. Choi, Y.S. Jang, and H.S. Yoo, *J Vet Sci* 12: 401-3 (2011)

[50] H. Shams, *Vet J* 170: 289-299(2005)

[51] E.N. Meeusen, J. Walker, A. Peters, P.P. Pastoret, and G. Jungersen, *Clin Microbiol Rev* 20: 489-510 (2007)

[52] H.A. Kim, H.S. Yoo, M.S. Yang, S.Y. Kwon, J.S. Kim, P.S. Choi, *J Plant Biotechnol* 37: 313-318 (2010).

[53] M. Herman, *J. AgroBiogen* 6(2):113-125 (2010).

[54] European Parliament, *Official J. Euro Commun* 106:1-39 (2001).

[55] T.E, Nickson, *Plant Phys*.147:494-502 (2008).

[56] Government of the Republic of Indonesia, at 2010 http://www.indonesiabch.org/z_peraturan_en.php (2010).

[57] Government of the Republic of Indonesia, at 1999 http://www.indonesiabch.org/z_peraturan_en.php (1999).

[58] Government of the Republic of Indonesia, at 2005 http://www.indonesiabch.org/docs/gr21-2005.pdf

[59] C. James, at 2001 http://www.isaaa.org/resources/publications/ briefs/43/highlights/default.asp

[60] N. A. Belson, *AgBioForum*. 3: 268-280 (2000).

In: Cultivars
Editor: Katya Carbone

ISBN: 978-1-62417-101-7

Chapter 11

INFLUENCE OF NITRATE ON NODULE STRUCTURE AND NITRATE REDUCTASE ACTIVITY IN A PEANUT CULTIVAR

Cecilia Belgoff, María del Carmen Tordable and Stella Castro
Departamento de Ciencias Naturales, Facultad de Ciencias Exactas,
Físico-Químicas y Naturales, Universidad Nacional de Río Cuarto, Córdoba, Argentina

ABSTRACT

It is well documented that the addition of nitrate to legume prevents nodulation and results in senescence nodule. In the symbiotic association TEGUA peanut cultivar with *Bradyrhizobium* sp. SEMIA 6144 strain, the addition of nitrate (5 mM) caused a delay in the formation of nodules and a decrease in number and dry weight; however, the parameters related to nitrogen fixation (nitrogen content, shoot and root dry weight) remained unchanged at 45 days post-inoculation. In peanut nodules, histological examination displayed some changes in the cytology of the infected cells associated with a lower cytoplasmic density in the presence of nitrate indicating senescence nodule. It is known that nitrate assimilation is an energy-demanding process and conversion of nitrate into nitrogenous compounds proceeds via nitrate and ammonia, and the main enzyme involved in this process is the nitrate reductase (NR) which is an inducible enzyme. In this chapter, we demonstrated that the activation state of the nitrate reductase of bacteroids was similar with and without nitrate supply while it was slightly higher in nodule cytosol with nitrate. These findings suggest that nitrogen fixation and nitrate reduction may contribute significantly to the nitrogen nutrition of the TEGUA peanut plant in presence of nitrate.

INTRODUCTION

Peanut (*Arachis hypogaea* L.) is an agriculturally valuable plant with widespread distribution in the world, serving as a subsistence food crop as well as a source of various food products. It is a unique leguminous plant for its characteristic behavior to produce the pods underground in direct contact with soil. This large-seeded oilseed crop is cultivated extensively for production of edible oil and protein-rich grains. Furthermore, peanut seed contains many compounds of nutritional benefit for human health. Ahmed et al. and Awad et al. indicated peanuts as a good source of β-sitosterol, a plant sterol, believed to have anticancer properties [1,2].

Peanuts are also a source of plant proteins, dietary fiber, antioxidant vitamins, minerals, and bioactive compounds such as the flavonoids, resveratrol, and plant sterols that may further benefit human health [1,2,3]. Resveratrol is a phytoalexin and antifungal compound with human health benefits for reducing atherosclerosis and carcinogenesis [4]. Resveratrol synthesis in peanut plants increased with induced biotic and abiotic stresses such as UV light, wounding, paraquat and hydrogen peroxide addition [5]. Chen et al. indirectly associated the resveratrol level in peanut roots with their antioxidant activity in pork fat patties [6].

Like all legumes, peanut can meet its nitrogen (N) demand from different sources, such as nitrate from the soil or fertilizer and N_2 by symbiotic association with rhizobia. Peanut is usually nodulated with rhizobia belonging to the *Bradyrhizobium* genus [7,8]. The symbiotic association peanut-rhizobia is especially interesting because there are no cell-to-cell infection threads as in other legume-rhizobia symbiosis such as *Sinorhizobium meliloti*-alfalfa and *Rhizobium tropici*-bean. Instead, infection through the epidermis involves intercellular penetration (crack entry), i.e., entry at the point of emergence of lateral roots. Thus, several of the phenomena that are essential in the infection thread mode do not occur in the crack entry infection process. Rhizobia appear to enter the cortical cells through the structurally altered cell walls. Within cortical cells, rhizobia multiply rapidly, and the invaded cells divide repeatedly to form a determinate nodule [7,9].

The peanut plant has a distinct main stem and a variable number of lateral branches. The carriage of the laterals determines the growth habit of the plant. Two distinct forms of growth habit, postrate and erect, have long been recognized and have provided the basis for both agronomic and taxonomic classifications. In earlier classifications, the postrate bunch form was associated with the alternate branching pattern (subspecies *hypogaea*) and the erect habit with sequential branching pattern (subspecies *fastigiata*) [10]. Peanut cultivars differ botanically (Virginia, Spanish and Valencia types), with each type characterized by a different growth habit and branching pattern structures. The new cultivars that have been obtained from crossing these botanical types have no specific distribution and have been developed to meet environmental demands, such as a short cycle, resistance to drought, high oleic acid content, among others. Differences among cultivars in terms of nodulation (nodule weight and numbers) and the ability of symbiotic N_2 fixation have been detected [11].

Delfini et al. have investigated the response of two peanut cultivars (TEGUA and UTRE) with different growth habit and branching pattern structures, using as nitrogen sources the N-fertilizer or symbiotic N_2 fixation as well as the pattern of nitrate reductase (NR) activity distribution [12]. With regard to the contribution of nitrogen to the plant, TEGUA cultivar showed a better response to inoculation with *Bradyrhizobium* sp. SEMIA 6144, while UTRE cultivar presented a better response to nitrate (5 mM KNO_3). Both cultivars proved to have differences in the NR activity in the analyzed organs (leaves, roots and nodules) depending on supplied N source. Considering that little is known about peanut nodulation and nitrogen fixation in the presence of nitrate, the aim of this chapter is to contribute to knowledge of the effects caused by the nitrate on the anatomical and histological structures of nodules as well as the role of nitrate reductase enzyme in the TEGUA peanut cultivar.

INDUCTION OF SENESCENCE IN THE PEANUT NODULES BY EFFECT OF NITRATE

Legume nodules are unique symbiotic organs that develop on the roots and, in a few species, also on the stems, after infection with rhizobia. The plant provides sucrose to nodule host cells, where it is oxidized to dicarboxylic acids and used as energy source by the bacteroids to fix atmospheric nitrogen. Nodules have a high potential for the production of reactive oxygen species (ROS) such as singlet oxygen ($^1O_2$), superoxide anion ($O_2^{\cdot-}$), hydrogen peroxide (H_2O_2), and hydroxyl radical ($OH^{\cdot}$). This is due, among other factors, to the acid pH (5.5–6.5) and high leghemoglobin (Lb) concentration (1–2 mM) in the infected cells, the elevated rates of bacteroid respiration, and the abundance of oxygen-labile proteins [13].

The ROS molecules are highly toxic, since they are able to modify all the primary constituents of the cell such as lipids, DNA, carbohydrates, and proteins [14] leading to senescence and cell death [15,16]. However, nodules possess an array of antioxidant metabolites and enzymes that prevent the generation of highly oxidizing radicals and hence the damage of molecules. In addition, antioxidants regulate the intracellular concentrations of $O_2^{\cdot-}$, H_2O_2, which serve useful purposes in signalling stressful situations and in activating defense genes [17]. ROS are also involved in the nodule senescence process, and a model in which ROS and antioxidants interact with hormones such as abscisic acid in orchestrating nodule senescence has been proposed [18,19]. Nodules have high levels of the redox buffers glutathione and ascorbate, which are important in the nodule senescence. These metabolites decline with nitrogen fixation as the nodule ages, but the resultant decrease in redox buffering capacity does not necessarily lead to enhanced ROS [19].

Legume nitrogen fixation is particularly sensitive to environmental perturbations such as nitrate fertilization where the initial decrease of nitrogenase activity is associated with a decline in the O_2 concentration reaching the infected cells and bacteroids [20,21]. Prolongation of stress induces premature nodule senescence, which shares some features with natural senescence (nodule aging), such as the loss of nitrogen fixation, the increase in lytic activities, and the formation of green pigments from Lb [22].

There are several reports on the effect of nitrate in different legumes. In lupine, nitrogenase activity was inhibited by 35% after 3 to 5 days with 20 mM nitrate, and structural degradation of nodules was only recognizable after 10 days [23]. In bean and pea, nitrogenase activity was inhibited by approximately 85% after only 2 days with 10 mM nitrate. Structural data revealed differences in the progression of senescence between pea and bean nodules. In pea nodules the plasmolysis of host cells and the disruption of symbiosome membranes and bacteroids were already evident after 2 days, while in bean nodules, nitrate had little effect on the shape or poly-β-hydroxybutyrate content of bacteroids even after 4 days. These data suggest that pea nodules were particularly sensitive to nitrate. On the other hand, nitrate had a two-stage effect on bean nodules. In the first stage (1–2 days), there was a major decline in the nitrogenase activity, an increase in the oxygen-diffusion barrier (ODB) resistance and in the carbon cost of nitrogenase, but only moderate or no changes in the nodule content of carbohydrates, antioxidants and pyridine nucleotides. Furthermore, there was a significant increase in Lb, soluble protein and ascorbate. At this early stage, there were no detectable changes in the nodules at the ultrastructural level. In contrast, in the second stage (4 days), there was an almost complete loss of both nitrogenase activity and amount, along with a marked decline in Lb, soluble protein, sugars, antioxidants (except ascorbate), and nucleotides. There was also a general deterioration of the nodule ultrastructure, in particular of symbiosome membranes, and an accumulation of oxidized proteins [24].

Puppo et al. and Cermola et al. observed that in bean nodules with different degrees of senescence the density of bacteroids in no senescent cells was much greater than in senescent cells [19,25]. However, Müller et al. found no difference in the amount of bacteroids in senescent nodules in comparison with young nodules [26]. Trichant et al. observed a lower symbiosome density in alfalfa senescent nodules [27]. It was also found that the mitochondria of the cells in senescent nodules were grouped and their internal structure was destroyed. Lorenzo et al. showed ultrastructural alterations consisting in the loss of the outer layer of the peroxisome in senescent nodules of lupinus white (*Lupinus albus* L. cv. Multolupa) [28]. On the other hand, a decrease in the ribosome number and chromatin content of the nucleus was observed as well as changes in the rough endoplasmic reticulum, the dictyosomes of the Golgi apparatus, mitochondria and cytoplasm in senescent nodules of lupinus white (*Lupinus albus* L. cv. Stara) and yellow lupine (*Lupinus luteus* L. cv. Fakel) [29]. Vacuoles, more than any other organelle, synthesized and released endoproteases, which were mainly responsible for the degradation of cytoplasmic proteins and organelles during senescence [30]. In the process of nodule senescence the plastids were the last organelles destroyed [31].

Khetmalas and Bal distinguished different stages of *Arachis pintoi* nodule development, using three criteria: color, size, and day-post-inoculation (DPI) [32]. Nodules at different developmental stages (immature/small/white:1.0–1.5 mm in diameter, mature/ medium/ pink:1.5–2.0 mm in diameter and mature/large/red:>2.0 mm in diameter) were found from two selected plants at weekly intervals up to 10 weeks post-inoculation. In 70 day-old nodules undergoing senescence, oleosomes reappeared in the infected cells, and the senescent nodule appeared pink. The spherical bacteroids at this stage seemed to disappear and large vacuoles were found in the infected cells, which contained rod-shaped rhizobia. The senescent cells showed disruption of the tonoplasts. Both transmission and scanning electron microscopic observations support the presence of rod-shaped rhizobia during senescence. The prematurely induced senescence also showed the mixed population of spherical rod-forms along with the presence of oleosomes in the infected cells. The light microscopic observations in squash preparation of senescent nodules revealed both rod and spherical forms with some forms intermediate between them. In contrast, *Arachis. hypogaea* L. senescent nodules did not reveal such change in bacteroid forms.

In TEGUA peanut cultivar, the effect of 5 mM nitrate on the anatomical and histological structures of nodules at different stages of ontogeny (15, 30 and 45 DPI) was analyzed. The results obtained showed that exomorphological observations of the different peanut plant developmental stages revealed only presence of nodules in the roots of control plants at 15 DPI (Figure 1A-B).

After 30 days of post-inoculation, the nodules showed similar size and shape in both control and nitrate treatments (Figure 1C-D). At 45 DPI, the size of the nodules of control plants was greater than those formed in plants grown in the presence of nitrate (Figure 1E-F). At 15 DPI, the anatomical and histological structures of control roots showed the presence of lateral root primordium. Very close to these roots, the first ontogenetic stages of nodule formation consisted of parenchymatous tissue in which was not found site of infection. With the addition of nitrate, the root structures revealed the presence of lateral root in differentiation without development of the nodule. At 30 DPI, in the roots of both treatments, nodules were found with infected tissue in the center and surrounded by several layers of cortex on which there were vascular tissues immersed (Figure 2A-B). Furthermore, the parenchyma cells of the cortex had plenty of starch granules, especially the zone adjacent to the infected area (Figure 2C-D).

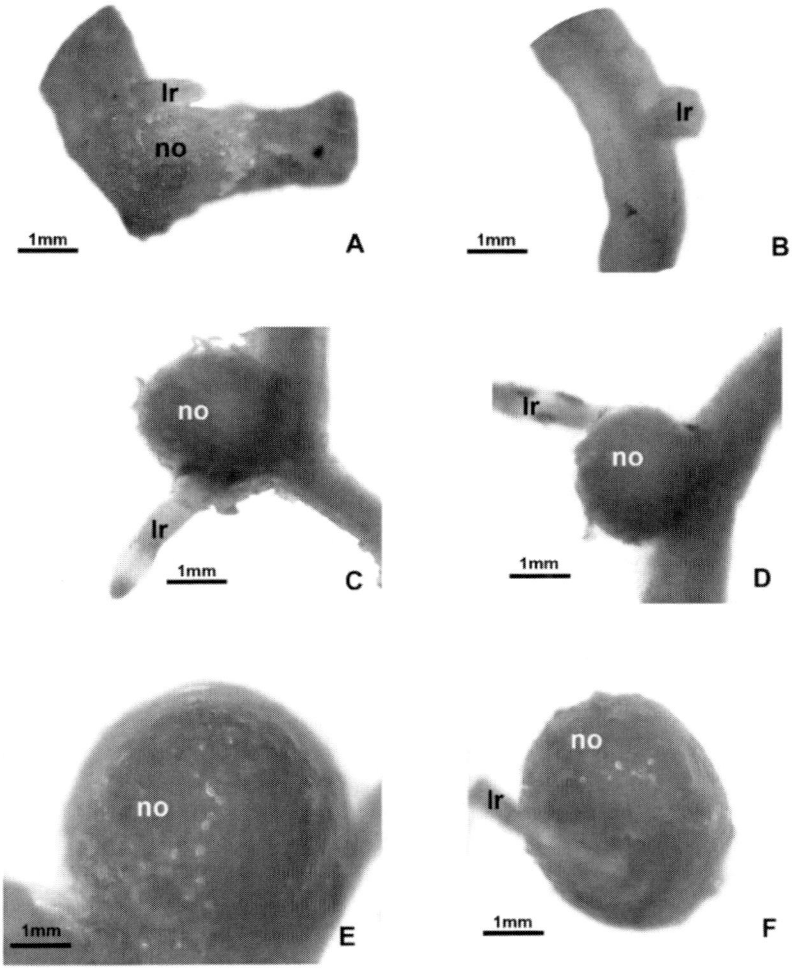

Figure 1. Exomorphological view of roots and nodules. A, C and E – control. B, D and F -nitrate. A-B, 15 DPI. C-D, 30 DPI. E-F, 15 DPI. Abbreviations: lr: lateral root; no: nodule.

In the control plants, the nodule infected area displayed cells with major cytoplasmic density and low intercellular spaces between them (Figure 2C). In contrast, in plants grown in the presence of nitrate, the nodule infected area showed a higher amount of uninfected cells and defined intercellular spaces, as well as less cytoplasmic dense in the infected cells (Figure 2D). At 45 DPI, the nodule structure and the infected area between treatments was similar to the stage of 30 DPI (Figure 2E-F). The size of the infected cells of nodules was about 25 μm for both treatments at 30 and 45 DPI.

The anatomical and histological studies of peanut nodules at 30 and 45 DPI indicated that they had similar anatomical structure, i.e., an infected central zone, active fixation, surrounded by a zone cortical with tissue vascular, in the two treatments (control and nitrate). However, histological examination showed some changes in the cytology of the infected cells associated with a lower cytoplasmic density in the presence of nitrate. This structure of the nodule could be related to the presence of senescent infected cells similar to what were found in *Arachis pintoi* nodules induced by nitrate [32]. Also, Skeokand et al. reported that nitrate caused a steady increase in the disorder in infected cells and the ratio of number of uninfected cells to infected cells in chickpea nodules [33].

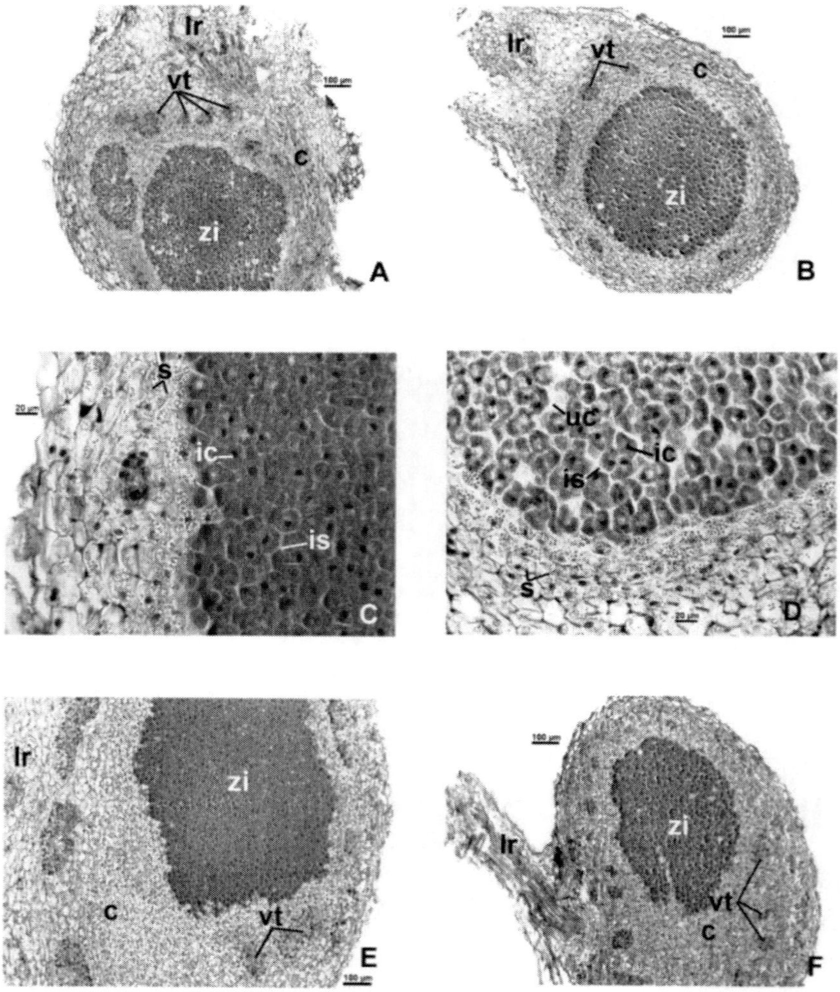

Figure 2. Light microscopy photographs of cross sections of nodules. A-control and B-nitrate, 30 DPI. C-control and D-nitrate, sector in detail of nodules. E-control and F- nitrate, 45 DPI. Abbreviations: c: cortex; ic: infected cell; is: intercellular space; lr: lateral root; s: starch; uc: uninfected cell; vt: vascular tissues; zi: zone of infection.

It is well documented that the addition of nitrate to legume prevents nodulation and results in nodule senescence and inhibition of nitrogen fixation [34,35]. However, there are differences between species and cultivars of legumes in their sensitivity to nitrate. Thus, it was found that degradative processes occurred within nodules after exposure to nitrate resulting in loss of both nodule dry weight and bacteroid protein. Nodule dry weight on a plant basis declined to a minimum value when 10 mM nitrate was applied, remaining fairly constant upon further increase of nitrate concentration even up to 200 mM nitrate. The decline of bacteroid protein content was another clear indicator of senescence. However, this decrease was only detected at concentrations above 20 mM nitrate. Decreases in nitrogenase activity, total soluble protein and Lb contents of legume nodules are generally considered as indicators of nodule senescence.

Such changes have been described for nodules of plants following application of combined nitrogen [36] or foliage removal [37], and also for nodules of plants reaching physiological maturity [38]. In order to determine if the addition of nitrate affects the growth and symbiotic N_2 fixation in TEGUA cultivar, 5 mM nitrate was added in the symbiotic association peanut-*Bradyrhizobium* sp SEMIA 6144.

The results obtained revealed a delay in the formation of nodules and a decrease in the number and dry weight; however, shoot and root dry weight as well as nitrogen fixation measured as nitrogen content remained unchanged at 45 DPI (Table 1). Thus, it is possible to infer that this concentration of nitrate was able to affect the nodulation process without changes in nitrogen metabolism in peanut plants.

Nitrogen is the major limiting nutrient for most plant species and nitrate is often the major source of inorganic N available from soil or fertilizer. Nitrate reduction has three different purposes: the utilization of nitrate as a nitrogen source for growth (nitrate assimilation), the generation of metabolic energy by using nitrate as a terminal electron acceptor (nitrate respiration), and the dissipation of excess in reducing power for redox balancing (nitrate dissimilation) [39]. Nitrate assimilation is an energy-demanding process and conversion of nitrate into nitrogenous compounds proceeds via nitrate and ammonia, and the main enzyme involved in this process is the nitrate reductase (NR), which is an inducible enzyme whose activity depends on the availability of nitrate and light [40].

Table 1. Estimation of physiological and symbiotic parameters in *Bradyrhizobium* sp. SEMIA6144 - peanut symbiosis

Nitrate (mM)	Shoot dry weight [mg plant^{-1}]	Root dry weight [mg plant^{-1}]	Nitrogen content [mg plant^{-1}]	Nodule numbers [plant^{-1}]	Nodule dry weight [mg plant^{-1}]
0	422.22±40.24a	190.10±68.19a	15.89±1.55a	38.69±1.43a	31.40±7.53a
5	478.20±64.47a	174.12±58.57a	14.32±2.14a	22.75±2.97b	12.66±1.24b

Data are mean ± standard error, n=10. Data followed by the same letter in each column do not differ significantly at $P < 0.05$, according to Duncan´s test.

REDUCTION OF NITRATE IN THE PEANUT NODULES

It is known that legumes differ considerably in their tolerance to nitrate. Some varieties of soybean were capable of fixing nitrogen despite the presence of relatively high nitrate concentration. The nitrogenase activity in yellow lupine nodules increased when plants were grown in the presence of nitrate for 40 days [41]. It was also reported that nitrate might have accessed to the inside of bacteroids as a strong induction of NR activity independently of the type of *Rhizobium* strains used as inoculant in lucerne (*Medicago sativa* cv Aragón) plants [42]. The majority of legume plants respond to the presence of nitrate in medium by loss of their protein content and nodule dry weight. The presence of active NR, however, may alter the scale of this process. In yellow lupine infected with three *Bradyrhizobium* sp. (*Lupinus*) strains, the nodule dry weight produced by the strains with NR activity (either inducible or constitutive) was higher than the nodule dry weight produced by the NR-negative strain [43]. This effect was positively correlated with the nitrogen content in plant shoots. Other symbiotic associations also showed some degree of resistance to nitrate such as root nodules initiated by *Sinorhizobium melilloti* which were capable of keeping a constant protein concentration in the presence of 5 mM nitrate [44]. These examples prompted some authors to claim that nitrate alone was unable to upset both the symbiosis itself and its initiation.

It is documented that NR activity allows nitrite to accumulate inside bacteroids only for a short time and in low quantity. Also, it was found that the different isoenzymes of NR may use its substrate in an assimilatory way helping nitrogenase to provide nitrogen-rich compounds to the legume. While others may possibly cooperate in a denitrification process to get rid of nitrogenous toxic products [45]. Delgado et al. showed that NR activity was detected in the cytosol and membranes from bacteroids of *Bradyrhizobium japonicum* L-236 strain isolated from soybean nodules that were only N_2-dependent [46]. The molecular mass (M) of the NR, estimated by gel filtration, was about 76 kDa in both (bacteroid cytosol and membranes). Treatment of plants with 20 mM nitrate induced NR activity in the bacteroid cytosol, but did not affect NR activity in the membranes. Thus, the M of the cytosolic enzyme was 160 kDa, while that of the NR from the membranes remained unchanged (76 kDa).

Modulation of NR activity involves first phosphorylation of a conserved serine residue, and a subsequent binding of proteins of the 14-3-3 family which cause inactivation of the enzyme [47]. The activation state of NR was determined as ratio of NR activity assayed *in vitro* in the presence of Mg^{2+} to total NR activity measured with EDTA (Table 2). In the presence of Mg^{2+}, phosphorylated NR is kept in its inactive form (complexed with inhibitory 14-3-3 proteins). Thus, the NR activity measured reflects the amount of dephosphorylated NR present in the extract, whereas in the absence of Mg^{2+} (chelated by EDTA), the total extractable NR is revealed. Our findings demonstrated that the activation state of NR in bacteroids was similar with and without nitrate supply, while the percentage of NR active was slightly higher in nodule cytosol in the presence of nitrate than the same without nitrate.

Table 2. Nitrate reductase (NR) activity (nmol NO_2^- min^{-1} mg^{-1} protein) in response to the addition of 5 mM of nitrate. The activation state of the NR is defined as the ratio of activity assayed in the presence of Mg^{2+} to activity in the presence of EDTA

	Bacteroids		Cytosol	
	Control	Nitrate	Control	Nitrate
NRA with EDTA	22.69±3.42a	36.02±2.52b	1.80±0.65a	2.74± 0.47a
NRA with Mg^{2+}	17.94±2.92a	28.84±1.27a	0.89±0.43a	1.92± 0.65a
Activation state (%)	79	78	50	70

Data are mean ± standard error, n=5. Data followed by the same letter in each column do not differ significantly at $P < 0.05$, according to Duncan´s test.

Our previous work on the pattern of NR activity distribution in TEGUA cultivar showed a higher NR activity in the nodules compared with leaves and roots [12]. These results agreed with those reported by Polcyn and Lucinski who found that NR activity in nodules of yellow lupine was higher than in other parts of the plant [48]. Chamber-Pérez et al. demonstrated a high positive correlation between NR activity and the majority of the parameters related to nitrogen fixation (shoot dry and fresh weight, total nitrogen content and nitrogenase activity), suggesting that NR activity may play functions complementary to those of nitrogenase in lupine nodules [49]. In this work, our findings revealed the presence of an active NR in the bacteroids of peanut nodules which may contribute significantly to the nitrogen nutrition of the plant by reducing nitrate.

CONCLUSION

The influence of nitrate on the symbiotic association peanut-*Bradyrhizobium* sp. SEMIA6144 showed a delay in the formation of nodules correlated with a decrease in the number of nodules. However, it did not affect plant growth and biological nitrogen fixation. Furthermore, an active bacteroidal nitrate reductase was found in nodules which allowed the survival of bacteria inside them and nitrogen fixation process. These findings suggest that nitrogen fixation and nitrate reduction may contribute significantly to the nitrogen nutrition of the TEGUA peanut plant in presence of nitrate.

REFERENCES

[1] E. M. Ahmed and C. T. Young, *Peanut Science and Technology,* Texas (1982).

[2] A B. Awad, K. C. Chan, A. C. Downie and C. S. Fink, *Nut. Cancer*, 36, 238 (2000).

[3] T. H. Sanders, R. W. McMichael and K.W. Hendrix, *J. Agric. Food Chem,* 48, 1243 (2000).

[4] G. J. Soleas, E. P. Diamondis and D. M. Goldberg, *Clin. Biochem*, 30, 91 (1997).

[5] M. Chung, M. R. Park, J. C. Chun and S. J. Yun, *Plant Sci,*164, 103 (2003).

[6] R. S. Chen, P. L. Wu and Y. Y. Chiou, *J. Agric. Food Chem,* 50, 1665 (2002).

[7] F. Boogerd and D. Van Rossum, *Microbiol. Rev*, 21, 5 (1997).

[8] T. Taurian, O. M. Aguilar and A. Fabra, *Symbiosis* 33, 59 (2002).

[9] A. Fabra, S. Castro, T. Taurian, J. Angelini, F. Ibañez, M. Dardanelli, M. Tonelli, E. Bianucci and L. Valetti, *Crit. Rev. Microbiol,* 3, 179 (2010).

[10] A. Krapovickas, Evolution of the genus *Arachis,* Jerusalem (1973), pp131–151.

[11] J. C. Wynne, G. H. Elkan, T. G. Isleib and T. J. Schneeweis, *Peanut Sci,* 10, 110 (1983).

[12] R. Delfini, C. Belgoff, E. Fernández, A. Fabra and S. Castro, *Plant Growth Regul*, 61, 153 (2010).

[13] D. A. Dalton, S. A. Russell, F. J. Hanus, G. A. Pascoe and H. J. Evans, *PNAS,* 83, 3811 (1986).

[14] M. Moller, P. E. Jensen, and A. Hansson, *Annu. Rev. Plant Biol*, 58, 459 (2007).

[15] K. Overmyer, M. Brosché and J. Kangasjärvi, *Trends Plant Sci*, 8, 335 (2003).

[16] R. Rivero, M. Kojima, A. Gepstein, H. Sakakibara, R. Mitler, S. Gepstein and E. Blumwald, *PNAS*, 104, 19631 (2007).
[17] R. Mittler, S. Vanderauwera, M. Gollery and F. Van Breusegem, *Trends Plant Sci*, 9, 490 (2004).
[18] K. Groten, H. Vanacker, C. Dutilleul, F. Bastian, S. Bernard, R. Carzaniga and C. H. Foyer, *Plant Cell Environ*, 28, 1293 (2005).
[19] A. Puppo, K. Groten,F. Bastian, R. Carzaniga, M. Soussi, M. M. Lucas, M. R. de Felipe, J. Harrison, H. Vanacker and C. Foyer, *New Phytol*, 165, 683 (2005).
[20] D. B. Layzell, S. Hunt and G. R. Palme, *Plant Physiology*, 92, 1101 (1990).
[21] P. R. Escuredo, F. R. Minchin, Y. Gogorcena, I. Iturbe-Ormaetxe, R. V. Klucas and M. Becana, *Plant Physiol*, 110, 1187 (1996).
[22] G. Sarath, N. E. Pfeiffer, C. S. Sodhi and F. W. Wagner, *Plant Physiol*, 82, 346 (1986).
[23] K. Swaraj, J. S. Laura and N. R. Bishnoi, *J. Plant Physiol*, 141, 202 (1993).
[24] M. A. Matamoros, L. Baird, P. Escuredo, D. A. Dalton, F. Minchin, I. Iturbe-Ormaetxe, M. C. Rubio, J. Morán, A. Gordon and M. Becana, *Plant Physiol*, 121, 97 (1999).
[25] M. Cermola, E. Fedorova, R. Taté, A. Riccio, R. Favre and E. J. Patriarca, *MPMI*, 13, 733 (2000).
[26] J. Müller, A. Wimken, T. Boller, *J. Exp. Bot,* 52, 2181 (2001).
[27] J.C. Trinchant and J. Rigaud, *Arch. Microbiol.* 124, 49 (1980).
[28] C. Lorenzo, M. M. Lucas, A. Vivo and M. R. de Felipe, *J. Exp. Bot,* 41, 1573 (1990).
[29] N. I. Andreeva, G. M. Kozharinova and S. F. Izmailov, S. F. *Russian J. Plant Physiol,* 45, 101 1998.
[30] R. C. Huffaker, *New Phytol,* 116, 199 (1990).
[31] M. M. Lucas, G. Van de Sype, D. Hérouart, M. J. Hernández, and A. Puppo and M. R. de Felipe, *Protoplasma,* 204, 61 (1998).
[32] M. Khetmalas and A. Bal, *Plant Sci*, 168, 557 (2005).
[33] S. Sheokand, S. Babber and K. Swaraj, *Biol. Plantarum*, 41, 435 (1998).
[34] M. Becana and J. Sprent, *Physiol. Plantarum*, 70, 757 (1987).
[35] J. G. Streeter, *Crit. Rev. Plant Sci,* 7, 1 (1988).
[36] P. C. Chen and D. A. Phillips, *Plant Physiol*, 59, 440 (1977).
[37] C. P. Vance, G. H. Heichel, D. K. Barnes, J. W. Bryan and L. E. Johnson, *Plant Physiol*, 64, 1 (1979).
[38] N. E. Pfeiffer, C. M. Torres and F. W. Wagner, *Plant Physiol*, 71, 797 (1983).
[39] C. Moreno-Vivián, P. Cabello, M. Martínez-Luque, R. Blasco and F. Castillo, *J. Bacteriol*, 181, 6573 (1999).
[40] W. H.Campbell, *Annu. Rev. Plant Physiol. Plant Mol. Biol*, 50, 277 (1999).
[41] P. Lang, R. Martin and M. P. Golvano, *Plant Physiol. Biochem*, 31, 639 (1993).
[42] C. Arrese-Igor, J. Garacía-Plazaola, A. Hernández and P. M. Aparicio-Tejo, *Physiol. Plantarum*, 80, 185 (1990).
[43] A. Serrano and M. Chamber, M. *J. Plant Physiol*, 136, 240 (1990).
C. Arrese-Igor, F. R. Minchin, A. J. Gordon and A. K. Nath, *J. Exp. Bot,* 48, 905 (1997).
[44] M. Chamber-Pérez, M. Camacho, A. Burgos, M. Lucas, M. Pascual, J. Manclús and M. de Felipe, *J. Plant Physiol*, 159, 525 (2002).
[45] M. Delgado, M. Fernández- López and E. Bedmar, *Plant Physiol. Biochem*, 36, 279 (1998).
[46] W. M. Kaiser and S. C. Huber, *J. Exp. Bot,* 52,1981 (2001).

[47] W. Polcyn and R. Lucinski, *Physiol. Plantarum*, 158, 829 (2001).
[48] M. A. Chamber-Pérez, M. Camacho-Martínez and J. J. Soriano-Niebla, *J. Plant Physiol*, 150, 92 (1997).

In: Cultivars
Editor: Katya Carbone

ISBN: 978-1-62417-101-7

Chapter 12

FATTY ACID UN-SATURATION IN THE RESPONSE OF TOMATO TO TEMPERATURE STRESS

Anita Maienza[1], Ramireddy Sanampudi Venkata[2], Silvia Rita Stazi[1]**, Stefano Grego[2] and Andrea Mazzucato[2]

[1]Department for Innovation of Biological Systems, Agrifood and Forestry (DIBAF), Tuscia University, Viterbo, Italy
[2]Department of Science and Technology for Agriculture, Forestry, Nature and Energy (DAFNE), Tuscia University, Viterbo, Italy

ABSTRACT

Plant survival under abiotic stresses, like thermal stress, requires the retaining of structural and functional composition of cell membranes. Among the others factors, maintaining biological membranes fluidity is thought to help improving the resistance to these kinds of stresses. In fact, adjustment of membrane fluidity makes it possible to maintain an environment suitable for the functioning of critical integral proteins during stress. Cold tolerance has been correlated with the degree of unsaturation of fatty acids in the phospholipids and glycolipids composing the membranes. Indeed, plants respond to cold temperatures by releasing α-linolenic acid (18:3) from membranes and significantly increasing tri-unsaturated linolenic acid at low temperature exposition. This observation suggests that possibly the degree of unsaturation of membrane lipid fatty acids is a factor helping also the resistance at high temperatures. However, heating stress has been less investigated and the role of fatty acids in preserving plants' functionality has not been confirmed. Our experiments confirm that the higher concentrations of saturated fatty acids help improving high temperature tolerance of plants. In the case of tomato plants, this can be obtained by silencing the *Fatty Acid Desaturase 7* (*FAD7*) gene.

[*]E-mail address: srstazi@unitus.it

INTRODUCTION

Plants often are subjected to stresses due to low or high temperatures, and high salt and heavy metal concentration in soils. These stresses can significantly reduce the productivity of food and fiber crops that, together with other sources of abiotic stresses (as those caused by global climate changes, etc.), makes it very important to understand how plants can resist to them. Biomembranes provide the major sites for the perception and propagation of different abiotic stress signals, leading to alleviation of deleterious stress effects [1]. Plant cell membranes contain different type of lipid molecules as building blocks. In this chapter, we show that it is the change of concentration of the different types of lipid molecules determining the tolerance of plants to some form of abiotic stresses.

At first, we summarize the characteristics of the relevant types of lipids, namely glycolipids and phospholipids. These two classes of lipids share the same general structure, consisting of a hydrophilic head and a hydrophobic end corresponding to the associated fatty acid. The contemporary presence of a hydrophilic head and a hydrophobic end gives to lipids an amphiphilic nature. The structure of the associated fatty acid determines the functional characteristics of a lipid. The hydrophobic chain of lipids present in membranes of plants typically contains up to 20 carbon atoms. If these carbon atoms are all single-bonded the chain is said saturated. If they are bonded by one or more double bonds, they are said monounsaturated or polyunsaturated, respectively. Mono and polyunsaturated fatty acids can present cis/trans isomerism.

The nomenclature of fatty acids follows the pattern of A:BωC, where "A" denotes the total number of carbon atoms in the acid, "B" the number of double bonds from the aliphatic (ω) end of the molecule, and "C" the carbon atom from the aliphatic end before the double bond. This is followed by a "c" for *cis* or a "t" for *trans* configuration. The prefixes "i", "a", or "br" indicate *iso, anteiso,* or branched (position undetermined). Mid-chain branching is denoted by "me", and cyclopropyl fatty acids are indicated with the prefix "cy", with the ring position indicated from the aliphatic end of the molecule [2]. Examples of fatty acids, together with their nomenclature, are reported in Table1 and Figure 1.

Lipid molecules and their derivatives play the role of bioactive agents to mediate stress signal in plant cells [1]. The presence of mono/polyunsatured acids helps keeping membrane fluidity at an optimal value by introducing bends or kinks in the fatty acids chains, resulting in a higher tolerance to temperature changes [3]. Plant cell membranes can change the lipid and fatty acid composition (and saturation degree) according to the environmental conditions [4, 5], wounding [6], and pathogen invasion [7]. Indeed, the growth of plants at high temperature is thought to be influenced by the fatty acid composition of membrane lipids [8]. Changes in their saturation degree are mediated by changes in unsaturated fatty acid concentration, a function provided, in part, by the regulatory activity of fatty acid desaturases. In particular, release of alfa-linolenic acid (18:3ω3) from membrane lipids in response to abiotic stresses is regulated by lipase activity [9]. The main objective of our research is, therefore, to further investigate this subject by studying the possibility to use genetic engineering to influence the level of unsaturation of fatty acids, validating their ability to improve plants' adaptability to high temperature conditions.

This chapter focuses on changes in fatty acid saturation of cellular membrane of *Solanum lycopersicum* L. as a response to high temperature stresses. Moreover, it will be discussed a case study where the activity of a fatty acid desaturase gene has been silenced in order to confer to the plant increased tolerance to heat.

Table 1. Main fatty acids present in plant cells

	Saturated	
10:0	Caprinic acid	$C_{10}H_{20}O_2CH_3$ $(CH_2)_8COOH$
12:0	Lauric acid	$C_{12}H_{24}O_2CH_3$ $(CH_2)_{10}COOH$
14:0	Myristic acid	$C_{14}H_{28}O_2CH_3$ $(CH_2)_{12}COOH$
16:0	Palmiltic acid	$C_{16}H_{32}O_2CH_3$ $(CH_2)_{14}COOH$
17:0	Margaric acid	$C_{17}H_{34}O_2CH_3$ $(CH_2)_{15}COOH$
18:0	Stearic acid	$C_{18}H_{36}O_2CH_3$ $(CH_2)_{16}COOH$
	Monounsaturated	
16:1ω7	Palmitoleic acid	$C_{16}H_{30}O_2$ $CH_3(CH_2)_5CH{=}CH(CH_2)_7COOH$
18:1ω9	Oleic acid	$C_{18}H_{34}O_2$ $CH_3(CH_2)_7CH{=}CH(CH_2)_7COOH$
20:1ω9	Gadoleic acid	$C_{20}H_{38}O_2$ $CH_3(CH_2)_7CH{=}CH(CH_2)_9COOH$
22:1ω11	Cetoleic acid	$C_{22}H_{42}O_2$ $CH_3(CH_2)_9CH{=}CH(CH_2)_9COOH$
22:1ω9	Erucic acid	$C_{22}H_{42}O_2$ $CH_3(CH_2)_7CH{=}CH(CH_2)_{11}COOH$
	Polyunsaturated	
18:2ω6	Linoleic acid	$C_{18}H_{32}O_2$
18:3ω3	Linolenic acid	$C_{18}H_{30}O_2$
18:4ω3	Stearidonic acid	$C_{18}H_{28}O_2$

Figure 1. Above are given examples of omega-9 (n-9), omega-6 (n-6) or omega-3 (n-3) fatty acids. Nomenclature of polyunsaturated fatty acids as a function of the position of the first double bond in the molecule.

PLANT RESISTANCE MECHANISM TO HEAT STRESS

Plants react to heat stress by several mechanisms, including synthesis of specialized proteins, change of membrane fluidity, osmoprotection, production of late embryogenesis abundant proteins [10]. Additional factors involve signaling cascades and transcriptional control, which can counteract the stress effects [11].

Membrane lipid saturation is considered an important element in high temperature tolerance. Promoting the change of the unsaturation level of fatty acids could represent a useful method for enhancing the tolerance of plants to environmental stresses, especially temperature [12]. This mechanism is sketched in Figure 2 [9].

Trienoic acids (TAs), hexadecatrienoic acid (16:3ω3) and linolenic acid (18:3ω3) are the principal polyunsaturated fatty acids present in membrane lipids. Plastid omega 3 fatty acid desaturase (FAD7) catalyzes the conversion of dienoic fatty acids (16:2 and 18:2) to trienoic fatty acids (16:3ω3 and 18:3ω3) of glycerolipids, which are the main constituents of chloroplast membranes. The reactions catalyzed by FAD7 have been hypothesized to be the rate-limiting step of the production of trienoic fatty acids TAs [10]. Therefore, the control of the expression of FAD genes by genetic engineering is one of the most effective methods to modify the content of dienoic and trienoic fatty acids and to regulate the response of the plant to heat stress.

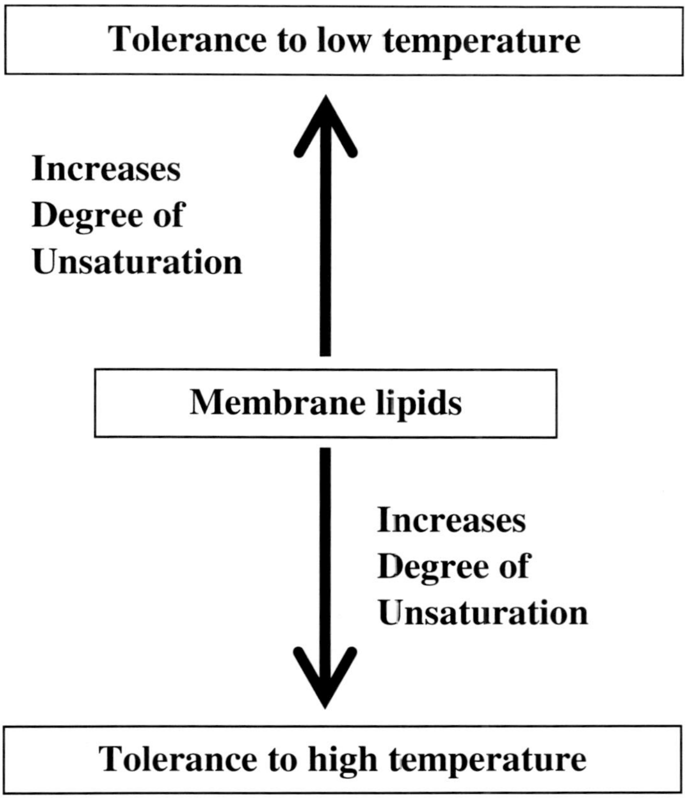

Figure 2. Schematic representation of the influence of saturation levels of membrane lipids on the temperature sensitivity of plants [9].

EFFECT OF HIGH TEMPERATURE ON TOMATO PRODUCTION

Tomato (*Solanum lycopersicum* L.) belongs to the Solanaceae family and is an important vegetable crop all over the world. It is used in various forms, in fresh salads, cooked foods and in processed forms like ketchup, paste etc. In terms of human diet, tomato is a major component of daily meals in many countries and constitutes an excellent source of health-promoting compounds thanks to its balanced mixture of minerals and antioxidants, including vitamins C and E, lycopene, β-carotene, lutein and flavonoids such as quercetin [13].

In addition to genetic determinants, the quality of tomato fruits depends on environmental factors, such as light and temperature, composition of the nutrient solution and crop management.

Tomato thrives well in tropical and subtropical regions. In general, it can grow vigorously and is highly productive in the 18–28°C temperature range [14]. In the Mediterranean regions, the control of climatic conditions in glasshouses in summer is problematic because of high light intensity, elevated temperatures and high air vapor pressure deficit. In fact, the summer production of fresh tomatoes grown in glasshouses in the Mediterranean basin usually shows a yield depression, together with an increased occurrence of in blossom-end-rot and cracking. Nevertheless, an increase in dry matter and sugar content promoting eating quality was also observed [15].

Tomato plants have, within certain limits, the ability to counterbalance the effect of thermal stresses. As a result, plants exposed to a fluctuating temperature regime often suffer no overall loss of yield [11, 16] or change in the dry matter partitioning [17] when compared with those grown in a constant regime having the same mean temperature. However, temperature fluctuations may affect the pattern of crop yield (i.e. the rate of developmental events), such as fruit ripening, which has been shown to be determined largely by temperature [18]. For example, commercial crop yield presents large weekly fluctuations. This is, probably, the result of change of ripening time of fruits [19]. The major factors affecting production are day and night temperatures [20], daytime temperatures resulting the most critical factor. Fruit number, fruit weight and number of seeds per fruit at a mean daytime temperature of 29°C are markedly lower than the corresponding values at 25°C. Consistent results have been found by other authors [21].

It is worth mentioning that temperature also affects other factors, such as the rate of fruit growth [22] and the partitioning of assimilates between vegetative and generative parts. This latter effect may, in turn, indirectly affect the rate of fruit development or abortion, and the fruit set. In particular, moderate increase of average temperature is likely to cause a significant reduction in tomato fruit set, which is attributed to the disruption of specific physiological and molecular events during the narrow window of male reproductive development [23]. In the following paragraphs we will analyze the effect of temperature on several aspects of tomato life cycle.

Effect of Temperature on Flower Drop

High temperatures stress induces flower abscission by limiting the supply of nutrients to the sink tissues, leading to a reduced fruit set in *S. lycopersicum*. An additional possible reason for the flower abscission is an increased production of ethylene by reproductive organs [24].

According to Roberts [25], high temperature induced flower abscission is prevented or reduced by the action of auxin diffusion down to the pedicel. In fact, this prevents the separation of cells of the abscission layer.

Effect of Temperature on Anther Dehiscence, Pollination and Fertilization

High temperature stress disturbs the process of pollen release, the process of transfer of pollen to the stigma surface and, finally, the process of fertilization affecting the growth of pollen tubes. Moreover, high temperature stress also affect the number pollen grain produced, the number of pollen tubes penetrating the stigma [26] and the rigidity of the pollen tube wall. The exposure of tomato flowers to 40°C for 18 h after pollination, leads to ovary abortion and endosperm degeneration due to the inhibition of pollen tube growth and failure of fertilization [27]. High temperatures negatively regulates the activity of S-adenosyl methionine decarboxylase (SAMDC), which is involved in the biosynthesis of polyamines, like spermidine and spermine, by reducing their availability to germinating pollen grains [28]. An additional effect of high temperatures, shown by *in vitro* incubation of pollen grains, is a reduced respiratory activity over time [29].

Effect of Temperature on Fruit Set

The fruit set process is extremely important in most of the crops because it is directly related to their yield and productivity. The fruit set process is influenced by a number of biotic and abiotic stresses, among which high temperatures is the most important. It has been shown that thermal stress is of particular importance for tomato [30]. Sato et al. [31] showed that a mean daily temperature of 29°C reduces the fruit number, fruit weight and number of seeds per fruit in comparison to normal mean daily temperature of 25°C. At high daytime temperatures, tomato shows the phenotype of increased style exertion as a dominant factor, which affects the fruit set by reducing the chances of perfect pollination and fertilization [32]. High night temperatures reduce fruit set by showing detrimental effect on functionality of female reproductive organs [21]. A lower percentage of fruit set of tomato under higher temperatures is also caused by degeneration of endosperm between 24 and 96 h after pollination [27].

High Temperature Tolerance Mechanism Associated to Fatty Acid Composition in Leaves of Transgenic Plants of Solanum lycopersicum L - A Case Study

Literature indicates that mutations in the *FAD7* gene sequence produce no specific effects on the length of carbon chain (16- or 18-carbon), glycerol backbone (sn-1 or sn-2), or lipid head group [33]. Therefore, the observed increased tolerance of plants to high temperatures with mutated *FAD7* alleles cannot be due to any of these effects. Previous investigations on tobacco plants [34] have shown that a reduction in the amount of TAs results in an improved resistance to high temperatures. The reduction of TAs has been obtained by gene silencing of the chloroplast ω-3 desaturase in transgenic plants. Our aim is to extend this approach to tomato and try to identify the mechanism underneath the enhanced tolerance of this plant to temperature stress. The hypothesis discussed in this chapter is that transcription reduction, obtained by silencing the *FAD7* gene, can affect the degree of saturation of fatty acids of membrane lipids, and that it is the changed degree of saturation of membrane fatty acids that improves plant tolerance to high temperatures.

To check whether tomato plants genetically engineered as described above improve their resistance to thermal stress, and validate the hypothesis on the mechanism of this improved resistance, we have grown control and silenced plants at high temperature (T = 32°C). In both cases we measured the degree of unsaturation of fatty acids in the lipidic membrane. In practice, tomato cotyledons (cv Microtom) were transformed by co-cultivation with a strain of *Agrobacterium tumefaciens* containing the plasmid vector pBI121 with portions of the tobacco *FAD7* gene cloned in sense and antisense orientation. Figure 3 shows the semi-quantitative reverse transcription-polymerase chain reaction (RT-PCR) analysis of *FAD7* gene expression, which confirmed the complete silencing of the gene in transformed plants.

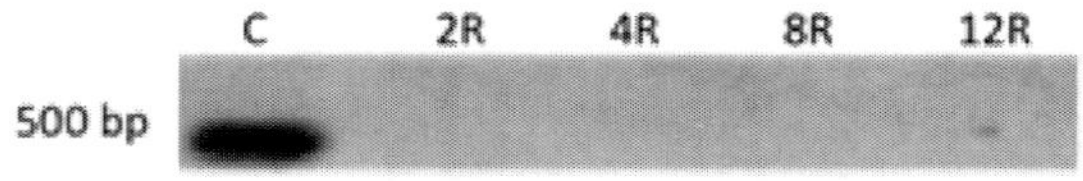

Figure 3. Amplification of FAD7 in the control but not in the transgenic plants.

Plants transformed as described above were grown under thermal stress in a growth chamber at 35°C with a photoperiod light/dark of 16/8 hours. To assess the effect of the transgene in seed germination and seedling growth, seeds from transgenic and non-transgenic plants were treated for 48 h at a temperature of 37 °C in the dark and subsequently placed in standard germination conditions. Below we report the experimental protocol for fatty acids extraction and analysis.

Fatty Acid Extraction and Analysis

We collected 1 g of tissue from four genotypes of FAD7-silenced plants and untransformed control plants undergone to the treatment described above. Plant tissues were grounded in liquid nitrogen to a fine powder using a mortar and pestle. The powdered tissue was treated for the extraction of the trienoic linolenic acid from the total plant lipids. FA extraction was done using a modified version of the procedure suggested by Bligh and Dyer [35]. The lipids were extracted from the leaves by adding a solution of phosphate-buffer (pH 7.4), methanol, chloroform in the ratio 0.8:2:1 (v/v), incubated for a minimum of 2 h and then extracted again with the same solution but with different concentrations (buffer, methanol, chloroform 1.8:2:2). Samples were then incubated overnight. After that, the samples were centrifuged for 20 min at 4000 rpm to separate the solvent from the leaves, and using a separating funnel to separate the phases. The organic phases of the samples were then dried with anhydrous sodium sulfate, filtered (Whatman 41) and dried again by rotary evaporation prior to storage at -20°C. Total lipids were fractionated into neutral lipids, glycolipids and phospholipids by silicic acid column solid-phase extraction (SPE) (3ml/500 mg silica Sep-pak VacTM, Waters) with acetone-chloroform-methanol solutions. The three portions were dried by rotary evaporation and then stored at -20°C. The phospholipids were then subjected to a mild alkaline methylation to trans-esterify fatty acids into methyl esters. The organic phases were analyzed for fatty acid methyl esters (FAME).

FAME were separated and quantified by gas chromatography-mass spectrometry (GC-MS). Samples were dissolved in dichloromethane, containing nonadecanoic acid methyl ester (C19:0) as internal standard and injected into a GC-MS (Shimadzu GCMSQP5050), equipped with a VF-5ms column, (30 m length, 0.25 mm internal diameter), with helium as carrier gas. The oven was programmed from 89°C (hold for 2 min) to 280°C (hold for 5 min), at a temperature rate of 6.0°C/min. Fatty acids were identified by relative retention times compared with standards (FAME 37 47885-4; Supelco, Inc.). Single FAME were identified and expressed as percentage over the total amount of fatty acids (% abundance)

Results and Discussion

Productivity

We compared the productivity of transformed and control plants in terms of fruit number per 12 plants from the first three trusses, which were grown under high temperature (Table 2).

Table 2. Productivity of *FAD7* silenced and untransformed plants

Plant [a]	Fruit Number [b]
C	25 (0.96)
2R	76 (1.21)
4R	32 (0.69)
8R	30 (0.90)
12R	43 (0.86)

[a] 2R, 4R, 8R and 12R are four samples of primary transformants taken out of 12 plants undergone to the same treatment. [b] Standard errors of the mean (SEM) are given in brackets.

Table 3. Seed germination percentage of transformed and control plants at 35°C

Plant	Seed germination[a] (%)
Control	66.66 (1.45)
2R	51.66 (1.66)
4R	93.33 (0.66)
8R	98.33 (0.33)
12R	100.00 (0.00)

[a]Standard errors of the mean (SEM) given in brackets.

Table 4. Mean hypocotyl length in control (C) and transformed plants (2R to 12R) after germination and seedling growth under heat stress

Plant	H.L.[a] (cm)
	32°C
C	3.5 (0.5)
2R	2.3 (0.1)
4R	5.1 (0.1)
8R	4.0 (0.2)
12R	5.5 (0.2)

[a]Standard error of the mean (SEM) is given in brackets.

We found that the productivity of silenced plants was significantly higher than that of control ones. We also found that the rate of seed germination in transformed plants under heat stress was higher than in untransformed control plants (Table 3), even though the variation of this trait among transformed plants is relatively high. Also the length of hypocotyl of transformed plants was longer than that of control plants (Table 4, Figure 4).

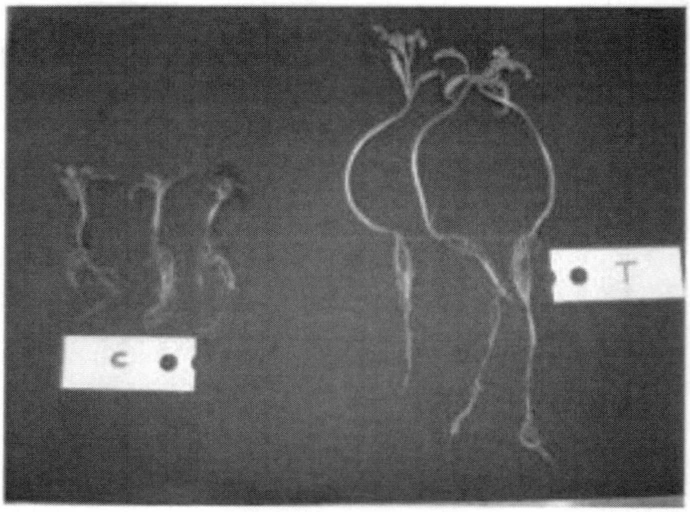

Figure 4. Seedling growth of plants transformed with *FAD7* (T) and control (C) plants at 35°C.

Our results are in agreement with Corbineau et al. [36]. Similar results were also found in different plant species like amaranth [37], cocklebur [38], tobacco [39] rapeseed and flax [40].

Fatty Acids Measurement

Our experiments show that with the exception of 8R, the silencing of *FAD7* produces an increase of stearic acid (18:0) in tomato lipid membranes. This increase is larger in the 4R and 12R samples. These two samples show also a sizable reduction of C18:3 (about 40%, Table 5). This reduction is in agreement with the observed phenotype of an increased tolerance to high temperatures stress.

At the same time we observe an increase in C16:0 (with the exception of 2R). Thus, we observed an overall increase in saturation of the lipid membrane fatty acids in the plants genetically engineered to silence *FAD7*. These results, in conjunction with those on plant productivity, support the hypothesis that saturation of lipid membrane fatty acids is the principal mechanism of plant tolerance to high temperatures stress [41].

Table 5. Percentage of various fatty acid in control plants (C) and plants silenced for FAD7 gene grown at 35°C[a]

PLANT	% 16:0	% 18:0	% 18:2	% 18:3
C	35.0[b]	1.4[c]	19.3[a]	44.2[a]
2R	34.7[b]	3.9[b]	12.7[a]	40.8[a]
4R	54.9[a]	5.5[a]	15.7[a]	23.7[b]
8R	42.4[b]	2.7[b]	18.3[a]	36.4[a]
12R	54.6[a]	4.8[a]	16.4[a]	24.1[b]

[a]Statistical analysis was performed using the statistical package SPSS 17.0.1. The effect of treatments on total fatty acid content was assessed by analysis of variance (ANOVA). Letter denote significant differences ($P \leq 0.05$) between fatty acids

CONCLUSION

High temperatures affect growth and productivity of plants: vegetative growth, flowering, pollen development, fruit set and, finally, total yield of crop plants.

However, there are plants that resist better to thermal stresses. Understanding the mechanism of temperature tolerance of these plants, and inducing analogous mechanisms in other plants, can represent a viable approach to increase the productivity of high value crops and/or counterbalancing the effect of global warming [42]. This approach has already been applied with some successes to a number of species [43, 44]. For agricultural crops, durable stress resistance will likely require the directed regulation of multiple genes. Thus, the identification of stress resistance-genes relationships is a crucial step forward.

In this chapter, we have presented results of our experiments on tomato plants. We have shown that silencing of the *FAD7* gene results in an increase of plant growth and productivity under high temperature conditions. Our results on the composition of lipid membrane fatty acids support the hypothesis that the improved tolerance to high temperatures is due to an increased level of their saturation.

ACKNOWLEDGMENT

The authors acknowledge Dr. Simone Meloni for critical reading of the manuscript, Dr Tatsuro Hamada for kindly providing the construct for silencing the FAD7 gene, Mr. Maurizio Enea Picarella for technical assistance and Dr. Daniela Dalmonech for fruitful discussions.

REFERENCES

[1] W. Wang, B. Vinocur, O. Shoseyov, A. Altman, *Trends Plant Sci.* 9, 244 (2004).
[2] Ringelberg, J.D. Davis, G.A. Smith, S.M. Pfiffner, P.D . Nichols, J.S. Nickels, J.M. Henson, J.T. Wilson, M. Yates, D.H. Kampbell, H.W. Read, T.T. Stocksdale & D.C. White, *FEMS Microbiol. Ecol.* 62, 39 (1989).
[3] A.L. Lehninger, Biochemistry 2nd ed Worth Publ, New York (1977).
[4] Y. Murakami, M. Tsuyama, Y. Kobayashi, Y.H. Kodama, K. Iba, *Science.* 287, 476 (2000).
[5] N. Murata, O. Ishizaki-Nishizawa, S. Higashi, H. Hayashi, Y. Tasaka, *Nature.* 356, 710 (1992).
[6] N. Takumi, K. Hiroaki, Y. Shuichi, K. Iba, *Plant Physiol.* 121, 1239 (1999).
[7] T. Yaeno, O. Matsuda, K. Iba, *Plant J.* 40, 931 (2004).
[8] M. Horikawa, K. Sakamoto, *Biochem. Bioph. Res. Co.* 390, 1402 (2009).
[9] A. Grover, M Agarwal, S. Katiyar-Agarwal, C. Sahi, S. Agarwal. *Curr. Sci. India.* 79, 557 (2000).
[10] K. Iba, *Annu. Rev. Plant. Biol.* 53, 225 (2002).
[11] R.G. Hurd, C.J. Graves, *Acta Hort.* 148, 547 (1984).
[12] C. Somerville and J. Brows, *Science.* 252, 80 (1991).
[13] M. Dorais, D. L. Ehret, A.P. Papadopoulos, *Phytochemistry Reviews.* 7, 231 (2008).
[14] A. Saeed, K. Haya, A.A. Khan, A. Iqbal, *Int J Agr Biol.* 4, 649 (2007).
[15] G.W. Winsor, P. Adams, *Crops Res Inst Annu.* 134 (1976).
[16] A.N.M de Koning, *Sci Hortic. Amsterdam.* 45, 117 (1988).
[17] E. Heuvelink, *Sci Hortic. Amsterdam.* 64, 193 (1995).
[18] R.G. Hurd, C.J. Graves, *Acta Hort.* 148, 547 (1984).
[19] S.R Adams K.E. Cockshull R.J. Cave, *Ann. Bot.* 88, 869 (2001)
[20] E.L. Moore, W.O. Thomas, *Proceedings of the American Society for Horticultural Science.* 60, 289 (1952).
[21] M.M. Peet and M. Bartholomew, *J. Am. Soc. Hortic. Sci.* 121, 514 (1996).
[22] B.D. Pearce, R.I. Grange, K. Hardwick, *J. Am. Soc. Hortic. Sci.* 68, 1 (1993).
[23] S. Sato, M. Kamiyama, T. Iwata, N. Makita, H. Furukawa, H. Ikeda, *Ann. Bot.* 97, 731 (2006).
[24] B. Aloni, L. Karni, Z. Zaidman, Y. Riov, M. Huberman and K. Goren, *J. Hort. Sci.* 69, 923 (1994).
[25] J.A. Roberts, K.A. Elliott, Z.H. Gonzalez-Carranza, *Ann. Rev. Plant Biol.* 53, 131 (2002).
[26] Y. Gross and J. Kigel, *Field Crops Res.* 36, 201 (1994).
[27] S. Iwahori , *J. JPN Soc. Hortic. Sci.* 35, 379 (1966).
[28] J. Song, J. Win, M. Tian, S. Schornak, F. Kaschani, M. Ilyas, A. L. Renier, H. van der Hoorn and K.Sophien, *Proc. Natl. Acad. Sci USA.* 106, 1654 (2009).
[29] I.C. Karapanos, *Ph D Thesis, Agricultural University of Athens.* 22, (2007).
[30] I. Rylski, *Sci. Hortic. Amsterdam.* 10, 27 (1979).
[31] S. Sato, M.M. Peet, J.F. Thomas, *J. Exp. Bot.* 53, 1187 (2002).
[32] R. Fernandez-Munoz and J. Cuartero, *J. Hortic. Sci.* 66, 395 (1991).

[33] H. Kodama, T. Hamada, G. Horiguchi, M. Nishimura and K. Iba, *Plant Phys.* 105, 601 (1994).

[34] R.G. Upchurch, *Biotechnol. Lett* 30, 967 (2008).

[35] E.G. Bligh and W.J. Dyer, *Can J. Biochem. Physiol.* 37, 911 (1959).

[36] F. Corbineau, R.M. Rudnicki, D. Come, *Plant Growth Regul.* 7, 157 (1988).

[37] B. Bialeca, J. Kepczynski, *Plant Growth Regul.* 41, 7 (2003).

[38] M. Nojavan-Asghari and K. Ishizava, *Plant Growth Regul.* 17, 13 (1998).

[39] C.A. Preston, H. Betts and I.T. Baldwin, *Chem. Ecol.* 28, 2343 (2002).

[40] R.W. Wilen, B. E. Ewan and L.V. Gusta, *Can. J. Bot.* 72, 1009 (1994).

[41] R.S. Venkata, *Ph D Thesis, Agricultural University of Tuscia*, (2011).

[42] FAO, *Session Report, 26-30 March, Rome, Italy.* (2001)

[43] J. Xing, C.K. Chin, *Physiol. Mol. Plant Pathol.* 57, 217 (2000).

[44] T. H. Chen and N. Murata, *Curr. Opin. Plant Biol.* 5, 250 (2002).

INDEX

A

B

C

D

F

G

H

K

L

M

N

O

P

R

S

T

V

W